AF352883

Dynamics and Control of Robot Manipulators

Dynamics and Control of Robot Manipulators

Editors

Marco Carricato
Edoardo Idà

MDPI • Basel • Beijing • Wuhan • Barcelona • Belgrade • Manchester • Tokyo • Cluj • Tianjin

Editors
Marco Carricato
University of Bologna
Bologna, Italy

Edoardo Idà
University of Bologna
Bologna, Italy

Editorial Office
MDPI
St. Alban-Anlage 66
4052 Basel, Switzerland

This is a reprint of articles from the Special Issue published online in the open access journal *Actuators* (ISSN 2076-0825) (available at: https://www.mdpi.com/journal/actuators/special_issues/robot_manipulator).

For citation purposes, cite each article independently as indicated on the article page online and as indicated below:

LastName, A.A.; LastName, B.B.; LastName, C.C. Article Title. *Journal Name* **Year**, *Volume Number*, Page Range.

ISBN 978-3-0365-8426-3 (Hbk)
ISBN 978-3-0365-8427-0 (PDF)

Cover image courtesy of Edoardo Idà

Contents

About the Editors

Marco Carricato

Prof. Marco Carricato received an M.Sc. degree (with honors) in Mechanical Engineering in 1998, and a Ph.D. degree in Mechanics of Machines in 2002. He has been working with the University of Bologna since 2004. He is currently a Full Professor in the Department of Industrial Engineering, where he is the head of the IRMA L@B (Industrial Robotics, Mechatronics & Automation Lab @ Bologna). His research interests include robotic systems, servo-actuated automatic machinery and the theory of mechanisms, with a particular emphasis on parallel manipulators (displacement analysis, kinematics, dynamics, synthesis, gravity compensation, cable drives), mobile collaborative robotics, efficiency and optimization of servomechanisms, and the theory of 'screws'. His work in the aforementioned areas has been the subject of a number of scientific publications in international conferences and journals.

Edoardo Idà

Dr. Edoardo Idà received B.Sc. and M.Sc. degrees (with honors) in Mechanical Engineering at the University of Bologna, and a Ph.D. degree in Mechanics and Advanced Engineering Science at the same university in 2021. He is currently a Junior Assistant Professor in the Department of Industrial Engineering, where he is also the head of operations at IRMA L@B (Industrial Robotics, Mechatronics & Automation Lab @ Bologna). His research focuses on cable-driven robotic systems, continuum robots, automation, and mechanism design.

 *actuators*

Article

Pattern-Moving-Based Partial Form Dynamic Linearization Model Free Adaptive Control for a Class of Nonlinear Systems

Xiangquan Li [1,2] **and Zhengguang Xu** [1,*]

[1] School of Automation and Electrical Engineering, University of Science and Technology Beijing, and The Key Laboratory of Knowledge Automation for Industrial Processes of Ministry of Education, Beijing 100083, China; b20180295@xs.ustb.edu.cn
[2] School of Information Engineering, Jingdezhen University, Jingdezhen 333000, China
[*] Correspondence: xzg@ies.ustb.edu.cn; Tel.: +86-133-1118-7553

Abstract: This work addresses a pattern-moving-based partial form dynamic linearization model free adaptive control (P-PFDL-MFAC) scheme and illustrates the bounded convergence of its tracking error for a class of unknown nonaffine nonlinear discrete-time systems. The concept of pattern moving is to take the pattern class of the system output condition as a dynamic operation variable, and the control purpose is to ensure that the system outputs belong to a certain pattern class or some desired pattern classes. The P-PFDL-MFAC scheme mainly includes a modified tracking control law, a deviation estimation algorithm and a pseudo-gradient (PG) vector estimation algorithm. The classification-metric deviation is considered as an external disturbance, which is caused by the process of establishing the pattern-moving-based system dynamics description, and an improved cost function is proposed from the perspective of a two-player zero-sum game (TP-ZSG). The bounded convergence of the tracking error is rigorously proven by the contraction mapping principle, and the validity of the theoretical results is verified by simulation examples.

Keywords: pattern moving; partial form dynamic linearization (PFDL); nonlinear system; two-player zero-sum game; model free adaptive control (MFAC)

Citation: Li, X.; Xu, Z. Pattern-Moving-Based Partial Form Dynamic Linearization Model Free Adaptive Control for a Class of Nonlinear Systems. *Actuators* **2021**, *10*, 223. https://doi.org/10.3390/act10090223

Academic Editor: Marco Carricato and Edoardo Idà

Received: 4 July 2021
Accepted: 2 September 2021
Published: 5 September 2021

Publisher's Note: MDPI stays neutral with regard to jurisdictional claims in published maps and institutional affiliations.

1. Introduction

In the process of industrial production, there is a range of complex equipment, such as sintering machines, rotary kilns, blast furnaces, and so on. Due to the increase in complexity, such as nonlinearity, high order, large delay, time-varying, and parameter perturbation, it is very difficult to establish an accurate mathematical model [1]. To a certain extent, this kind of production system is mainly governed by the law of statistical moving rather than the existing Newton's law of mechanics. A group of the same or similar system working conditions can produce the corresponding products with the same or similar quality index parameters [2].

A feasible method of system modeling and control is the pattern recognition technology for these considered systems [3], and most researchers' practice is to design the corresponding model and controller according to the different pattern classes of the system working condition [4,5]. Different from the previous multi-controller model design method based on pattern classes, a novel pattern-moving-based system dynamics description method was proposed in [6]. Its basic idea is to take the pattern class as a moving variable, and this variable is mapped to a computable space by class centers [7], interval numbers [8], and cells [9] due to its lack of arithmetic operation attribute. One advantage of the system dynamics description method introduced in [6] is that it is robust to system parameter disturbance and measurement noise. Regarding robust control, a well-known method is sliding mode control [10–12], which has a good ability to deal with external disturbances and system uncertainties. In recent years, a series of important research achievements have been made in sliding mode control, and many improved methods have been proposed, such as global sliding mode control [13] and terminal sliding mode control [14]. Different from the methods proposed in [10–14], the pattern-moving-based system dynamics description method is able to eliminate the system disturbance in the process of pattern classification, as long as the influence of the disturbance on the output does not change the pattern class to which the output belongs. In the case of various metric methods of pattern class, the linear autoregressive model with exogenous input (ARX) or interval ARX (IARX) model has been established, and the minimum-variance-based controller [6], optimal controller [15], predictive controller [16], and state-feedback-based [7] controller have been designed. However, it is well known that it is not easy to identify the system model order and parameters. In addition, even if a pattern-moving-based mathematical prediction model such as ARX or IARX is proposed, it is always an approximation of the real plant, and the unmodeled dynamics of the system are inevitable. Therefore, it is of significance to propose a pattern-moving-based data-driven control (DDC) method and design a controller whose parameters are adjusted by adopting the online input/output (I/O) data and the offline historical data simultaneously.

The data-driven controller is designed directly depending on the offline or/and online I/O data, instead of the explicit mathematical model of the controlled plant [17]. Generally, DDC can be almost cataloged into the following classes according to the different ways in which the data are used: (1) adaptive dynamic programming [18] and iterative learning control [19] based on offline and online data; (2) iterative feedback tuning [20] and virtual reference feedback tuning [21] based on offline data; (3) traditional MFAC [17,22–24] based on online data. The traditional MFAC method does not use the state space model but puts forward new concepts such as pseudo-gradient (PG) vector or pseudo-partial derivative (PPD) to capture the dynamic characteristics of the controlled plant, and it designs the controller through the dynamic linearization data model of the controlled plant at each operating point. Thus far, three equivalent dynamic linearization data models have been proposed, i.e., PFDL, compact-form dynamic linearization (CFDL), and the full-form dynamic linearization (FFDL) data model. By setting input correlation and output correlation components with different memory lengths, the three kinds of data models are different equivalent descriptions of system evolution, and they have different dynamic description capabilities for the controlled plant. Recently, due to many advantages of the MFAC method, such as the fact that establishing a controller merely depends on the measurement I/O data, the monotonic convergence of tracking error, and the bounded-input bounded-output stability of the closed-loop system, it has achieved many application results in many fields, and a few examples are as follows: the MFAC-based fault-tolerant control [25]; sensorless brushless direct current motor based on MFAC [26]; multi-agent systems tracking control [27]; MFAC-based sliding mode control [28]; chemical process based on MFAC [29], etc.

However, although the traditional MFAC algorithms have good control qualities for single-input single-output (SISO), multiple-input single-output, and multiple-input multiple-output time-varying structures and parameters in nonlinear discrete-time systems, there are few reports on MFAC for single-input multiple-output (SIMO) nonlinear systems or systems where the desired exact value of the output target cannot be determined exactly. In view of this kind of nonlinear system, a P-PFDL-MFAC method is proposed in this work and it considers that the difference in the output between next time and the current time is related to the differences in inputs in a time window between the current time and a specific previous time. The length of the time window corresponds to the number of PG vector elements, which is also called the pseudo-order of the equivalent PFDL data model. This is the most significant difference between the method proposed here and the pattern-moving-based CFDL-MFAC (P-CFDL-MFAC) scheme in [30], which considered that the output difference between next time and the current time is only related to the input difference between the current time and the previous time. The control purpose of this kind of system is to make the system outputs belong to one or some specific pattern classes. The first contribution of this work is to combine the pattern-moving-based system dynamics description with the traditional PFDL-MFAC method, and to design a control law algorithm based on two-player zero-sum game and saddle point theory [31,32] under the condition of classification-metric deviation. Another major contribution is that the bounded convergence of the tracking error dynamics of the closed-loop control system is rigorously proven by using the contraction mapping principle.

The remainder of this work is organized as follows. Section 2 introduces the preliminary of the work. Section 3 presents the problem formulation and designs a pattern-moving-based PFDL-MFAC scheme. The bounded convergence of the closed-loop system's tracking error is proven in Section 4. Section 5 presents two simulation examples to demonstrate the correctness and efficiency of the proposed algorithms. A conclusion is given in Section 6.

Notation: $\mathbb{R}$ denotes the real number domain; $\mathbb{Z}_+$ denotes the positive integer domain; $\mathbb{R}^n$ is the real n-dimensional space; $[\cdot]^T$ is the transpose of $[\cdot]$; $\|\cdot\|$ is the Euclidean norm, and $\|\cdot\|_v$ is the consistent matrix norm.

2. Preliminary

Consider a class of SIMO nonaffine nonlinear discrete-time systems with unknown structure, order and parameters.

$$
\begin{cases}
y_1(k+1) = f_1(y_1(k), \cdots, y_1(k-n_1), u(k), \cdots, u(k-m_1)) + d_1(k), \\
y_2(k+1) = f_2(y_2(k), \cdots, y_2(k-n_2), u(k), \cdots, u(k-m_2)) + d_2(k), \\
\quad \vdots \\
y_q(k+1) = f_q(y_n(k), \cdots, y_n(k-n_q), u(k), \cdots, u(k-m_q)) + d_q(k),
\end{cases}
\tag{1}
$$

where $q > 1$; $y_i(k)$ denotes the output of $f_i(\cdot)$ and it satisfies $y_i(k) \in \mathbb{R}$; $u(k)$ is the whole system input and it satisfies $u(k) \in \mathbb{R}$; m_i , n_i represent the unknown input and output orders, respectively, and they satisfy that $m_i \in \mathbb{Z}_+$, $n_i \in \mathbb{Z}_+$; $d_i(k)$ is the weak output measurement noise; $f_i(\cdot)$ denotes an unknown nonlinear discrete-time function; $i \in \{1, \cdots, q\}$.

Assumption 1. *The input of this kind of system* (1) *is bounded, i.e., a constant* M_1 *exists and satisfies that* $|u(k)| \leq M_1$.

A pattern-moving-based system dynamics description [6–9,30] that corresponds to system (1) is proposed in the following steps.

(1) Feature extraction $(T(\cdot))$. A large number of inputs and outputs are collected offline, and the input data set $\{u(k)\}$ and q-dimensional output vector set $\{[y_1(k), \cdots, y_q(k)]\}$ are obtained. Through the principal component analysis (PCA) feature extraction [33] of the output data, the first principal component information is obtained, and then the one-dimensional principal component information set $\{y(k)\}$ will be obtained.

(2) Classification $(M(\cdot))$ and hybrid metrics $(D(\cdot), \bar{D}(\cdot))$. Using pattern classification technology to classify the first principal component information, the number of pattern classes (N), the class center value (s_i), and the class radius (r_i) of each pattern class (dx_i) can be obtained, $i = [1, \cdots, N]$. Since the pattern class does not have the arithmetic operation attribute, the pattern class variable needs to be measured. Because the pattern class is a collection of pattern samples with the same or similar attributes, the method of combining the class center explicit metric $D(\cdot)$ and implicit metric $\bar{D}(\cdot)$ is adopted, i.e., $s_i = D(dx_i)$ and $\bar{dx}_i = \bar{D}(dx_i)$. The implicit metric values are unknown, but there is a definite relationship between an implicit metric value and a class center explicit metric value, such as $|s_i - \bar{dx}_i| \leq r_i$. The class center explicit metric represents the statistical attribute of the pattern class, while the implicit metric denotes the difference in each pattern sample in one pattern class.

(3) Establishing the pattern-moving-based system dynamics equations. The inputs $\{u(k)\}$, implicit metric values $\{\bar{dx}(k)\}$, and class center explicit metric values $\{s(k)\}$ are employed to construct the following dynamics equations.

$$\bar{dx}(k+1) = f(\bar{dx}(k), \ldots, \bar{dx}(k-n), u(k), \ldots, u(k-m)), \tag{2}$$

$$s(k+1) = D(M(\bar{dx}(k+1))) = \begin{cases} s_1, \ \bar{dx}(k+1) \in [s_1 - r_1, s_1 + r_1], \\ s_2, \ \bar{dx}(k+1) \in (s_2 - r_2, s_2 + r_2], \\ \vdots \\ s_N, \ \bar{dx}(k+1) \in (s_N - r_N, s_N + r_N], \end{cases} \tag{3}$$

where $f(\cdot)$ is an unknown SISO nonlinear discrete-time system function; m, n denote the input and output orders of system (2), respectively.

By choosing a reasonable classification method, such as a modified quantized control classification [34], it can be obtained that $C_i = s_i + r_i = s_{i+1} - r_{i+1}$, which is named the class threshold. It exits a classification-metric deviation $e(k+1)$ between the $\bar{dx}(k+1)$ and $s(k+1)$, and $|e(k+1)| = |s(k+1) - \bar{dx}(k+1)| \le r_i$, while $s(k+1) = s_i$. Let $r_{max} = max_{i \in [1,N]}\{r_i\}$, then $|e(k)| \le r_{max}$.

Remark 1. *As mentioned in the Introduction, the description of system dynamics based on pattern moving was first proposed in [6], and further studied in [7–9,30]. The basic idea is to treat the pattern class as a moving variable. Since this variable does not have the attribute of arithmetic operation, it is necessary to measure it into a computable space, and then construct the corresponding dynamic equation in this space. Obviously, the SISO nonlinear system or linear time-varying system can also be treated by the dynamic description method proposed in this section, but the feature extraction $(T(\cdot))$ process is not required.*

Remark 2. *The ultimate goal of classifying and measuring the first principal component information is to obtain a SISO system dynamics description in a computable space. From the perspective of pattern recognition technology, when the contribution rate of the first principal component obtained after feature extraction is more than 85%, it is considered that the first principal component information does not lose the original information or it loses very little. If the contribution rate of the first principal component information does not reach 85%, more principal component information should be considered. Then, after classification and class center explicit metric, the metric result of each pattern class variable is a vector. A pattern-moving-based SIMO system dynamics description is to be constructed in a computable space, but the output dimension may be less than that of the original system. For the pattern-moving-based SIMO system, its control method remains to be studied in the future. In this work, we only consider the case in which the contribution rate of the first principal component information is greater than 85%.*

3. Problem Formulation and Control Scheme

3.1. Problem Formulation

Through the above system dynamics description method, the model free adaptive tracking control problem of system (1) is transformed into the corresponding control problem of system (2) and (3). In order to carry out our next analysis, the following assumptions and lemma are proposed first.

Assumption 2. *The partial derivatives of nonlinear system function $f(\cdot)$ with respect to all variables of the system (2) exist and are continuous.*

Assumption 3. *The system (2) satisfies the generalized Lipschitz condition, i.e.,*

$$|\bar{dx}(k_1+1) - \bar{dx}(k_2+1)| \le b\|U_l(k_1) - U_l(k_2)\|,$$

where $U_l(k) = [u(k), \cdots, u(k-l+1)]^T \in \mathbb{R}^l$, l denotes the input pseudo-order, which satisfies $l > 1$, and b is a positive constant.

Lemma 1 ([22,23])**.** *For the considered system (2) satisfying Assumptions 2 and 3, there must exist a time-varying parameter vector $\varphi_{f,l}(k)$ which is called a pseudo-gradient (PG) vector. If $\|\Delta U_l(k)\| \neq 0$, the system (2) can be described as the following PFDL data model.*

$$\Delta \bar{d}x(k+1) = \varphi_{f,l}^T(k)\Delta U_l(k), \tag{4}$$

where $\|\varphi_{f,l}(k)\| \leq b$; $\Delta \bar{d}x(k+1) = \bar{d}x(k+1) - \bar{d}x(k)$; $\varphi_{f,l}(k) = [\varphi_1(k), \cdots, \varphi_l(k)]^T$; $\Delta U_l(k) = U_l(k) - U_l(k-1)$.

Because the implicit metric values $\{\bar{d}x(k)\}$ are not available, the traditional MFAC methods can not be directly used in such systems. Therefore, this work will focus on the design of a new control scheme that merely depends on the obtained data $\{s(k)\}$, $\{u(k)\}$ and the performance analysis of the closed-loop control system.

3.2. The P-PFDL-MFAC Scheme

It can be seen from the system dynamics Equations (2) and (3) that there is a classification-metric deviation $e(k+1)$ between the initial predicted output $\bar{d}x(k+1)$ and the final output $s(k+1)$ of the system, and this deviation $e(k+1)$ is always considered as a bounded external disturbance [12] in this work. Based on the saddle point theory of TP-ZSG proposed in [30–32], an improved cost function is designed in order to obtain a deviation estimation algorithm and an adaptive tracking control law, which aims to find an equilibrium point between the classification-metric deviation difference and the input difference. The basic idea is that even under large deviation fluctuation, a small input variation value can be found to optimize the loss function.

$$\begin{aligned} J(\Delta u(k), \Delta e(k+1)) = &|s^*(k+1) - s(k+1)|^2 + \lambda|u(k) - u(k-1)|^2 \\ &- \gamma^2|e(k+1) - e(k)|^2, \end{aligned} \tag{5}$$

where $\Delta e(k+1) = e(k+1) - e(k)$; λ is utilized to limit the variation in the control input difference, which satisfies $\lambda > 0$; $s^*(k+1)$ denotes the desired class center explicit metric value at time instant $k+1$; γ is employed to limit the difference change in classification-metric deviation, which satisfies $\gamma > 1$; $\Delta u(k) = u(k) - u(k-1)$.

By solving the following equations

$$\frac{\partial J(\Delta u(k), \Delta e(k+1))}{\partial \Delta u(k)} = 0,$$

and

$$\frac{\partial J(\Delta u(k), \Delta e(k+1))}{\partial \Delta e(k+1)} = 0,$$

one has the optimal results, such as

$$\Delta e(k+1) = \frac{1}{1-\gamma^2}\left(s^*(k+1) - s(k) - \varphi_{f,l}^T(k)\Delta U_l(k)\right), \tag{6}$$

and

$$\Delta u(k) = \frac{\varphi_1(k)\rho_1(s^*(k+1) - s(k) - \Delta e(k+1))}{\lambda + |\varphi_1(k)|^2} - \frac{\varphi_1(k)\sum_{i=2}^{l}\rho_i\varphi_i(k)\Delta u(k-i+1)}{\lambda + |\varphi_1(k)|^2}, \tag{7}$$

where $\gamma > 1$; $\lambda > 0$; ρ_i is a step-size, which satisfies $\rho_i \in (0,1]$ and makes the control algorithm more general; $i \in \{1, \cdots, l\}$.

In order to estimate the PG vector, the following objective function is designed.

$$J(\varphi_{f,l}(k)) = \left| s(k) - s(k-1) - \varphi_{f,l}^T(k)\Delta U_l(k-1) \right|^2 + \mu\|\varphi_{f,l}(k) - \varphi_{f,l}(k-1)\|^2, \quad (8)$$

where μ is a weight factor and it satisfies $\mu > 0$.

By letting

$$\frac{\partial J(\varphi_{f,l}(k))}{\partial \varphi_{f,l}(k)} = 0,$$

one can obtain the estimation algorithm of the PG vector as follows:

$$\varphi_{f,l}(k) = \varphi_{f,l}(k-1) + \frac{\eta\Delta U_l(k-1)\left(\Delta s(k) - \varphi_{f,l}^T(k-1)\Delta U_l(k-1)\right)}{\mu + \|\Delta U_l(k-1)\|^2}, \quad (9)$$

where $\Delta s(k) = s(k) - s(k-1)$; η is a step-size that satisfies $\eta \in (0,2]$ and makes the estimation algorithm more general; $\mu > 0$.

Combining the above algorithms (6), (7), and (9), and proposing a reset algorithm of the PG estimation vector and a limitation mechanism of classification-metric deviation, the P-PFDL-MFAC scheme can be obtained.

$$\hat{\varphi}_{f,l}(k) = \hat{\varphi}_{f,l}(k-1) + \frac{\eta\Delta U_l(k-1)\left(\Delta s(k) - \hat{\varphi}_{f,l}^T(k-1)\Delta U_l(k-1)\right)}{\mu + \|\Delta U_l(k-1)\|^2}, \quad (10)$$

$$\hat{e}(k+1) = \hat{e}(k) + \frac{1}{1-\gamma^2}\left(s^*(k+1) - s(k) - \hat{\varphi}_{f,l}^T(k)\Delta U_l(k)\right), \quad (11)$$

$$\begin{aligned}
u(k) = u(k-1) &- \frac{\hat{\varphi}_1(k)\sum_{i=2}^{l}\rho_i\hat{\varphi}_i(k)\Delta u(k-i+1)}{\lambda + |\hat{\varphi}_1(k)|^2} \\
&+ \frac{\hat{\varphi}_1(k)\rho_1(s^*(k+1) - s(k) - \Delta\hat{e}(k+1))}{\lambda + |\hat{\varphi}_1(k)|^2},
\end{aligned} \quad (12)$$

$$\hat{\varphi}_1(k) = \hat{\varphi}_1(1), \text{if } \|\hat{\varphi}_{f,l}(k)\| \le \varepsilon, \text{or } \|\Delta U_l(k-1)\| \le \varepsilon, \text{or sign}(\hat{\varphi}_1(k)) \ne \text{sign}(\hat{\varphi}_1(1)), \quad (13)$$

$$\hat{e}(k) = \begin{cases} r_j, & \text{if } \hat{e}(k) > r_j, s(k) = s_j \\ -r_j, & \text{if } \hat{e}(k) \le -r_j, s(k) = s_j. \end{cases} \quad (14)$$

where $\eta \in (0,2], \mu > 0, \gamma > 1, \lambda > 0, \rho_i \in (0,1], i \in \{1,\cdots,l\}, j \in \{1,\cdots,N\}$; $\hat{\varphi}_{f,l}(k)$ is the estimation vector of PG $\varphi_{f,l}(k)$; ε denotes a small positive constant; $\hat{\varphi}_1(1)$ is the initial value of $\hat{\varphi}_1(k)$; the algorithm (13) is the reset algorithm of the PG estimation vector, and the algorithm (14) denotes the limitation mechanism of classification-metric deviation.

It is known from the above algorithms that the PG estimation vector directly affects the quality of the control scheme. In order to enhance the time-varying parameters' tracking ability for the PG estimation (10), it is necessary to add the reset algorithm (13). The limitation mechanism (14) is added to ensure that the deviation within one pattern class is not greater than the corresponding pattern class radius. The pseudo-order l is supposed to be less than or equal to the sum of the input and output orders $(m + n)$. A large number of experiments show that the lower the system complexity, the smaller the value of l can be. On the contrary, the higher the system complexity, the greater the l should be. It is obvious that the proposed P-PFDL-MFAC algorithms in this work degenerate to the P-CFDL-MFAC algorithms designed in [30] when $l = 1$.

4. Performance of the Closed-Loop System

The focus of this section is to analyze the performance of the closed-loop tracking control system, i.e., to prove the tracking error bounded stability of the closed-loop control system. Before this, the following assumptions and lemmas are proposed.

Assumption 4. *Considering the nonlinear system* (2), *for any desired bounded output* $\bar{d}x^*(k+1)$, *a bounded input* $u^*(k)$ *always exists and it can make the system output equal to* $\bar{d}x^*(k+1)$.

Assumption 5. *The signal of the first element of the PG vector* $\varphi_{f,l}(k)$ *is assumed to be known and unchanged at any time k with* $\|\Delta U_l(k)\| \neq 0$, *i.e.,* $\varphi_1(k) \geq \epsilon > 0$ *(or* $\varphi_1(k) \leq \epsilon < 0$), ϵ *is a small positive constant. In this work, in order to simplify the derivation of the conclusion, it is always assumed that* $\varphi_1(k) \geq \epsilon > 0$ *without loss of generality.*

Lemma 2 ([22]). *Let*

$$A = \begin{bmatrix} a_1 & a_2 & \cdots & a_p \\ 1 & 0 & \cdots & 0 \\ & \ddots & \ddots & \vdots \\ & & 1 & 0 \end{bmatrix}_{(p \times p)}.$$

If $\sum_{i=1}^{p} |a_i| < 1$, *then* $s(A) < 1$, *where* $s(A)$ *is the spectral radius of A.*

Lemma 3 ([17]). *Let* $A \in \mathbb{R}^{p \times p}$. *For any given* $\varepsilon > 0$, *there exists an induced consistent matrix norm such that* $\|A\|_v \leq s(A) + \varepsilon$, *where* $s(A)$ *has the same meaning as Lemma 2.*

It is known to all that Assumption 4 is a necessary condition for the design and solution of the control problem, and it also shows that the output of the system (2) is controllable. Many plants satisfy the condition of Assumption 5 to some extent, and its actual physical background is also very clear, i.e., the plant's output increasing or decreasing corresponds to the control input increasing or decreasing. Next, our main results will be proven.

Lemma 4. *For the system* (2) *and* (3) *using the P-PFDL-MFAC scheme* (10)–(14) *under Assumptions 2–5,* $\|\hat{\varphi}_{f,l}(k)\|$ *is bounded.*

Proof of Lemma 4. When $\|\Delta U_l(k-1)\| \leq \varepsilon$, it is obvious that $\hat{\varphi}_{f,l}(k)$ is bounded from the reset algorithm (13) of the P-PFDL-MFAC scheme. When $\|\Delta U_l(k-1)\| > \varepsilon$, subtracting $\varphi_{f,l}(k)$ in both sides of Equation (10) obtains

$$
\begin{aligned}
\tilde{\varphi}_{f,l}(k) =& \hat{\varphi}_{f,l}(k-1) - \varphi_{f,l}(k) + \varphi_{f,l}(k-1) + \frac{\eta \Delta U_l(k-1)\Delta s(k)}{\mu + \|\Delta U_l(k-1)\|^2} \\
& - \frac{\eta \Delta U_l(k-1)\hat{\varphi}_{f,l}^T(k-1)\Delta U_l(k-1)}{\mu + \|\Delta U_l(k-1)\|^2} \\
=& \left[I - \frac{\eta \Delta U_l(k-1)\Delta U_l^T(k-1)}{\mu + \|\Delta U_l(k-1)\|^2} \right] \hat{\varphi}_{f,l}(k-1) - \varphi_{f,l}(k) + \varphi_{f,l}(k-1) \\
& + \frac{\eta \Delta \hat{e}(k)\Delta U_l(k-1)}{\mu + \|\Delta U_l(k-1)\|^2},
\end{aligned}
\tag{15}
$$

where $\tilde{\varphi}_{f,l}(k) = \hat{\varphi}_{f,l}(k) - \varphi_{f,l}(k)$.

Taking the norm on both sides of (15) and using Lemma 1, $|\hat{e}(k)| \leq r_{max}$ yields

$$
\|\tilde{\varphi}_{f,l}(k)\| \leq 2b + 2\eta r_{max} + \left\| \left[I - \frac{\eta \Delta U_l(k-1)\Delta U_l^T(k-1)}{\mu + \|\Delta U_l(k-1)\|^2} \right] \hat{\varphi}_{f,l}(k-1) \right\|.
\tag{16}
$$

Square the first term on the right of (16) and obtain the following inequality:

$$
\left\| \left[I - \frac{\eta \Delta U_l(k-1)\Delta U_l^T(k-1)}{\mu + \|\Delta U_l(k-1)\|^2} \right] \tilde{\varphi}_{f,l}(k-1) \right\|^2 \leq \left\| \tilde{\varphi}_{f,l}(k-1) \right\|^2 +
$$
$$
\left(-2 + \frac{\eta \|\Delta U_l(k-1)\|^2}{\mu + \|\Delta U_l(k-1)\|^2} \right) \frac{\eta \left(\tilde{\varphi}_{f,l}^T(k-1)\Delta U_l(k-1) \right)^2}{\mu + \|\Delta U_l(k-1)\|^2}.
\tag{17}
$$

Since $\mu > 0$ and $\eta \in (0,2]$, it can be obtained that $-2 + \frac{\eta \|\Delta U_l(k-1)\|^2}{\mu + \|\Delta U_l(k-1)\|^2} < 0$, and it is obvious that $\frac{\eta \left(\tilde{\varphi}_{f,l}^T(k-1)\Delta U_l(k-1) \right)^2}{\mu + \|\Delta U_l(k-1)\|^2} > 0$. Thus, there must exist a constant $0 < d_1 < 1$ that satisfies $\left\| \left[I - \frac{\eta \Delta U_l(k-1)\Delta U_l^T(k-1)}{\mu + \|\Delta U_l(k-1)\|^2} \right] \tilde{\varphi}_{f,l}(k-1) \right\| \leq d_1 \left\| \tilde{\varphi}_{f,l}(k-1) \right\|$. It can be further deduced that

$$
\begin{aligned}
\|\tilde{\varphi}_{f,l}(k)\| &\leq d_1 \|\tilde{\varphi}_{f,l}(k-1)\| + 2b + 2\eta r_{max} \\
&\leq d_1^2 \|\tilde{\varphi}_{f,l}(k-1)\| + d_1(2b + 2\eta r_{max}) + 2b + 2\eta r_{max} \\
&\leq \cdots \leq d_1^{k-1} \|\tilde{\varphi}_{f,l}(1)\| + \frac{(2b + 2\eta r_{max})(1 - d_1^{k-1})}{1 - d_1}.
\end{aligned}
\tag{18}
$$

In view of (18), $\|\tilde{\varphi}_{f,l}(k)\|$ is bounded, since $\|\varphi_{f,l}(k)\|$ is bounded; thus, $\|\hat{\varphi}_{f,l}(k)\|$ is bounded. $\square$

Theorem 1. *For system (2) and (3) using the P-PFDL-MFAC scheme (10)–(14) under Assumptions 3–6 with the desired signal $s^*(k+1) = s^* = const$, if the controller parameters meet the following conditions*

(1) *letting $\bar{\rho}_1 = \frac{\gamma^2 \rho_1}{\gamma^2 - 1 + \rho_1}$ and $\bar{\rho}_1 \in (0,1]$;*

(2) *letting $\bar{\rho}_i = \frac{(\gamma^2 - 1)\rho_i + \rho_1}{\gamma^2 - 1 + \rho_1}$ and $\bar{\rho}_i \in (0,1], i = 2, \cdots, l$;*

(3) *letting $\bar{\lambda} = \frac{(\gamma^2 - 1)\lambda}{\gamma^2 - 1 + \rho_1}$, and there exists a $\bar{\lambda}_{min}$ such that $\bar{\lambda} > \bar{\lambda}_{min}$,*

then the closed-loop control system guarantees that

$$
\lim_{k \to \infty} |s^* - s(k+1)| \leq M,
$$

where M is a constant and $M > 0$.

Proof of Theorem 1. Substituting the classification-metric deviation estimation algorithm (11) into control algorithm (12), one has

$$
u(k) = u(k-1) + \frac{\frac{\gamma^2 \rho_1}{\gamma^2 - 1 + \rho_1}\hat{\varphi}_1(k)(s^* - s(k))}{\frac{(\gamma^2 - 1)\lambda}{\gamma^2 - 1 + \rho_1} + |\hat{\varphi}_1(k)|^2} - \frac{\hat{\varphi}_1(k)\sum_{i=2}^{l}\frac{(\gamma^2 - 1)\rho_i}{\gamma^2 - 1 + \rho_1}\hat{\varphi}_i(k)\Delta u(k - i + 1)}{\frac{(\gamma^2 - 1)\lambda}{\gamma^2 - 1 + \rho_1} + |\hat{\varphi}_1(k)|^2}.
\tag{19}
$$

Given $\bar{\rho}_1, \bar{\rho}_i, \bar{\lambda}$, Equation (19) can be written as

$$
u(k) = u(k-1) + \frac{\bar{\rho}_1 \hat{\varphi}_1(k)(s^* - s(k))}{\bar{\lambda} + |\hat{\varphi}_1(k)|^2} - \frac{\hat{\varphi}_1(k)\sum_{i=2}^{l}\bar{\rho}_i \hat{\varphi}_i(k)\Delta u(k - i + 1)}{\bar{\lambda} + |\hat{\varphi}_1(k)|^2},
\tag{20}
$$

where $\bar{\rho}_i \in (0,1], i = 1, \cdots, l$.

Since $\gamma > 1$, $\lambda > 0$ and $\rho_1 \in (0,1]$, thus $\bar{\lambda} > 0$. It is known from Lemma 4 that $\|\hat{\varphi}_{f,l}(k)\|$ is bounded and noted that $\|\hat{\varphi}_{f,l}(k)\| \leq b_1$; here, b_1 is a positive constant. Given $\|\hat{\varphi}_{f,l}(k)\| \leq b_1$, $\|\varphi_{f,l}(k)\| \leq b$, $\gamma > 1$, $\lambda > 0$, $\rho_i \in (0,1]$, $\bar{\rho}_i \in (0,1]$, $\bar{\lambda} > 0$, there exist bounded constants W_i, $i \in \{1, 2, 3, 4, 5\}$ such that the following inequalities (21)–(25) hold when $\bar{\lambda} > \bar{\lambda}_{min}$.

Letting $\bar{\lambda} > \bar{\lambda}_{min} \geq b^2$ and using inequality $x^2 + y^2 \geq 2xy$, one obtains

$$\left| \frac{\hat{\varphi}_1(k)}{\bar{\lambda} + |\hat{\varphi}_1(k)|^2} \right| \leq \left| \frac{\hat{\varphi}_1(k)}{2\sqrt{\bar{\lambda}}|\hat{\varphi}_1(k)|} \right| < \left| \frac{1}{2\sqrt{\bar{\lambda}_{min}}} \right| = W_1 < \frac{0.5}{b}, \tag{21}$$

$$0 < W_2 \leq \left| \frac{\hat{\varphi}_1(k)\varphi_i(k)}{\bar{\lambda} + |\hat{\varphi}_1(k)|^2} \right| \leq b \left| \frac{\hat{\varphi}_1(k)}{2\sqrt{\bar{\lambda}}|\hat{\varphi}_1(k)|} \right| < 0.5, \tag{22}$$

$$W_1 \| \varphi_{f,l}(k) \| = W_3 < 0.5. \tag{23}$$

From the inequalities (22) and (23), it is deduced that

$$W_2 + W_3 < 1. \tag{24}$$

Letting $\left\{ \sum_{i=2}^{l} \left| \frac{\hat{\varphi}_1(k)\hat{\varphi}_i(k)}{\bar{\lambda} + |\hat{\varphi}_1(k)|^2} \right| \right\}^{\frac{1}{l-1}} \leq W_4$ and choosing $\bar{\rho}_{max} = \max_{i=1,\dots,l} \bar{\rho}_i$, one has

$$\sum_{i=2}^{l} \bar{\rho}_i \left| \frac{\hat{\varphi}_1(k)\hat{\varphi}_i(k)}{\bar{\lambda} + |\hat{\varphi}_1(k)|^2} \right| \leq \bar{\rho}_{max} \sum_{i=2}^{l} \left| \frac{\hat{\varphi}_1(k)\hat{\varphi}_i(k)}{\bar{\lambda} + |\hat{\varphi}_1(k)|^2} \right| \leq \bar{\rho}_{max} W_4^{l-1} = W_5 < 1. \tag{25}$$

Defining tracking error $w(k) = s^* - s(k)$ and letting

$$A(k) = \begin{bmatrix} -\frac{\bar{\rho}_2\hat{\varphi}_1(k)\hat{\varphi}_2(k)}{\bar{\lambda}+|\hat{\varphi}_1(k)|^2} & -\frac{\bar{\rho}_3\hat{\varphi}_1(k)\hat{\varphi}_3(k)}{\bar{\lambda}+|\hat{\varphi}_1(k)|^2} & \cdots & -\frac{\bar{\rho}_l\hat{\varphi}_1(k)\hat{\varphi}_l(k)}{\bar{\lambda}+|\hat{\varphi}_1(k)|^2} & 0 \\ 1 & 0 & \cdots & 0 & 0 \\ 0 & 1 & \cdots & 0 & 0 \\ \vdots & \vdots & \vdots & \vdots & \vdots \\ 0 & 0 & \cdots & 1 & 0 \end{bmatrix}, \tag{26}$$

the control algorithm (12) can be written as

$$\begin{aligned} \Delta U_l(k) &= [\Delta u(k), \cdots, \Delta u(k-l+1)]^T \\ &= A(k)[\Delta u(k-1), \cdots, \Delta u(k-l)]^T + \frac{\bar{\rho}_1\hat{\varphi}_1(k)}{\bar{\lambda} + |\hat{\varphi}_1(k)|^2} Cw(k), \end{aligned} \tag{27}$$

where $C = [1, 0, \cdots, 0]^T \in \mathbb{R}^l$. The secular equation of $A(k)$ is

$$z^l + \frac{\bar{\rho}_2\hat{\varphi}_1(k)\hat{\varphi}_2(k)}{\bar{\lambda} + |\hat{\varphi}_1(k)|^2} z^{l-1} + \cdots + \frac{\bar{\rho}_l\hat{\varphi}_1(k)\hat{\varphi}_l(k)}{\bar{\lambda} + |\hat{\varphi}_1(k)|^2} z = 0.$$

From Lemma 2 and inequality (25), one has $|z| < 1$ and obtains

$$|z|^{l-1} \leq \sum_{i=2}^{l} \bar{\rho}_i \left| \frac{\hat{\varphi}_1(k)\hat{\varphi}_i(k)}{\bar{\lambda} + |\hat{\varphi}_1(k)|^2} \right| \leq \bar{\rho}_{max} W_4^{l-1} < 1.$$

Further, it can be deduced that $|z| \leq \bar{\rho}_{max}^{\frac{1}{l-1}} W_4$. From Lemma 3, one can obtain $\|A(k)\|_v \leq s(A(k)) + \varepsilon \leq \bar{\rho}_{max}^{\frac{1}{l-1}} W_4 < 1$. According to the definition of $U_l(k)$, it is clear that $\Delta U_l(0) = 0$. Letting $d_2 = \bar{\rho}_{max}^{\frac{1}{l-1}} W_4$ and taking the norm on both sides of (27), one obtains

$$\|\Delta U_l(k)\| \leq \|A(k)\|_v \|\Delta U_l(k-1)\| + \bar{\rho}_1 \left| \frac{\hat{\varphi}_1(k)}{\bar{\lambda} + |\hat{\varphi}_1(k)|^2} \right| |w(k)|$$

$$\leq d_2 \|\Delta U_l(k-1)\| + \bar{\rho}_1 W_1 |w(k)| \leq \cdots = \bar{\rho}_1 W_1 \sum_{i=1}^{k} d_2^{k-i} |w(k)|. \tag{28}$$

From Lemma 1 and Equation (27), one has

$$
\begin{aligned}
w(k+1) &= s^* - s(k+1) = s^* - \bar{d}x(k+1) - e(k+1) \\
&= w(k) - \Delta e(k+1) - \varphi_{f,l}^T(k)\Delta U_l(k) \\
&= \left[1 - \frac{\bar{\rho}_1 \hat{\varphi}_1(k)\varphi_1(k)}{\bar{\lambda} + |\hat{\varphi}_1(k)|^2} \right] w(k) - \varphi_{f,l}^T(k)A(k)\Delta U_l(k-1) - \Delta e(k+1).
\end{aligned}
\tag{29}
$$

Choosing a reasonable $\bar{\rho}_1$, one can obtain

$$\left| 1 - \frac{\bar{\rho}_1 \hat{\varphi}_1(k)\varphi_1(k)}{\bar{\lambda} + |\hat{\varphi}_1(k)|^2} \right| = \left| 1 - \left| \frac{\bar{\rho}_1 \hat{\varphi}_1(k)\varphi_1(k)}{\bar{\lambda} + |\hat{\varphi}_1(k)|^2} \right| \right| \leq 1 - \bar{\rho}_1 W_2 = d_3 < 1. \tag{30}$$

From the above inequality and $|e(k)| \leq r_{max}$, taking the norm on both sides of the Equation (29), one obtains

$$|w(k+1)| < d_3|w(k)| + d_2\|\varphi_{f,l}(k)\|\|\Delta U_l(k-1)\| + 2r_{max} < \cdots$$

$$< d_3^k|w(1)| + d_2 \sum_{i=1}^{k-1} d_3^{k-1-i}\|\varphi_{f,l}(i+1)\|\|\Delta U_l(i)\| + 2r_{max} \sum_{i=1}^{k-1} d_3^{k-1-i}$$

$$< d_3^k|w(1)| + 2r_{max} \sum_{i=1}^{k-1} d_3^{k-1-i} + d_2 \sum_{i=1}^{k-1} d_3^{k-1-i}\|\varphi_{f,l}(i+1)\|\bar{\rho}_1 W_1 \sum_{j=1}^{i} d_2^{i-j}|w(j)|. \tag{31}$$

Letting $d_4 = \bar{\rho}_1 W_3$, it is clear that $d_4 < 1$. The inequality (31) can be recorded as

$$|w(k+1)| < d_3^k|w(1)| + d_2 d_4 \sum_{i=1}^{k-1} d_3^{k-1-i} \sum_{j=1}^{i} d_2^{i-j}|w(j)| + \frac{2r_{max}(1 - d_3^{k-1})}{1 - d_3}. \tag{32}$$

Letting

$$g(k+1) = d_3^k|w(1)| + d_2 d_4 \sum_{i=1}^{k-1} d_3^{k-1-i} \sum_{j=1}^{i} d_2^{i-j}|w(j)|,$$

it is obvious that $g(2) = d_3|w(1)|$. One can see that if $g(k+1)$ is bounded, then $w(k)$ is bounded.

Next, the boundedness of $g(k+1)$ will be proven.

$$
\begin{aligned}
g(k+2) &= d_3^{k+1}|w(1)| + d_2 d_4 \sum_{i=1}^{k} d_3^{k-i} \sum_{j=1}^{i} d_2^{i-j}|w(j)| \\
&= d_3 g(k+1) + d_4 d_2^k|w(1)| + \cdots + d_4 d_2^2|w(k-1)| + d_4 d_2|w(k)| \\
&< d_3 g(k+1) + d_4 d_2^k|w(1)| + \cdots + d_4 d_2 g(k) \\
&\quad + d_4 d_2^2|w(k-1)| + d_4 d_2 \frac{2r_{max}(1 - d_3^{k-2})}{1 - d_3}.
\end{aligned}
\tag{33}
$$

Note that $\bar{h}(k) = d_3 g(k+1) + d_4 d_2^k |w(1)| + \cdots + d_4 d_2^2 |w(k-1)| + d_4 d_2 g(k)$. Since $d_3 = 1 - \bar{\rho}_1 W_2 > \bar{\rho}_1 (W_2 + W_3) - \bar{\rho}_1 W_2 = \bar{\rho}_1 W_3 = d_4$, one obtains

$$
\begin{aligned}
\bar{h}(k) &< d_3 g(k+1) + d_4 d_2^k |w(1)| + \cdots + d_4 d_2^2 |w(k-1)| + d_3 d_2 g(k) \\
&< d_3 g(k+1) + d_4 d_2^k |w(1)| + \cdots + d_4 d_2^2 |w(k-1)| \\
&\quad + d_3 d_2 \left[d_3^{k-1} |w(1)| + d_2 d_4 \sum_{i=1}^{k-2} d_3^{k-2-i} \sum_{j=1}^{i} d_2^{i-j} |w(j)| \right] \\
&= d_2 g(k+1).
\end{aligned}
\tag{34}
$$

From the inequalities (33) and (34), one has

$$
g(k+2) \le (d_2 + d_3) g(k+1) + d_4 d_2 \frac{2 r_{max} (1 - d_3^{k-2})}{1 - d_3}.
$$

Since $d_2 + d_3 = 1 - \bar{\rho}_1 W_2 + \bar{\rho}_{max}^{\frac{1}{l-1}} W_4$, by choosing the reasonable $\bar{\rho}_i, i = 1, \cdots, l$, it exits $d_2 + d_3 = d_5 \in (0,1)$ and one obtains

$$
g(k+2) \le d_5 g(k+1) + d_4 d_2 \frac{2 r_{max}}{1 - d_3} \le \cdots \le d_5^k g(2) + d_4 d_2 \frac{2 r_{max}}{1 - d_3} \frac{1 - d_5^k}{1 - d_5}.
\tag{35}
$$

It is clear that $g(k)$ is bounded convergent; thus, the tracking error $w(k)$ is bounded convergent, i.e., $\lim_{k \to \infty} |w(k)| \le M$, M is a positive constant. $\square$

Remark 3. *The contraction mapping principle is utilized to prove the bounded convergence in this work, and many inequalities are employed to handle the mapping relationships in Lemma 4 and Theorem 1. A critical technique is to let λ, γ, and ρ_i take reasonable values that can guarantee the existence of constants W_1, W_2, W_3, W_4, W_5, $\bar{\lambda}$, $\bar{\gamma}$, $\bar{\rho}_i$, d_1, d_2, d_3, d_4, and d_5 to make the inequalities used in the above derivations hold.*

Remark 4. *It is obvious that the desired tracking target is an arbitrary bounded constant s^* in Theorem 1. In fact, for the closed-loop control system based on pattern moving, the desired tracking target should be one or some specific pattern classes (dx_i), i.e., one or some specific pattern class centers $(s^* = s_i, i = 1, \cdots, N)$. Therefore, instead of focusing on each specific value of the system output, the P-PFDL-MFAC method focuses on whether the system outputs belong to one or some specific pattern classes, and this is the most significant difference between the method designed in this work and the model free adaptive quantization control method proposed in [35,36]. From this point of view, under the control input and output disturbance, even if the implicit metric value of the pattern class to which the system outputs belong satisfies $|\bar{dx}(k+1) - s^*| \le r_i$ when the desired target $s^* = s_i$, it is still considered that the system's tracking error is zero.*

Remark 5. *The designed P-PFDL-MFAC method is employed for the considered system (2) and (3), which corresponds to a practical SIMO system (1). When the system is under the control input $u(k)$ at time instant k, the output vector $[y_1(k+1), \cdots, y_l(k+1)]$ is obtained, and then $s(k+1)$ is obtained by feature extraction $T(\cdot)$, pattern classification $M(\cdot)$, the class center explicit metric $D(\cdot)$ with the real-time output data $[y_1(k+1), \cdots, y_l(k+1)]$, and a large amount of offline historical data. Generally speaking, the P-PFDL-MFAC method can be considered a novel data-driven method based on offline historical data and online real-time data, and this is a major difference from the traditional MFAC methods.*

5. Simulation

Two examples are given to demonstrate the feasibility and effectiveness of the achieved algorithms in this section. In the simulation example of reference [37], the speed control of a Stanford manipulator's joint 4 proposed in [38] was discussed. It considered that the controlled object is a discrete-time system with jump parameters while the load changes. In the first example below, this discrete-time system is also taken as the consideration object, and the designed P-PFDL-MFAC scheme is implemented. Example 2 is a SIMO nonlinear discrete-time numerical case. In this simulation case, the designed control scheme is adopted, and the control effects with different pseudo-orders are compared.

Example 1. *Consider a SISO discrete-time system with jump parameters*

$$y(k) = a_2(k)y(k-2) + b_0(k)u(k-1) + b_1(k)u(k-2) + g(k) + e(k), \tag{36}$$

where $y(k)$ is the system output, which denotes the speed of a Stanford manipulator's joint 4; $u(k)$ is the system input, which denotes the motor's voltage and satisfies $u(k) \in [0,10]$; $e(t)$ denotes the system random noise and it satisfies that $|e(k)| \leq 0.01$; $g(k)$ is considered as a constant and $g(t) = 0.25$; $b_1(k)$ is also a constant and $b_1(k) = 0.2$; the other two system jump parameters are as follows:

$$a_2(k) = \begin{cases} -0.9, k \leq 200; \\ -0.75, 200 < k \leq 400; \\ -0.9, 400 < k \leq 600, \end{cases}$$

and

$$b_0(k) = \begin{cases} 0.4, k \leq 200; \\ 0.35, 200 < k \leq 400; \\ 0.4, 400 < k \leq 600. \end{cases}$$

The control goal of our designed scheme is that the outputs belong to one or some special pattern classes, which is the most significant difference from the simulation in [37]. Firstly, a large number of outputs obtained under effective control inputs are divided into several pattern classes. Then, one or some desired pattern classes are taken as the targets of system control.

Step 1: Classification ($M(\cdot)$) and metrics ($D(\cdot), \bar{D}(\cdot)$) of massive offline data. Here, 600 evenly distributed inputs are taken and the corresponding outputs are obtained. A modified quantized control classification and class center explicit metric method ($M(\cdot)$, $D(\cdot)$) [34] is adopted and described as follows.

$$s(k) = D(M(y(k))) = \begin{cases} y_0(k), & \text{if } T1_i < y(k) \leq T2_i, \\ 0, & \text{if } -T_N < y(k) \leq T_N, \\ -y_0(k), & \text{if } -T2_i < y(k) \leq -T1_i, \end{cases} \tag{37}$$

where $T1_i = \frac{1}{1+\Delta}\kappa_i$; $T2_i = \frac{1}{1-\Delta}\kappa_i$; $T_N = \frac{1}{1+\Delta}\rho_0^N\kappa_0$; $y_0(k+1) = \frac{1+\rho_0}{4}\kappa_i(\rho_0^{i-1} + \rho_0^i)$; $\Delta = \frac{1-\rho_0}{1+\rho_0}$; $\kappa_i = \rho_0^i\kappa_0$; $\rho_0 \in (0,1)$; κ_0 is the maximum working range of $y(k)$ ($\kappa_0 \geq \max\{|y(k)|\}$); N denotes the number of pattern classes; $i = 1, 2, \cdots, N-1$.

Given the upper limit of the initial class radius r_0 at the working point 0 and other parameters such as ρ_0 and κ_0, one can obtain $L \geq \lceil \frac{\ln(r_0 \frac{(1+\Delta)}{\kappa_0})}{\ln \rho_0} \rceil$, and the output sequence $\{y(k)\}$ is divided into $2L+1$ segments. Furthermore, $N = 2L+1$, s_i, $r_i = \frac{1+\rho^2}{4\rho}$ and class threshold C_i can be obtained, respectively, $i = 1, \cdots, N$. The parameter settings of the adopted classification method are $\rho_0 = 0.4$, $\kappa_0 = 15$, $r_0 = 0.2$. The distribution curves of $\{u(k)\}$, $\{y(k)\}$, and $\{s(k)\}$ are shown in Figure 1. Table 1 shows the property values of each pattern class.

Table 1. Property values of pattern class.

Class No.	Class Center s_i	Class Radius r_i	Threshold C_i
1	−7.3500	3.1500	−4.2000
2	−2.9400	1.2600	−1.6800
3	−1.1760	0.5040	−0.6720
4	−0.4704	0.2016	−0.2688
5	−0.1882	0.0806	−0.1075
6	0	0.1075	0.1075
7	0.1882	0.0806	0.2688
8	0.4704	0.2016	0.6720
9	1.1760	0.5040	1.6800
10	2.9400	1.2600	4.2000
11	7.3500	3.1500	10.5000

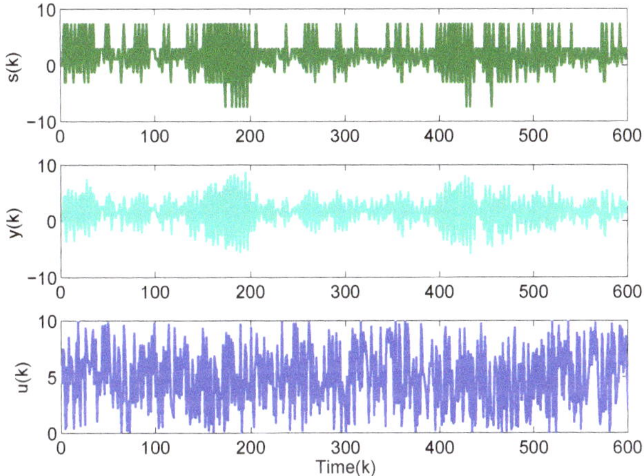

Figure 1. The curves of I/O data and class centers.

Remark 6. *To the best of our knowledge, there are many clustering and classification algorithms in statistical pattern recognition, such as ISODATA, K-means, C-means, and so on. A class center explicit metric and modified quantized control classification method is adopted in this work. As mentioned in [2], the product quality is directly related to the working conditions. Therefore, the parameter settings of condition classification are determined by the result of product quality clustering. Here, it is assumed that the first principal component information $y(k) \in (0.2688, 0.6720]$ corresponds to good product quality, so the initial parameters ($\rho_0 = 0.4$, $\kappa_0 = 15$, $r_0 = 0.2$) are configured to ensure that the working condition data $y(k) \in (0.2688, 0.6720]$ belong to one pattern class.*

Step 2: A pattern-moving-based system dynamics description is established with the obtained property values and data sets $\{u(k)\}$, $\{\bar{d}x(k)\}$, and $\{s(k)\}$.

$$
\begin{cases}
\bar{d}x(k) = f(\bar{d}x(k-1), \cdots, \bar{d}x(k-n_y), u(k-1), \cdots, u(k-n_u)), \\
s(k) = D(M(\bar{d}x(k))) = \begin{cases}
-7.3500, \bar{d}x(k) \in (-10.5, -4.2], \\
\quad\vdots \\
0.0000, \bar{d}x(k) \in (-0.1075, 0.1075], \\
\quad\vdots \\
7.3500, \bar{d}x(k) \in (4.2, 10.5],
\end{cases}
\end{cases}
\tag{38}
$$

where $f(\cdot)$ is an unknown nonlinear system function; n_u, n_y denote the unkown input and output orders of $f(\cdot)$, respectively.

Step 3: Application of the control scheme. Nine pattern classes are obtained and the designed P-PFDL-PMFAC scheme (10)–(14) is employed to track the following targets.

$$s^*(k) = 0.4704,$$

where $s^* = 0.4704$ denotes that the object is pattern class 8.

Set the initial conditions as $y(1:2) = 0$, $e(1:2) = 0$, $u(1:2) = 0$, $\hat{\varphi}_1(2) = 1$, $\hat{\varphi}_2(1:2) = 0$, $\varepsilon = 10^{-5}$, $s(1:2) = 0$. The controller parameters are set as $\gamma = 10$, $\lambda = 0.01$, $\mu = 1$, $\eta = 0.5$, $\rho_1 = \rho_2 = 0.5$, $l = 2$ and the resetting initial value is $\hat{\varphi}_1(1) = 0.5$. Figure 2 shows the system output process, and Figure 3 shows the curves of control input, PG estimation values, and deviation. From the controlled output of the system, it can be seen that although it has undergone drastic adjustment at the beginning, it can track the target quickly and achieve a good tracking effect.

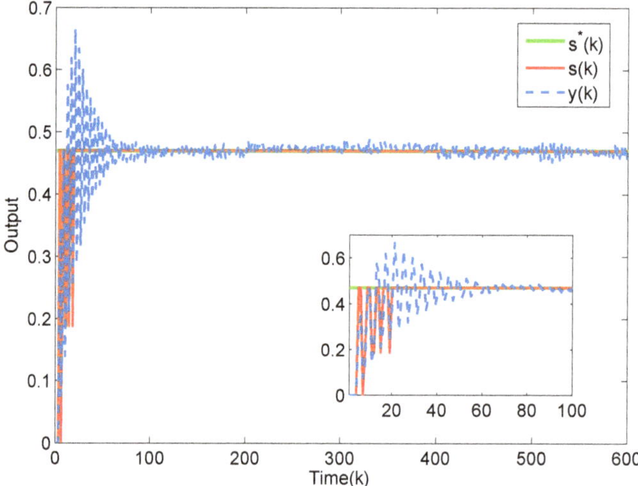

Figure 2. The curves of desired class center, original output, and its corresponding class center.

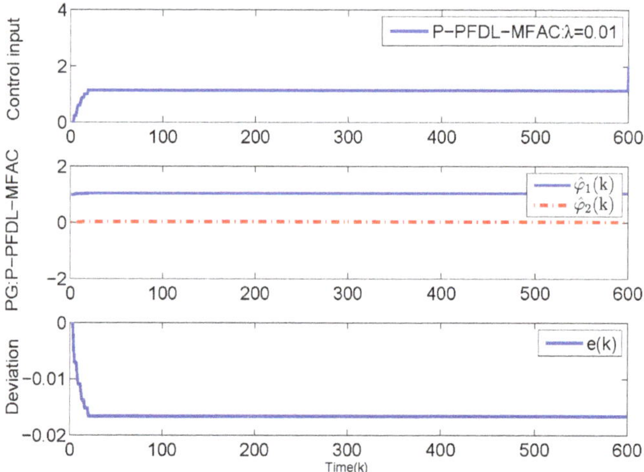

Figure 3. The curves of control input, PG estimation values, and deviation.

Example 2. *A single input and three outputs of the nonlinear discrete-time system are given as follows.*

$$\begin{cases} y_1(k+1) = 1.2\sin(0.5y_1(k)) + u^2(k) + \dfrac{u(k)}{1+u^2(k)} + u(k-1) + d(k), \\[2mm] y_2(k+1) = 1.3\sin(0.5y_2(k)) + 0.2y_2(k-1) + \dfrac{u(k)}{1+u^2(k)} + 0.5u(k-1) + d(k), \quad (39) \\[2mm] y_3(k+1) = 1.4\sin(0.5y_3(k)) + 0.5u^2(k) + \dfrac{u(k)}{1+u^2(k)} + u(k-1) + d(k), \end{cases}$$

where $y_i(k)$ denotes one of the three outputs, $i = 1, 2, 3$; $d(k)$ is the Gaussian white noise and $d(k) \sim \mathcal{N}(0, 0.01^2)$; $u(k)$ denotes the system input and $u(k) \in [-2, 2]$; the system is merely employed to produce the I/O data with unknown system structure, orders, and parameters.

Feature extraction ($T(\cdot)$), classification ($M(\cdot)$), and metrics ($D(\cdot), \bar{D}(\cdot)$) of massive offline data. Here, 1000 evenly distributed inputs are taken and the corresponding outputs are obtained. The outputs are normalized and the PCA technology is employed to deal with them. One can obtain the first principal component information $\{y(k)\}$ (the contribution rate: 85.4518% > 85%). The same classification-metrics method (37) as in Example 1 is adopted. The parameter settings of the adopted classification method are $\rho_0 = 0.4$, $\kappa_0 = 5$, $r_0 = 0.2$. The distribution curves of $\{u(k)\}$, $\{y_i(k)\}$, $\{y(k)\}$, and $\{s(k)\}$ are shown in Figure 4, $i = 1, 2, 3$. Table 2 shows the property values of each pattern class.

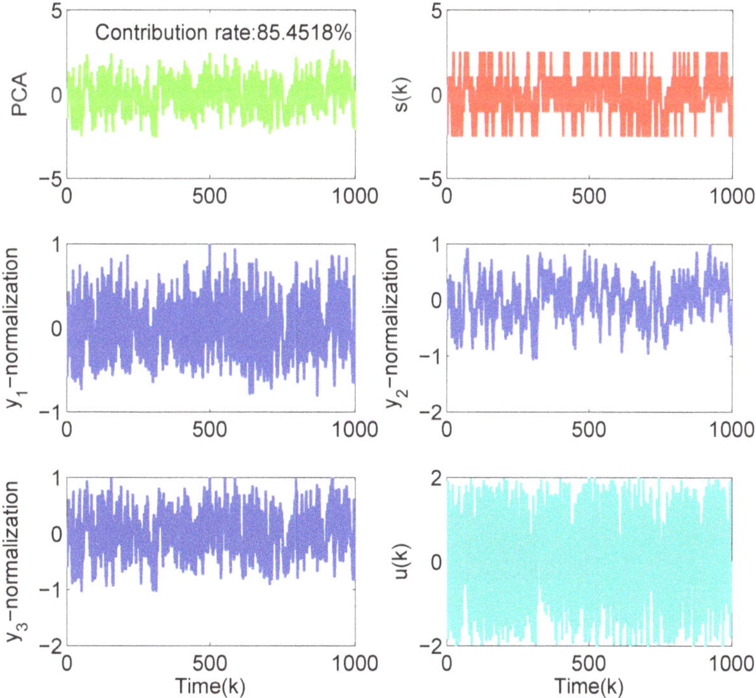

Figure 4. The curves of I/O data, PCA information, and class center.

Table 2. Property values of pattern class.

Class No.	Class Center s_i	Class Radius r_i	Threshold C_i
1	−2.4500	1.0500	−1.4000
2	−0.9800	0.4200	−0.5600
3	−0.3920	0.1680	−0.2240
4	−0.1568	0.0672	−0.0896
5	0	0.0896	0.0896
6	0.1568	0.0672	0.2240
7	0.3920	0.1680	0.5600
8	0.9800	0.4200	1.4000
9	2.4500	1.0500	3.5000

A pattern-moving-based system dynamics description is established as follows.

$$
\begin{cases}
\bar{d}x(k+1) = f(\bar{d}x(k), \cdots, \bar{d}x(k-n_y), u(k), \cdots, u(k-n_u)), \\
s(k+1) = D(M(\bar{d}x(k+1))) = \begin{cases} -2.4500, \bar{d}x(k+1) \in (-3.5, -1.4], \\ \quad \vdots \\ 0.0000, \bar{d}x(k+1) \in (-0.0896, 0.0896], \\ \quad \vdots \\ 2.4500, \bar{d}x(k+1) \in (1.4, 3.5], \end{cases}
\end{cases}
\tag{40}
$$

Nine pattern classes are obtained and the designed P-PFDL-PMFAC scheme (10)–(14) is employed to track the following targets.

$$
s^*(k) = \begin{cases} 0.000, & 0 < k \le 500; \\ 0.980, & 500 < k \le 1000, \end{cases}
$$

where $s^* = 0$, $s^* = 0.980$ denote that the object is pattern class 5 and 8, respectively.

Set the initial conditions as $y_1(1:4) = 0$, $y_2(1:4) = 0$, $y_3(1:4) = 0$, $e(1:4) = 0$, $u(1:4) = 0$, $\hat{\phi}_1(2:4) = 1$, $\hat{\phi}_2(1:4) = 0$, $\hat{\phi}_3(1:4) = 0$, $\varepsilon = 10^{-5}$, $s(1:4) = 0$. The controller parameters are set as $\gamma = 10$, $\lambda = 0.01$, $\mu = 1$, $\eta = 0.5$, $\rho_1 = \rho_2 = \rho_3 = 0.5$ and the resetting initial value is $\hat{\phi}_1(1) = 0.5$. Figures 5–7 correspond to the curves of system input, outputs, PG estimation values, and deviation when the pseudo-order l is 1, 2, and 3, respectively. When $l = 1$, the P-PFDL-PMFAC scheme degenerates to the P-CFDL-MFAC method designed in [30], and the PG vector becomes a PPD. All three figures show that the target trajectory $s^*(k) = 0.980$ is well tracked. However, Figure 5 shows that the tracking effect of target trajectory $s^*(k) = 0$ is poor. Figure 6 shows that the tracking effect of target trajectory $s^*(k) = 0$ is slightly better, but there are also many cases where the tracking can not be achieved. It can be seen from Figure 7 that the target object $s^*(k) = 0$ is well tracked. The simulation results confirm that the value of pseudo-order should correspond to the complexity of the system, and they show that a reasonable pseudo-order can improve the control effect of the system. This numerical example illustrates that the designed scheme is a very feasible method for a class of nonlinear discrete-time systems when the outputs only need to be controlled to one or some specific pattern classes.

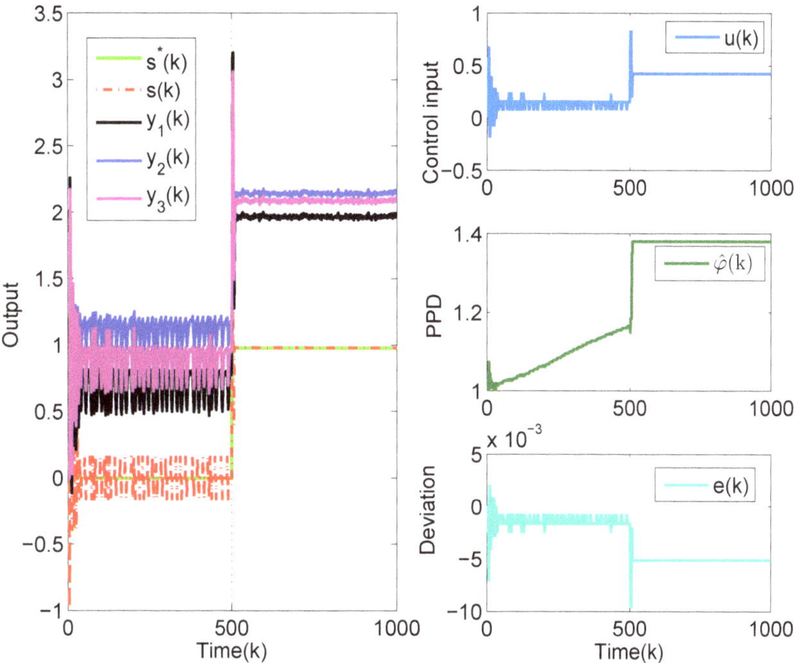

Figure 5. The curves of PPD estimation value, control input, deviation, and outputs with $l = 1$.

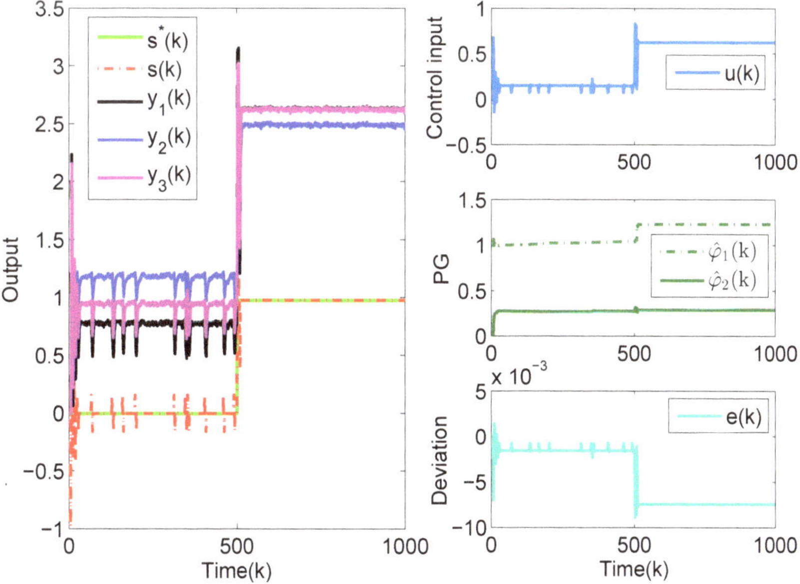

Figure 6. The curves of PG estimation values, control input, deviation, and outputs with $l = 2$.

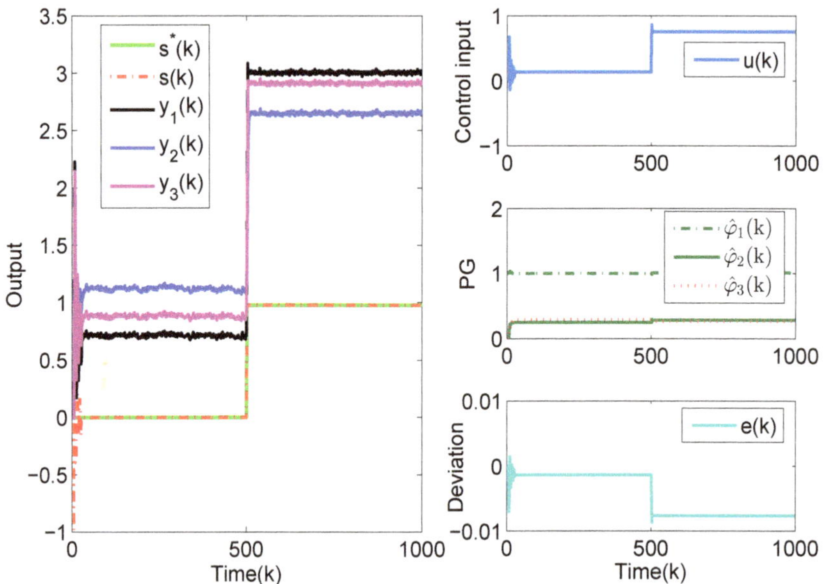

Figure 7. The curves of PG estimation values, control input, deviation, and outputs with $l = 3$.

6. Conclusions

A novel P-PFDL-MFAC scheme is proposed by combining the pattern-moving-based system dynamics description with the traditional PFDL-MFAC approach for a class of unknown practical SIMO nonaffine nonlinear discrete-time systems. Obviously, this scheme can also be applied to nonlinear or linear time-varying SISO systems, as long as the purpose of system control is to make all outputs belong to one or some pattern classes. Due to the existence of classification-metric deviation, an improved cost function for a deviation estimation algorithm and an adaptive tracking control law is designed based on the saddle point theory of TP-ZSG. The bounded convergence of the closed-loop system's tracking error has been proven and the effectiveness of the P-PFDL-MFAC scheme has been validated via two simulation examples.

Although it can be seen from the simulation results that the control strategy proposed in this work has a good effect on the output disturbance, the robustness of data-driven control should also include the ability to deal with data dropout, which may be caused by sensor fault, transmission network failure, or actuator damage. Therefore, the next topic that needs to be focused on is the robustness of pattern-moving-based model free adaptive control in the case of missing data.

Author Contributions: Conceptualization, X.L. and Z.X.; methodology, Z.X.; software, X.L.; validation, X.L.; formal analysis, X.L.; investigation, X.L.; resources, X.L.; data curation, Z.X.; writing—original draft preparation, X.L.; writing—review and editing, X.L.; visualization, X.L.; supervision, Z.X.; project administration, X.L.; funding acquisition, X.L. All authors have read and agreed to the published version of the manuscript.

Funding: This study was supported by the National Natural Science Foundation of China (62076025).

Institutional Review Board Statement: Not applicable.

Informed Consent Statement: Not applicable.

Data Availability Statement: Not applicable.

Conflicts of Interest: The authors declare no conflict of interest.

References

1. Yin, S.; Gao, H.; Kaynak, O. Data-driven control and process monitoring for industrial applications-Part I. *IEEE Trans. Ind. Electron.* **2014**, *61*, 6356–6359. [CrossRef]
2. Qu, S.D. Pattern recognition approach to intelligent automation for complex industrial processes. *J. Univ. Sci. Technol. Beijing* **1998**, *20*, 385–389.
3. Saridis, G.N. Application of pattern recognition method to control systems. *IEEE Trans. Autom. Control* **1981**, *26*, 638–645. [CrossRef]
4. Zhu, Q.; Onori, S.; Prucka, R. Pattern recognition technique based active set QP strategy applied to MPC for a driving cycle test. In Proceedings of the 2015 American Control Conference, Chicago, IL, USA, 1–3 July 2015.
5. Yu, B.; Zhang, X.; Wu, L.; Chen, X. A novel postprocessing method for robust myoelectric pattern-recognition control through movement pattern transition detection. *IEEE Trans. Hum.-Mach. Syst.* **2020**, *50*, 32–41. [CrossRef]
6. Xu, Z. Pattern Recognition Method of Intelligent Automation and Its Implementation in Engineering. Doctoral Dissertation, University of Science and Technology Beijing, Beijing, China, 2001.
7. Wang, M.; Xu, Z.; Guo, L. Stability and stabilization for a class of complex production processes via LMIs. *Optim. Control Appl. Methods* **2019**, *40*, 460–478. [CrossRef]
8. Sun, C.; Xu, Z. Multi-dimensional moving pattern prediction based on multi-dimensional interval T-S fuzzy model. *Control Decis.* **2016**, *31*, 1569–1576.
9. Guo, L.; Xu, Z.; Wang, Y. Dynamic modeling and optimal control for complex systems with statistical trajectory. *Discret. Dyn. Nat. Soc.* **2015**, *2015*, 1–8. [CrossRef]
10. Tayebi-Haghighi, S.; Piltan, F.; Kim, J.M. Robust composite high-order super-twisting sliding mode control of robot manipulators. *Robotics* **2018**, *7*, 13. [CrossRef]
11. Mobayen, S.; Tchier, F. A novel robust adaptive second-order sliding mode tracking control technique for uncertain dynamical systems with matched and unmatched disturbances. *Int. J. Control Autom. Syst.* **2017**, *15*, 1097–1106. [CrossRef]
12. Yeh, Y.L. A Robust Noise-Free Linear Control Design for Robot Manipulator with Uncertain System Parameters. *Actuators* **2021**, *10*, 121. [CrossRef]
13. Xi, X.; Mobayen, S.; Ren, H.; Jafari, S. Robust finitetime synchronization of a class of chaotic systems via adaptive global sliding mode control. *J. Vib. Control* **2018**, *124*, 3842–3854. [CrossRef]
14. Skruch, P. A terminal sliding mode control of disturbed nonlinear second-order dynamical systems. *J. Comput. Nonlinear Dyn.* **2016**, *11*, 054501. [CrossRef]
15. Xu, Z.; Wu, J.; Guo, L. Modeling and optimal control based on moving pattern. In Proceedings of the 32nd Chinese Control Conference, Xi'an, China, 26–28 July 2013.
16. Xu, Z.; Wu, J. Data-driven pattern moving and generalized predictive control. In Proceedings of the 2012 IEEE International Conference on Systems, Man and Cybernetics, Seoul, Korea, 14–17 October 2012.
17. Hou, Z.S.; Xiong, S.S. On model-free adaptive control and its stability analysis. *IEEE Trans. Autom. Control* **2019**, *11*, 4555–4569. [CrossRef]
18. Gao, W.; Jiang, Y.; Jiang, Z.; Chai, T. Output-feedback adaptive optimal control of interconnected systems based on robust adaptive dynamic programming. *Automatica* **2016**, *72*, 37–45. [CrossRef]
19. Ahn, H.; Chen, Y.; Moore, K. Iterative learning control: Brief survey and categorization. *IEEE Trans. Syst. Man. Cybern. Part C Appl. Rev.* **2007**, *37*, 1099–1121. [CrossRef]
20. Hjalmarsson, H.; Gevers, M.; Gunnarsson, S.; Lequin, O. Iterative feedback tuning: Theory and applications. *IEEE Control Syst. Mag.* **1998**, *18*, 26–41.
21. Campi, M.C.; Lecchini, A.; Savaresi, S.M. Virtual reference feedback tuning: A direct method for the design of feedback controllers. *Automatica* **2002**, *38*, 1337–1346. [CrossRef]
22. Hou, Z.S.; Jin, S.T. A novel data-driven control approach for a class of discrete-time nonlinear systems. *IEEE Trans. Control Syst. Technol.* **2011**, *19*, 1549–1558. [CrossRef]
23. Hou, Z.S.; Jin, S.T. *Model Free Adaptive Control: Theory and Applications*; CRC Press: Boca Raton, FL, USA, 2013.
24. Hou, Z.S.; Liu, S.D.; Tian, T.T. Lazy-learning-based data-driven model-free adaptive predictive control for a class of discrete-time nonlinear systems. *IEEE Trans. Neural Netw. Learn. Syst.* **2017**, *28*, 1914–1928. [CrossRef] [PubMed]
25. Wang, Z.S.; Liu, L.; Zhang, H.G. Neural network-based model-free adaptive fault-tolerant control for discrete-time nonlinear systems with sensor fault. *IEEE Trans. Syst. Man Cybernet Syst.* **2017**, *47*, 2351–2362. [CrossRef]
26. Li, H.T.; Ning, X.; Li, W.Z. Implementation of a MFAC based position sensorless drive for high speed BLDC motors with nonideal back EMF. *ISA Trans.* **2017**, *67*, 348–355. [CrossRef] [PubMed]
27. Bu, X.H.; Hou, Z.S.; Zhang, H.W. Data driven multiagent systems consensus tracking using model free adaptive control. *IEEE Trans. Neural Netw. Learn. Syst.* **2018**, *29*, 1514–1524. [CrossRef] [PubMed]
28. Treesatayapun, C. Varying-sliding condition adaptive controller for a class of unknown discrete-time systems with data-driven model. *Int. J. Model. Identif. Control* **2017**, *27*, 210–218. [CrossRef]
29. Zhu, Y.M.; Hou, Z.S.; Qian, F.; Du, W.L. Dual RBFNNs based model free adaptive control with aspen HYSYS simulation. *IEEE Trans. Neural Netw. Learn. Syst.* **2017**, *28*, 759–765. [CrossRef]

30. Li, X.Q.; Xu, Z.G.; Lu, Y.L.; Cui, J.R.; Zhang, L.X. Modified Model Free Adaptive Control for a Class of Nonlinear Systems with Multi-threshold Quantized Observations. *Int. J. Control. Autom. Syst.* **2021**. [CrossRef]
31. Sun, J.L.; Liu, C.S. Distributed zero-sum differential game for multi-agent systems in strict-feedback form with input saturation and output constraint. *Neural Netw.* **2018**, *106*, 8–19. [CrossRef]
32. Song, R.Z.; Zhu, L. Stable value iteration for twoplayer zero-sum game of discrete-time nonlinear systems based on adaptive dynamic programming. *Neurocomputing* **2019**, *340*, 180–195. [CrossRef]
33. Yang, J.; Zhang, D.; Frangi, A.F.; Yang, J. Two- dimensional PCA: A new approach to appearance-based face representation and recognition. *IEEE Trans. Pattern Anal. Mach. Intell.* **2008**, *26*, 131–137. [CrossRef]
34. Guo, G.J.; Chen, T.W. A new approach to quantized feedback control systems. *Automatica* **2008**, *44*, 534–542. [CrossRef]
35. Bu, X.H.; Qiao, Y.X.; Hou, Z.S.; Yang, J.Q. Model free adaptive control for a class of nonlinear systems using quantized informtion. *Asian J. Control* **2018**, *20*, 962–968. [CrossRef]
36. Bu, X.H.; Zhu, P.P.; Yu, Q.X.; Hou, Z.S.; Liang, J.Q. Model-free adaptive control for a class of nonlinear systems with uniform quantizer. *Int. J. Robust Nonlinear Control* **2020**, *30*, 6383–6398. [CrossRef]
37. Li, X.L.; Wang, S.N. Application of multimodel adaptive control algorithm in robotic manipulator control. *Robot* **2002**, *24*, 16–19.
38. Koivo, A.; Guo, T. Adaptive linear controller for robotic manipulators. *IEEE Trans. Autom. Control* **1983**, *28*, 162–171. [CrossRef]

Article

A New Method for Identifying Kinetic Parameters of Industrial Robots

Bin Kou [1,2], **Shijie Guo** [1,2,*] **and Dongcheng Ren** [1,2]

[1] Academy for Enigineering and Technology, Fudan University, Shanghai 200433, China; 18110860041@fudan.edu.cn (B.K.); 18110860025@fudan.edu.cn (D.R.)

[2] Guanghua Lingang Engineering Application and Technology R & D (Shanghai) Co., Ltd., Shanghai 201306, China

[*] Correspondence: guoshijie@fudan.edu.cn

Abstract: Identifying the kinetic parameters of an industrial robot is the basis for designing a controller for it. To solve the problems of the poor accuracy and easy premature convergence of common bionic algorithms for identifying the dynamic parameters of such robots, this study proposed simulated annealing with similar exponential changes based on the beetle swarm optimization (SEDSABSO) algorithm. Expressions for the dynamics of the industrial robot were first obtained through the SymPyBotics toolkit in Python, and the required trajectories of excitation were then designed to identify its dynamic parameters. Following this, the search pattern of the global optimal solution for the beetle swarm optimization algorithm was improved in the context of solving for these parameters. The global convergence of the algorithm was improved by improving the iterative form of the number N of skinks in it by considering random perturbations and the simulated annealing algorithm, whereas its accuracy of convergence was improved through the class exponential change model. The improved beetle swarm optimization algorithm was used to identify the kinetic parameters of the Zhichang Kawasaki RS010N industrial robot. The results of experiments showed that the proposed algorithm was fast and highly accurate in identifying the kinetic parameters of the industrial robot.

Keywords: industrial robot; kinetic parameter identification; beetle swarm optimization algorithm; stochastic perturbation

Citation: Kou, B.; Guo, S.; Ren, D. A New Method for Identifying Kinetic Parameters of Industrial Robots. *Actuators* **2022**, *11*, 2. https://doi.org/10.3390/act11010002

Academic Editors: Marco Carricato and Edoardo Idà

Received: 29 November 2021
Accepted: 21 December 2021
Published: 23 December 2021

Publisher's Note: MDPI stays neutral with regard to jurisdictional claims in published maps and institutional affiliations.

1. Introduction

Kinetic parameters are the main factor influencing the control of fast and highly precise movements of industrial robots [1]. The process of identifying of their kinetic parameters is usually divided into a number of steps, such as kinetic modeling, designing the excitation trajectory, data acquisition, and identifying and verifying the kinetic parameters [2]. Gautier et al. used a two degree-of-freedom (DOF) robot as the object of study and applied the extended Kalman filter and the least-squares method to identify its parameters. Memar used the SCHUNK Powerball LWA 4P as an experimental object, constructed a dynamics model for it, and implemented the least-squares method to identify the dynamic parameters of the industrial robot [3]. However, the least-squares method is susceptible to measurement noise that lowers its accuracy [4]. Fu et al. [5] used the particle swarm optimization (PSO) algorithm with least squares to identify the kinetic parameters of a seven-DOF collaborative robot in Xinsong, but PSO can easily fall into the local optimum owing to poor population diversity in the late stage of processing that reduces the accuracy of identification of the parameters. Ding et al. [6] identified the dynamics of the robot by using the genetic algorithm, but the process of coding of the algorithm is cumbersome.

Summarizing the existing research, it is found that the identification of robot dynamics parameters is a high-dimensional function problem. Common algorithms are either the PSO accuracy is not high enough, or the genetic algorithm coding is more complicated, so a simple programming, strong anti-interference ability, and convergence accuracy are needed. High algorithm. BSO is a bionic algorithm recently proposed, which mainly uses the principles of PSO and the Beetle Antenna Search (BAS) algorithm, so it has the advantage of simple programming [7–9]. This article proposes SEDSABSO on the basis of BSO.SEDSABSO combined random perturbation-based behavior and the simulated annealing (SA) algorithm with the global optimal solution of the BSO [10,11] to improve its ability to search for global optimal particles as well as the manner of changes in N particles through an exponential decay model. This improved the convergence of BSO without affecting its computational complexity. Therefore, using the SEDSABSO algorithm for robot dynamic parameter identification will have the advantages of simple programming, high convergence accuracy, and fast iteration speed.

The rest of the paper is organized as follows. Section 2 describes the D-H and the dynamic parameters of the RS010N robot and then linearized them through the SymPyBotics toolkit in Python. In Section 3, the genetic algorithm toolbox in MATLAB was used to design the excitation trajectories required to identify the dynamic parameters of the robot. Section 4 describes the principle of the SEDSABSO algorithm and its process for the identification of robot dynamics parameters. In Section 5, the minimum set of parameters for the dynamics of the RS010N robot was identified and was used to compare the per-formance of the SEDSABSO, BSO, and LDWPSO algorithms. The results verified the accuracy and effectiveness of the proposed algorithm [12].

2. Robotic arm Dynamics Model

2.1. RS010N Industrial Robots

In this paper, we used the Zhichang Kawasaki RS010N robot as the research object. Figure 1 shows its structural configuration, and Figure 2 shows the configuration of its DH coordinates. The RS010N robot is a typical six-DOF industrial robot. Because the kinetic parameters of its three rear joints are much smaller than those of its three front joints, the former have a smaller influence on the accuracy of control of the robot's motion. Thus, we considered only the first three joints of the RS010N industrial robot here [13].

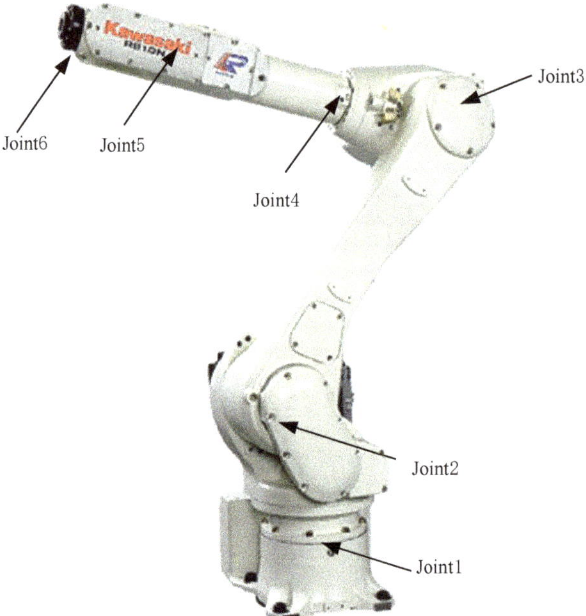

Figure 1. RS010N robot.

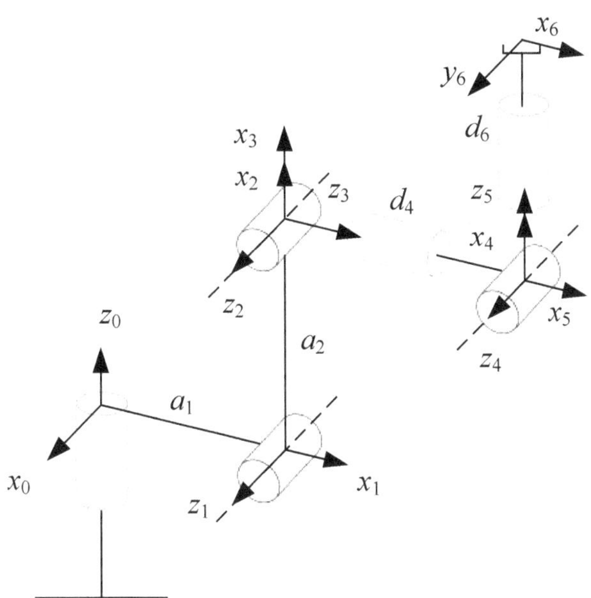

Figure 2. RS010N Robot DH coordinate system.

2.2. Dynamics Modeling

For an *n*-link robotic arm, the dynamics as modeled by the Newton-Euler method can be described as

$$\tau = D(q) \cdot \ddot{q} + C(q, \dot{q}) + G(q) \tag{1}$$

In Equation (1), τ denotes the driving moment, q denotes the vector of the position of the joints, and $\dot{q}$ and $\ddot{q}$ denote vectors of the velocity and acceleration of the joints, respectively. $D(q)$ denotes the inertial matrix, $C(q, \dot{q})$ denotes the Koch and centrifugal force terms, respectively, and $G(q)$ denotes the gravitational force term [14].

According to the improved Newtonian–Eulerian dynamics, the above equation can be transformed into

$$\tau = \Phi(q, \dot{q}, \ddot{q})P \tag{2}$$

where Φ is the observation matrix with a size of $n * 12n$, and P denotes the vector of the inertial parameters of the robot arm.

2.3. Minimum Set of Parameters for the Dynamical Model

The parameters of the first three joint linkages of the RS010N six-freedom robot are shown in Table 1 below. $a_0 = 350$ mm, $a_1 = 1035$ mm.

Table 1. RS010N connecting rod parameters.

i	α_{i-1}/rad	a_{i-1}/mm	d_i/mm	θ_i/rad
1	pi/2	a_0	0	θ_1
2	−pi/2	a_1	0	θ_2
3	0	0	0	θ_3

The kinetic parameters are shown in Table 2.

Table 2. Actual kinetic parameters.

Member Number	1	2	3
$I_{xx}/(\text{kg}\cdot\text{m}^2)$	36.33	-10.02	7.74
$I_{yy}/(\text{kg}\cdot\text{m}^2)$	40.61	7.05	1.02
$I_{zz}/(\text{kg}\cdot\text{m}^2)$	36.33	11.61	9.81
$I_{xy}/(\text{kg}\cdot\text{m}^2)$	0.00	6.14	3.34
$I_{xz}/(\text{kg}\cdot\text{m}^2)$	0.00	5.93	-3.42
$I_{yz}/(\text{kg}\cdot\text{m}^2)$	0.00	-7.82	-0.29
$m/(\text{kg})$	18.33	25.18	20.47
$x/(\text{mm})$	0.00	122.41	152.34
$y/(\text{mm})$	110.11	163.81	90.24
$z/(\text{mm})$	0.00	73.36	-34.21
F_c (N·m)	0.00	0.00	0.00
F_v (N·m)	0.00	0.00	0.00

We obtained the minimum set of parameters and related expressions for them through the SymPyBotics toolkit in Python.

$$
\begin{bmatrix}
I_{1xx} \\ I_{1yy} \\ I_{1zz} \\ I_{1xy} \\ I_{1xz} \\ I_{1yz} \\ m_1 \\ x_1 \\ y_1 \\ z_1 \\ F_{1c} \\ F_{1v} \\ I_{2xx} \\ I_{2yy} \\ I_{2zz} \\ I_{2xy} \\ I_{2xz} \\ I_{2yz} \\ m_2 \\ x_2 \\ y_2 \\ z_2 \\ F_{2c} \\ F_{2v} \\ I_{3xx} \\ I_{3yy} \\ I_{3zz} \\ I_{3xy} \\ I_{3xz} \\ I_{3yz} \\ m_3 \\ x_3 \\ y_3 \\ z_3 \\ F_{3c} \\ F_{3v}
\end{bmatrix}
\xrightarrow{\text{Linearization}}
\begin{bmatrix}
I_{1zz} + I_{2zz} + I_{3zz} + 2*a1*I_{1x} + m_1*a_1^2 - (m_2 + m_3)*(a_2{}^2 - a_1{}^2) \\
F_{1c} \\
F_{1v} \\
I_{2xx} - I_{2yy} - (m_2 + m_3)*a_2^2 \\
I_{2xy} \\
I_{2x} - a^2*(I_{2z} - I_{3y}) \\
I_{2yz} \\
I_{2zz} - (m_2 + m_3)*a_2^2 \\
I_{2x} + (m_2 + m_3)*a_2 \\
I_{2y} \\
F_{2c} \\
F_{2v} \\
I_{3xx} - I_{3zz} \\
I_{3xy} \\
I_{3xz} \\
I_{3yy} \\
I_{3yz} \\
I_{3x} \\
I_{3z} \\
F_{3c} \\
F_{3v}
\end{bmatrix}
\tag{3}
$$

According to Equation (3) above, the 36 kinetic parameters of the robot were linearized to 21, and the kinetic model can be written in the following form

$$\tau = \Phi_b(q, \dot{q}, \ddot{q}) p_b \tag{4}$$

where Φ_b is the full-rank observation matrix and P_b denotes the underlying parameter vector of the robotic arm.

3. Incentive Track Design

The commonly used excitation trajectory when identifying the kinetic parameters of industrial robots is the finite-term Fourier series [15,16]. It can be expressed in the following form:

$$\ddot{\theta}_i(t) = -\sum_{n=1}^{N} a_n n w_f \sin(n w_f t) - b_n n w_f \cos(n w_f t) \tag{5}$$

$$\dot{\theta}_i(t) = \sum_{n=1}^{N} a_n \cos(n w_f t) - b_n \sin(n w_f t) \tag{6}$$

$$\theta_i(t) = \theta_0 + \sum_{n=1}^{N} \frac{a_n}{n w_f} \sin(n w_f t) - \frac{b_n}{n w_f} \cos(n w_f t) \tag{7}$$

w_f denotes the fundamental frequency of the Fourier series. Each Fourier series contains a_n, b_n, and θ_0. N represents the number of harmonic terms of the Fourier series. Because each Fourier series has $2N + 1$ parameters, n represents the number of $1 \ldots 2N + 1$. The constraints on the RS010N robot are shown in Table 3.

Table 3. RS010N robot joint constraints.

Parameter	Joint i	Min	Max
	1	−180	180
Angle	2	−60	140
	3	−180	80
	1	−125	125
Angle velocity	2	−100	100
	3	−165	165
	1	−45	45
Angle acceleration	2	−40	40
	3	−75	75

When the robot moved, the observation matrix w was obtained by recording the angle, velocity, and acceleration of the joints of the robot in Φ_b. n denotes the nth sample.

$$w = \begin{bmatrix} \Phi_b(q_1, \dot{q}_1, \ddot{q}_1) \\ \ldots \\ \Phi_b(q_n, \dot{q}_n, \ddot{q}_n) \end{bmatrix} \tag{8}$$

Adaptation function of the incentive trajectory.

$$y = Cond(w) + P \tag{9}$$

where P denotes a penalty function and $Cond$ denotes the number of conditions of the acquisition matrix. When the trajectory of the industrial robot satisfies the above constraint, $P = 0$, and $P = 10^8$ otherwise. The excitation of the first three joints of the RS010N was obtained by the genetic algorithm toolbox in MATLAB as shown in Figure 3.

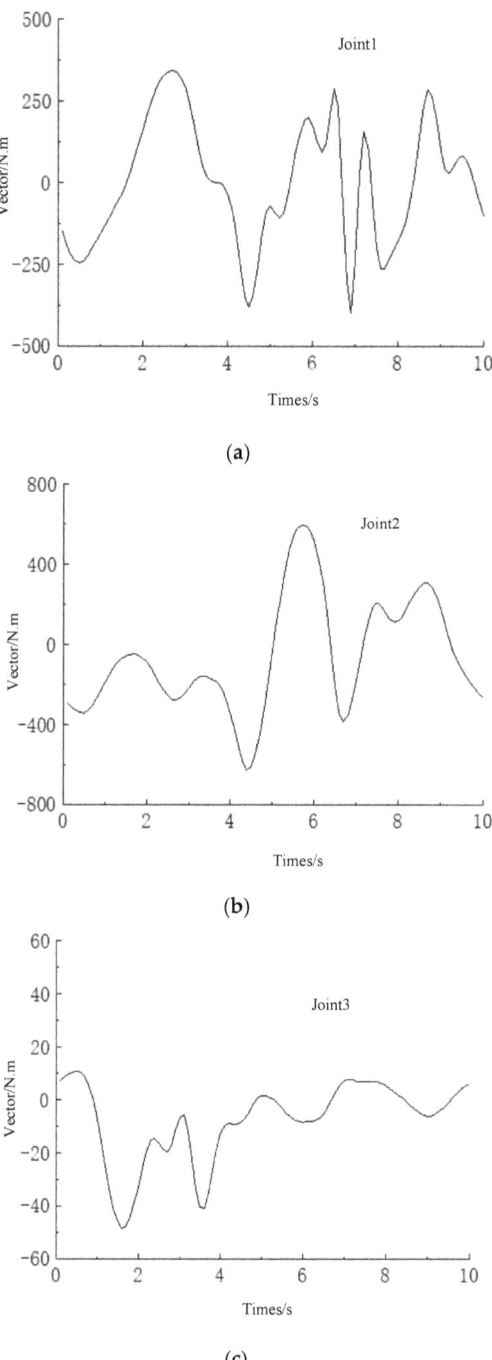

Figure 3. The excitation trajectory of the first three joints of the RS010N robot. (**a**) First joint motivation. (**b**) Second joint motivation. (**c**) Third joint motivation.

4. SEDSABSO for the Identification of Industrial Robot Dynamics Parameters

4.1. Improving Global Optimal Search

Owing to the large number of dynamic parameters of the industrial robot that needed to be identified, their solution was a high-dimensional function optimization problem with certain constraints. The BSO used to obtain the kinetic parameter identification of industrial robots easily falls into the premature phenomenon, so the global optimal solution search mode of BSO and the iterative change form of the number N of tennies in the iterative process of BSO are improved in this paper.

BSO

BSO was proposed by Wang et al. based on the PSO and the Aspen whisker algorithm. The position of a body in BSO represents a feasible solution to the optimization problem. The formula to update BSO is:

$$x_{ij}(t+1) = x_{ij}(t) + \lambda v_{ij}(t+1) + (1-\lambda)\xi_{ij}(t) \tag{10}$$

where t denotes the number of iterations of the skink, i denotes the number of skinks in the skink population, j denotes the dimensionality of the problem to be solved, λ denotes $x_{ij}(t + 1)$ denotes the position of skink i in the skink population at $t + 1$ iterations, $\xi_{ij}(t)$ denotes the position of the ith skink moving autonomously, and $v_{ij}(t + 1)$ denotes the velocity of the ith skink. It is expressed as:

$$v_{ij}(t+1) = w v_{ij}(t) + c_1 r_1 \left(p_{ij}(t) - x_{ij}(t)\right) + c_2 r_2 \left(g_{ij}(t) - x_{ij}(t)\right) \tag{11}$$

w denotes inertial weight, $p_{ij}(t)$ denotes the optimal position of individual i after t iterations of the Aspen population, $x_{ij}(t)$ denotes the given position of i after t iterations, $g_{ij}(t)$ denotes the optimal position of the Aspen population after t iterations, c_1 and c_2 denote adjustment factors, and r_1 and r_2 denote random numbers within the interval [0,1]

$$\xi_{ij}(t+1) = \delta(t) * V_{ij}(t) * sign\left(f\left(x_{lj}(t)\right) - f\left(x_{rj}(t)\right)\right) \tag{1}$$

In Equation (12), $sign(.)$ denotes the sign function, $\xi_{ij}(t + 1)$ denotes the $t + 1$th motion position searched by Aspen autonomously, $\delta(t)$ denotes the step size of the Aspen individual after t iterations, and $f(x_{lj}(t))$ and $f(x_{rj}(t))$ denote the fitness values of the left and right whiskers of the Aspen, respectively.

The positions of the left and right whiskers of the Aspen are indicated by:

$$\begin{aligned} x_{lj}(t + 1) &= x_{lj}(t) + V_{ij}(t)d_0/2 \\ x_{rj}(t + 1) &= x_{rj}(t) - V_{ij}(t)d_0/2 \end{aligned} \tag{13}$$

$x_{lj}(t + 1)$ denotes the position of the left whisker in iteration $t + 1$, $x_{rj}(t + 1)$ is the $t + 1$ right whisker position, and d_0 denotes the distance between the whiskers.

4.2. Improving BSO

4.2.1. Improved Global Optimal Solution

The global optimal particle P_g of BSO is randomly perturbed. Then, the new solution after perturbation is selected by the metropolis criterion of SA to improve the global convergence of BSO.

(1)   Simulated annealing algorithm

The simulated annealing algorithm controls the probability of jumping out of the local optimum by setting the temperature. The algorithm evaluates whether to jump out of the worse solution by the metropolis criterion. The procedure is as follows:

a.   The algorithm is initialized, and an initial solution x is randomly generated as the optimal solution.

b. A new solution x_t is obtained in the vicinity of the initial solution, denoted by $\Delta f = f(x_t) - f(x)$.

c. The new solution x_t is accepted according to min$\{1,$exp$(-\Delta f/T_k)\}$>random. T_k denotes the temperature and *exp* denotes the exponential function, with natural number e as the base.

(2) Random behavior

Stochastic behavior is a search behavior that can improve the diversity of algorithmic populations and is widely used in a variety of intelligent algorithms, such as in the foraging behavior within the fish swarm algorithm to search for food and the wandering behavior of the wolf pack algorithm to sense the scent of prey in the air in a random pattern [17,18]. Inspired by this, we incorporated stochastic behavior into BSO, and the optimal particles in it were randomly perturbed during each iteration to enhance its ability to escape from the local optimal solution to improve global convergence:

$$P_g{}^{new} = P_g * (1 + rand) \tag{14}$$

where $P_g{}^{new}$ denotes the new solution after the random perturbation of the optimal particle of the Aspen swarm, and *rand* represents a random number from zero to one. The fitness value of the optimal particle p_g of BSO was compared with that of the global optimal particle $P_g{}^{new}$ for random perturbation; then, the new solution was accepted according to the metropolis criterion of SA. BSO performed q random perturbations of its global optimal solution in each iteration.

4.2.2. Improved Iterative Approach to N Particles of the Aspen Swarm

Each skyline of BSO represents a potential solution to the problem. Assuming that the number of skylines in each iteration of BSO is N, the maximum number of its iterations is *Maxdt*, and the duration of the iteration of a single skyline is t, the total complexity of iterations of BSO is $N*Maxdt*t$ [19]. In solving a high-dimensional function problem, similar to determining the dynamic parameters of the industrial robot, the number of particles N at the moment of each iteration of BSO is constant. BSO is based on PSO, and a larger number of particles in PSO yields a higher accuracy of the convergence, but the time needed for convergence also increases.

For example, in previous work, we proposed a two-stage dynamic PSO that changed the number of particles in the swarm by linearly reducing the number of iterations. The experimental results showed that the accuracy of convergence of the algorithm was similar to that of the classical PSO algorithm, which also reflects a side-effect of the latter. The algorithm proposed here also had this property of a high initial accuracy of convergence [20]. Inspired by this, the number N of skinks in the skink swarm was changed dynamically. The number was larger in the early stage of the skink swarm algorithm and smaller in the later stage, such that the total number of skinks did not differ by much and the gap in the durations of their iterations was not large. We fully exploited the high efficiency of the early iterations of the skink swarm algorithm for the accuracy of its solution. The exponentially decreasing rate of curtailment was used for the number of cows **N** of the Aspen swarm algorithm by drawing on the idea proposed by Wang et al. [21]. That is, the particle swarm algorithm used a larger number of particles to search in the initial stage. With the increase of the number of iterations, the exponentially decreasing number of particle was used to reduce the Aspen number, so as to improve the search efficiency, and the improved iterative formula for the Aspen number is as follows:

$$N = round(e^{\alpha} * N_{start}) \tag{15}$$

$$\alpha = t\frac{Inf_{max} - Inf_{min}}{Maxdt} - Inf_{max} \tag{16}$$

where N_{start} denotes the initial number of skinks of the improved skink herd, i denotes the number of iterations of the skink herd algorithm, *Maxdt* denotes the maximum number of iterations, and f_{max} and f_{min} are the respective maximum and minimum values of the search factor set used to control the number of skinks N.

4.3. Robot Dynamic Parameter Identification Based on SEDSABSO

The SEDSABSO process is illustrated below:

From Figure 4, the flow of SEDSABSO for industrial robot dynamics parameter identification is:

(1) Initialization of the algorithm.
(2) Obtain SEDSABSO individual and group best-fit values.
(3) Update the position, velocity, and number of skinks N
(4) Perform a random perturbation search for the global optimal solution and then accept the searched solution with SA's Metropolis criterion and cycle through q searches.
(5) Compare with the global optimal solution obtained in step 3 after passing q times of search and proceed to the next step of the search by merit.
(6) Determine whether the algorithm ends, and if the termination condition is not satisfied, return to step 3. If it is satisfied, the global optimal solution is output. In turn, the parameters related to industrial robot dynamics are obtained.

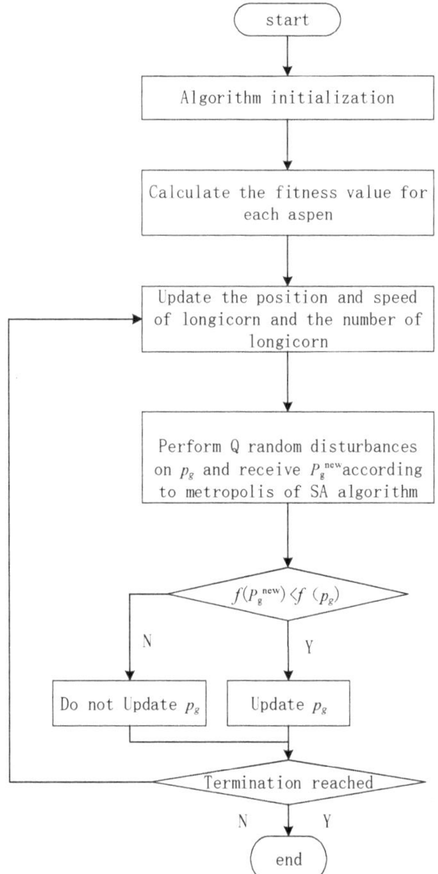

Figure 4. SEDSABSO algorithm flow chart.

5. Simulation Experiments and Results Analysis

5.1. Adaptation Function

The sum of the absolute values of the difference between the torque of joint recognition and the theoretical torque of the RS010N robot when sampled i times was used as the objective function, as in Equation (17)

$$\sum_{i=1}^{N} abs(f_k1(i) - f1(i)) + abs(f_k2(i) - f2(i)) + abs(f_k3(i) - f3(i)) \tag{17}$$

$i = 1, 2, \ldots, N$ denotes the *ith* sample of the robot, $f_k1(i), f_k2(i)$, and $f_k3(i)$ denote values of the first, second, and third moments of the joints values for the i robot recognition $f1(i)$, $f2(i), f3(i)$ denotes the theoretical value of the three joint moments sampled by the robot.

5.2. Analysis of Experimental Results

Experimental Parameter Setting

To improve the accuracy of identification of the kinetic parameters, the individual algorithms were initialized uniformly by using a small interval [22]. Then, the dynamic parameters of the robot were solved for by using the SEDSABSO algorithm, BSO algorithm, and linearly decreasing particle swarm algorithm (LDWPSO), where the number of particles N was set to 50 for the BSO and LDWPSO algorithms. The initial population N_{start} for SEDSABSO was set to 90, the learning factor was $c_1 = c_2 = 2.0$, and the inertial weight w was 0.8. The maximum number of iterations was set to 600. The initial temperature T for simulated annealing was set to 10,000, the decay scale of the annealing coefficient was set to 0.93, the step $\delta(t)$ was set to a constant value of 0.05, the adjustment factor λ was 0.9, and the ratio of the Aspen step to the distance between the whiskers, c, was set to two. After many trials, the optimal number of particle perturbations for the population was set to $q = 3, f_{max}$ was 0.9, and f_{min} was 0.3. The iteration curves of the Aspen number for SEDSABSO and BSO are shown below:

Figure 5 shows that the number of particles N was constant during iterations of the BSO algorithm. Assuming that the maximum number of iterations was $Maxdt$ and the duration of iteration of each particle was t, the total duration of iterations of BSO was $N*Maxdt*t$, and $N*Maxdt$ is equal to Area1+Area2 in the figure. For the SEDSABSO algorithm, the total iteration time was approximately the product of Figure 5. For the SEDSABSO algorithm, the total iteration time was approximately the product of the area of the class trapezoid enclosed by the dash and the horizontal axis and the time t, i.e., (Area3 + Area1) $* t$, and since the area of Area3 was not much different from that of Area2, the computational complexity of the improved Amanita group algorithm was approximately the same as that of the standard Amanita group algorithm.

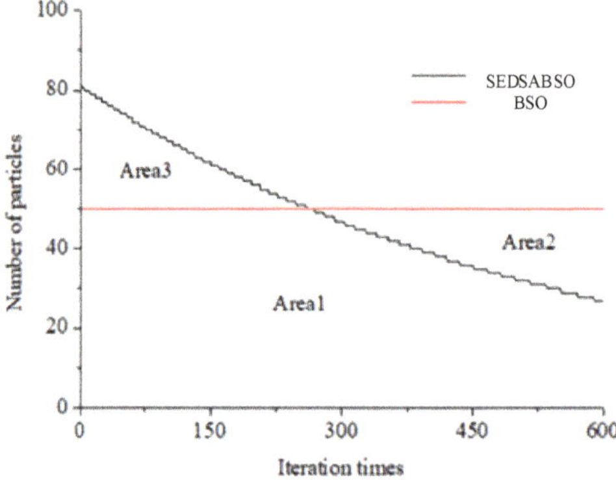

Figure 5. SEDSABSO and BSO Variation curve of number N of longicorn beetles.

Figure 6 below shows the adaptation iteration curves of each algorithm on an Intel(R) Core(TM) i7-8550U main frequency 4.00 GHz computer with 600 iterations through Matlab 9.1:

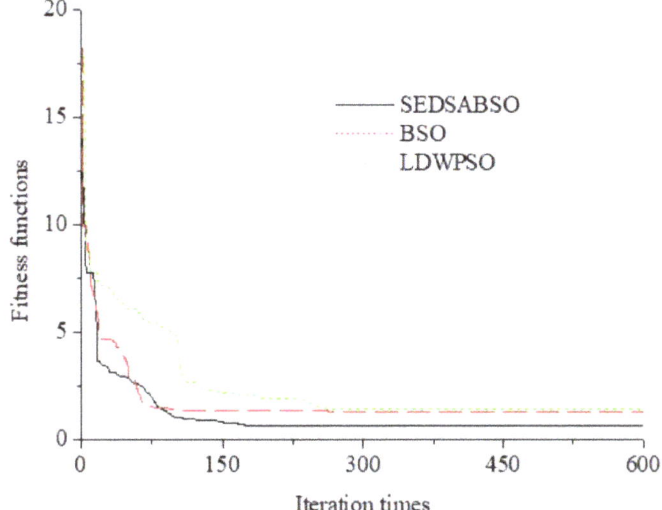

Figure 6. Iterative curve of the algorithm.

From Figure 6, we can see that the SEDSABSO algorithm proposed in this paper converges faster and has smaller fitness values than the BSO and LDWPSO algorithms. It is clear that the SEDSABSO algorithm proposed in this paper performs best.

In order to further verify the advanced level of the SEDSABSO algorithm, this paper ran each of the above three algorithms ten times and found the average fitness and average time as follows in Table 4:

Table 4. Comparison of average fitness values.

Algorithm	Average Fitness	Average Time/s
SDESABSO	0.95	16.93
BSO	1.12	17.31
LDWPSO	1.18	12.73

From Table 4, we can see that the average fitness values of the three algorithms after ten runs were only 0.95 for SEDSABSO, 1.12 for BSO, and 1.18 for LDWPSO, further confirming the stability and effectiveness of the proposed SEDSABSO algorithm. It can be seen that the average time of SEDSABSO was better than that of BSO, further proving the analysis in Figure 5. The difference with LDWPSO was not significant, and combined with the average fitness value, it can be seen that SEDSABSO performed optimally.

The linearized kinetic parameters of the RS010N industrial robot had a total of 21 minimum parameter sets, and the results of the kinetic parameters identified by SEDSABSO, BSO, and LDWPSO are shown in Table 5 below; the absolute errors between the identified kinetic parameters and the actual parameters were also obtained.

Table 5. Identification of dynamic parameters.

Dynamic Minimum Parameters	Theoretica Value	SEDSABSO Identification Value	BSO Identification Value	LDWPSO Identification Value	SEDSABSO Absolute Error	BSO Absolute Error	LDWPSO Absolute Error
$I_{1yy} + I_{2yy} + I_{3zz} + 2 * a_1 * I_{1x} + m_1 * a_1^2 - (m_2 + m_3) * a_2^2$	16.41	16.80	16.89	16.82	-0.39	-0.48	-0.41
F_{1c}	0	0.01	0.35	-0.20	-0.01	-0.35	0.2
F_{1v}	0	-0.07	-0.21	0.22	0.07	0.21	-0.22
$I_{2xx} - I_{2yy} + (m_2 + m_3) * a_2^2$	31.83	31.77	31.92	32.21	0.06	-0.09	-0.38
I_{2xy}	6.14	5.78	5.12	6.12	0.36	1.02	0.02
$I_{2x} - a_2 * (I_{2z} - I_{3y})$	5.76	5.54	6.22	5.83	0.22	-0.46	-0.07
I_{2yz}	-7.82	-7.84	-7.72	-8.10	0.02	-0.1	0.28
$I_{2zz} - (m_2 + m_3) * a_2^2$	-37.29	-37.11	-37.30	-37.94	-0.18	0.01	0.65
$I_{2x} + (m_2 + m_3) * a_2^2$	47.37	47.29	47.52	47.27	0.08	-0.15	0.1
I_{2y}	0.16	0.13	0.09	0.27	0.03	0.07	-0.11
F_{2c}	0	0.14	-0.02	-0.08	-0.14	0.02	0.08
F_{2v}	0	-0.21	-0.23	-0.03	0.21	0.23	0.03
$I_{3xx} - I_{3zz}$	-2.07	-2.20	-2.40	-1.51	0.13	0.33	-0.56
I_{3xy}	3.34	3.24	3.25	4.57	0.10	0.09	-1.23
I_{3xz}	-3.42	-3.34	-2.95	-4.10	-0.08	-0.47	0.68
I_{3yy}	1.02	0.77	1.12	-0.85	0.25	-0.1	1.87
I_{3yz}	-0.29	0.10	-0.34	-1.42	-0.39	0.05	1.13
I_{3x}	0.15	0.11	0.12	-0.04	0.04	0.03	0.19
I_{3z}	-0.034	0.00	-0.16	0.04	-0.034	0.126	-0.074
F_{3c}	0	0.18	-0.13	-1.20	-0.18	0.13	1.2
F_{3v}	0	0.06	-0.34	-0.1	-0.06	0.34	0.1

Taking the second parameter F_{1c} of the set of minimum parameters of the industrial robot in Table 5 as an example, the absolute errors identified by BSO and LDWPSO were -0.35 and 0.2, respectively, while the absolute error of SEDSABSO was -0.01. On the whole, the SEDSABSO algorithm yielded the smallest error in identifying the kinetic parameters of the industrial robot, while BSO and LDWPSO The kinetic parameters identified by the SEDSABSO algorithm deviated from the theoretical kinetic parameter values, which further reflects the superiority of the SEDSABSO algorithm. The maximum absolute error in the kinetic parameters as identified by the SEDSABSO algorithm was -0.39. The overall error in identification was thus small. The SEDSABSO algorithm can thus identify the kinetic parameters of the robot.

6. Conclusions

(1) In this paper, a new improved Beetle Antennae Search-QEDSABSO is proposed, which makes a class exponential change to the number of skinks in the iterative process of Beetle Antennae Search and effectively improved the utilization rate of Beetle Antennae Search skinks while the total number of skinks was basically unchanged. Simulation experiments showed that the proposed algorithm is more accurate and faster than the common particle swarm and Beetle Antennae Search in identifying the dynamics parameters of robots. The simulations showed that the proposed algorithm can identify the dynamical parameters of the robot with higher accuracy and faster speed than the common particle swarm and Beetle Antennae Search.

(2) The difficulty in identifying the kinetic parameters of industrial robots lies in the sheer number of variables that need to be determined and the selection of reasonable excitation trajectories. This paper designed the relevant excitation trajectories by using the genetic algorithm and linearized the kinetic parameters of the industrial robot to improve the accuracy of their recognition.

(3) The work provided the foundation for experiments compensating for the kinetic moments of the industrial robot. The minimum set of parameters of the kinetics could first be obtained by SymPyBotics. Then, the excitation trajectory of the industrial robot was designed by using the genetic algorithm, the data on its kinetic moments were collected, and the moments were identified by the SEDSABSO algorithm. Following this, the theoretical kinetic moments of the robot were calculated and compared with empirically sampled moments to obtain the error. Finally, this error was used to compensate for the kinetic moment of the robot.

Author Contributions: Methodology, B.K.; Software, B.K.; Validation, B.K.; Writing—original draft, B.K.; Writing—review & editing, B.K., S.G. article chart editor, B.K. and D.R. All authors have read and agreed to the published version of the manuscript.

Funding: This research was supported by the national key R & D plan project (2017YFB1301000).

Data Availability Statement: The data used in this study were self-tested and self-collected during the test. As the control method in this paper is still being further optimized, the data cannot be shared at present. Therefore, data sharing is not applicable to this article.

Conflicts of Interest: The authors declare no conflict of interest.

References

1. Liu, Z.; Zhao, B.; Zhu, H. Sorting Experimental Platform Research on Six-DOF Manipulator. *Mach. Des.Manuf.* **2013**, *27*, 210–213.
2. Wu, J.; Wang, J.; You, Z. An overview of dynamic parameter identification of robots. *Robot. Comput. Integr. Manuf.* **2010**, *26*, 414–419. [CrossRef]
3. Gautier, M.; Janot, A.; Vandanjon, P.O. A new closed-loop output error method for parameter identification of robot dynamics. *IEEE Trans. Control Syst. Technol.* **2013**, *21*, 428–444. [CrossRef]
4. Memar, A.H.; Esfahani, E.T. Modeling and Dynamic Parameter Identification of the SCHUNK Powerball Robotic Arm. In Proceedings of the ASME 2015 International Design Engineering Technical Conferences and Computers and Information in Engineering Conference, Boston, MA, USA, 2–5 August 2015; American Society of Mechanical Engineers: New York, NY, USA, 2015; p. VOSCT08A024.
5. Fu, X.; Yuan, J.; Wang, S.; Wang, N.; Zhang, W. Nonlinear Dynamic Identification of Robotic Manipulators Based on Particle Swarm Optimization Method. *Mech.-Electr. Integr.* **2017**, *023*, 3–8.
6. Ding, L.; Wu, H.; Yao, Y.; Li, Y.; Xie, B.H.; Chen, B. Parameters Identification of Industrial Robots Based on WLS-ABC Algorithm. *J. South China Unive. Technol. (Natl. Sci. Ed.)* **2016**, *44*, 90–95.
7. Jiang, X.; Li, S. BAS: Beetle antennae search algorithm for optimization problems. *arXiv* **2017**, arXiv:1710.10724. [CrossRef]
8. Chen, T.; Yin, H.; Jiang, H. Particle Swarm Optimization Algorithm Based on Beetle Antennae Search for Solving Portfolio Problem. *Comput. Syst. Appl.* **2019**, *28*, 171–176.
9. Wang, T.; Long, Y.; Qiang, L. Beetle Swarm Optimization Algorithm: Theory and Application. *arXiv* **2018**, arXiv:1808.00206. [CrossRef]
10. Zhou, T.; Qian, Q.; Fu, Y. Fusion Simulated Annealing and Adaptive Beetle Antennae Search Algorithm. *Commun. Technol.* **2019**, *52*, 1626–1631.
11. Khan, A.H.; Cao, X.; Li, S.; Katsikis, V.N.; Liao, L. BAS-ADAM:an ADAM based approach to improve the performance of beetle antennae search optimizer. *IEEE/CAA J. Autom. Sin.* **2020**, *7*, 461–471. [CrossRef]
12. Xu, C. *Research on Dynamic Parameter Identification And Feedforward Control of Articulated Robots*; Southeast University: Nanjing, China, 2017.
13. Zhang, J.; Duan, J. Robot Dynamic Parameter Identification Based on Improved Differential Evolution Algorithm. *J. Beijing Union Univ.* **2020**, *1*, 49–55.
14. Craig, J.J. *Introduction to Robotics Mechanics and Control*; Pearson Prentice Hall: Upper Saddle River, NJ, USA, 2005.
15. Sun, Y.; Zhou, B.; Meng, Z. Dynamic Parameter Identification of Industrial Robot Based on Genetic Algorithm. *Ind. Control Comput.* **2017**, *9*, 4–6.
16. Swevers, J.; Ganseman, C.; De Schutter, J.; Van Brussel, H. Experimental robot identification using optimised periodic trajectories. *Mech. Syst. Signal Process.* **1996**, *10*, 561–577. [CrossRef]
17. Jiang, M.; Yuan, D. Wavelet Threshold Optimization with Artificial Fish Swarm Algorithm. In Proceedings of the 2005 International Conference on Neural Networks, Beijing, China, 13–15 October 2005; IEEE Press: Piscataway, NJ, USA, 2005; pp. 569–572.
18. Liu, C.; Yan, X.; Liu, C.; Wu, H. The wolf colony algorithm and applications. *Chin. J. Electron.* **2011**, *20*, 212–216.
19. Wang, Q.; Li, L.; Lu, C.; Sun, F. Average Computational Time Complexity Optimized Dynamic Particle Swarm Optimization Algorithm. *Comput. Sci.* **2010**, *37*, 191.
20. Kou, B.; Guo, S.; Ren, D. Geometric parameter calibration of industrial robot based on improved particle swarm optimization. *J. Harbin Inst. Technol.* **2021**, *49*, 61–64.
21. Wang, Y.G.; Qu, T.T.; Li, S. Disruption particle swarm optimization algorithm based on exponential decay weight. *Appl. Res. Comput.* **2020**, *37*, 1020–1024.
22. Liu, Y.; Li, G.X.; Xia, D.; Xu, W.F. Identifying Dynamic parameters of a space robot based on improved genetic algorithm. *J. Harbin Inst. Technol.* **2010**, *42*, 1734–1739.

Article

Development of a Mechatronic System for the Mirror Therapy

Maurizio Ruggiu [†] and Pierluigi Rea [*,†]

Department of Mechanical, Chemical and Materials Engineering, University of Cagliari, Via Marengo, 2, 09123 Cagliari, Italy; maurizio.ruggiu@unica.it
* Correspondence: pierluigi.rea@unica.it
† These authors contributed equally to this work.

Abstract: This paper fits into the field of research concerning robotic systems for rehabilitation. Robotic systems are going to be increasingly used to assist fragile persons and to perform rehabilitation tasks for persons affected by motion injuries. Among the recovery therapies, the mirror therapy was shown to be effective for the functional recovery of an arm after stroke. In this paper we present a master/slave robotic device based on the mirror therapy paradigm for wrist rehabilitation. The device is designed to orient the affected wrist in real time according to the imposed motion of the healthy wrist. The paper shows the kinematic analysis of the system, the numerical simulations, an experimental mechatronic set-up, and a built 3D-printed prototype.

Keywords: parallel robots; mechatronics; motion simulation; mirror therapy

1. Introduction

Limb rehabilitation using robotic devices is an innovative form of rehabilitation based on interactions between the patient and the device. These systems augment the rehabilitation outcomes in neurologic disorders, such as stroke and multiple sclerosis. Robotic therapy provides high-intensity and repeated training and it can be used as an effective complement to standard rehabilitation from the beginning of a therapy [1,2]. The results show in all cases that patients who received the robotic therapy in addition to conventional therapy have greater reductions in motor impairment [3–6]. The robotic rehabilitation systems can be classified according to the design, depending on whether it operates as end-effector or as an exoskeleton. Another classification is between proximal and distal robots. Proximal robots are used to move the shoulder joint and the elbow joint; distal robots are rehabilitation robots for fine motor skills. They are used to train the hand and the fingers. Finally, rehabilitation robots can be classified as unilateral and bilateral. Unilateral devices use only the paralyzed limb for rehabilitation tasks, whereas bilateral robots use both limbs, the paralyzed and the healthy one [7].

There are numerous successful implementations of robotic rehabilitation devices. One of the first examples was the MIT-MANUS, which had a major impact on neurorehabilitation. The MIT-MANUS can move and guide the upper limb and record the trajectory, the velocity and the force of movement [8], indeed, this application can be classified as proximal and unilateral. A similar concept was described in [9]. An approach for robotic rehabilitation was used with MIME (Mirror Image Movement Enabler) [10], using an industrial robot to train the arm. A pneumatic system is the RUPERT (Robotic Upper Extremity Repetitive Therapy), which is an exoskeletal type. It consists of four pneumatic muscles, [11]. MACARM (Multi-Axis Cartesian-based Arm Rehabilitation Machine) is a planar cable-driven system designed for the rehabilitation of the upper limb [12]. The use of this kind of technology was further used in [13,14], although for correct operations, several issues for the design and modeling should be considered [15–17]. NeReBot and then MariBot are cable-driven systems with proximal, unilateral end-effector devices [18,19]. A cable-driven robot was proposed for mirror therapy in [20] and a real-

Citation: Ruggiu, M.; Rea, P. Development of a Mechatronic System for the Mirror Therapy. *Actuators* **2022**, *11*, 14. https://doi.org/10.3390/act11010014

Academic Editors: Marco Carricato and Edoardo Idà

Received: 1 December 2021
Accepted: 3 January 2022
Published: 5 January 2022

Publisher's Note: MDPI stays neutral with regard to jurisdictional claims in published maps and institutional affiliations.

time two-axis mirror robot system was developed for functional recovery of hemiplegic arms in [21].

In this paper, we present the design, analysis and experimental set-up of a system based on the mirror therapy concept suited to wrist rehabilitation. Mirror therapy is an effective occupational therapy for functional recovery of a hemiplegic arm after stroke [22–24]. It can facilitate brain neuroplasticity through activation of the sensorimotor cortex. The standard approach uses a simple mirror and the individual sits orthogonal to the mirror. The affected limb is positioned behind the mirror, so that it blocks the view and shows the non-paralyzed limb. Watching the mirror, the motion of the master limb is ideally projected to the paralyzed limb (the slave). The mirroring creates the illusion where it looks like the paralyzed limb would do the same movement as the non-paralyzed one. With this visual illusion, damaged nerve connections in the brain should be stimulated to make reconnections [25]. Moreover, mirror therapy has been used for chronic pain [26,27], and, as it was introduced by Vilayanur S. Ramachandran [28], as a therapy against phantom pain.

In our implementation, the system is designed to orient the affected wrist (slave) in real time according to the imposed motion of the healthy wrist (master).

The rest of the paper is organized as follows: In Section 2 a description of the system is provided, Section 3 presents the kinematic analysis of the mechanisms, Section 4 shows the numerical results obtained from the simulations whilst in Section 5 we describe the experimental set-up. Finally, in Section 6 the conclusions are drawn.

2. Description of the System

The system is basically composed of two units, which are called master and slave, the master is interacting with the full functional upper limb, while the slave is devoted to the affected upper limb. The two mechanisms are different because they must have different functions. The master is designed and built as a three-axis gimbal. Figure 1 illustrates a view of the master. In a typical gimbal, there are two or three motors on the system with the aim to prevent or eliminate vibration or locate an end-effector in space [29,30]. The basic aim of a gimbal is to minimize the vibration in video recording devices, and creating a reverse motion in the opposite direction of the vibration. The reverse motion can be provided by using an inertial measuring unit (IMU) sensor, which is placed on the camera and detects the camera movements and reports the motion to three servomotors positioned in line with the camera lens. The IMU detects the relative pose of the camera according to the ground, and based on the predetermined optimum position, the deviation between the two is evaluated. Then, an electronic board receives and processes data from the IMU and then transmits the information to the servomotors of the gimbal, which provides smooth motion. Thus, the servomotors that produce the opposite movement of the camera allows obtaining a smooth image. We have used the same concept to design a gimbal for tracking the orientation of the full-functional upper limb. Instead of using the IMU, we have chosen encoders. The motivation is that the orientation of the hand must be tracked and measured, instead of its angular velocity and acceleration. Nevertheless, in future applications, we may consider the use of an IMU.

The slave is a spherical parallel mechanism [31,32]. Figure 2 illustrates a view of the slave. It consists of a fixed base and an end-effector, hereafter referred to as the joystick, connected to each other by three identical limbs, each with an RRUR kinematic chain (R stands for revolving joint, U stands for universal joint formed by two concurrent R joints, P stands for prismatic joint and S stands for spherical joint) [33]. In each limb the first three R joints have parallel axes forming a planar chain whilst the last two R joints are perpendicular to each other, intersecting in the center of rotation of the joystick. Only the R pairs connected to the frame are actuated. The motors torques allow the joystick to fully rotate about the center of rotation. The mechanism is decoupled and not-overconstrained.

In literature, there are numerous applications of the parallel mechanisms for rehabilitation tasks. Some ankle rehabilitation devices proposed have the 3-RRR [34], 2-RRR/UPRR [35], 2-UPS/RRR [36], 3-RRS [37] geometries and to fit more closely the ankle motion Zhang et al. exploited a more complex parallel geometry [38]. There are, also, parallel geometries exploited for upper limb, wrist rehabilitation devices as those proposed in [39,40].

In summary, the choice of the three-axis gimbal geometry is motivated by its simplicity. In fact, the simple kinematics allows us to obtain the joystick orientation straight from the encoders measurements. Additionally, there are no strict requirements in terms of dynamics as the motion is imposed by the healthy hand. Conversely, the slave mechanism must reduce the inertial effects as the affected hand follows the driven motion and it has to be accurate and repeatable in posing the joystick. For these reasons, the parallel architecture appeared to be appropriate as it presents light moving links with the motors fixed at the base. Furthermore, the closed chain geometry may allow us to have high accuracy and·repeatability.

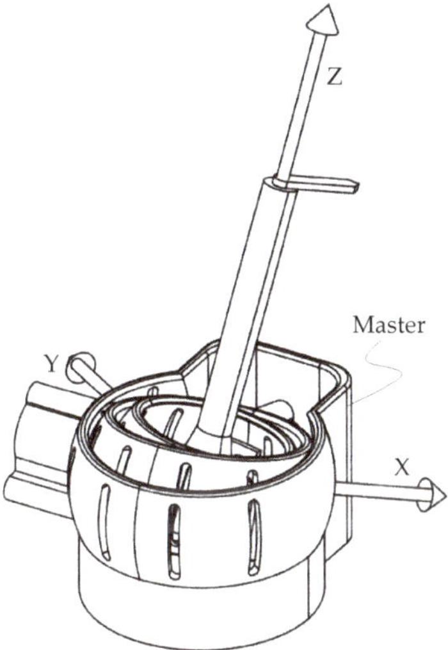

Figure 1. The master mechanism.

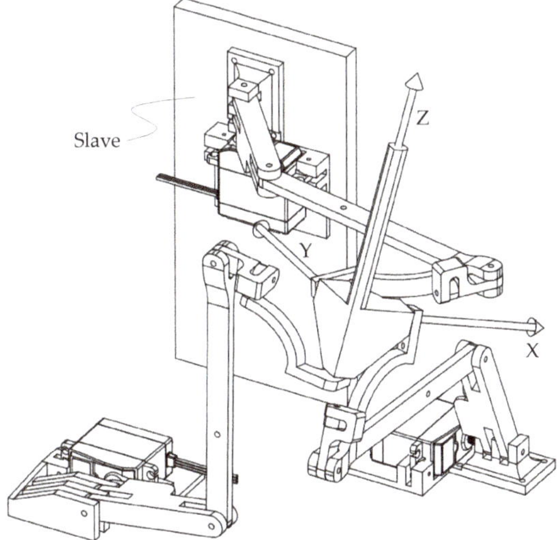

Figure 2. The slave mechanism.

3. Kinematics

The master mechanism is moved by the patient healthy arm with the rotations measured by encoders. A forward position kinematics allows us to obtain the orientation of the joystick. Because of the numerical efficiency, a quaternion parametrization was used such that the rotation of the joystick can be expressed as the compositional rotation of the intrinsic Tait–Bryan angles, ψ, θ, ϕ:

$$q = q_3 \otimes q_2 \otimes q_1, \tag{1}$$

with

$$q_1 = \cos\left(\frac{\psi}{2}\right) + \sin\left(\frac{\psi}{2}\right)\mathbf{k}, \quad yaw$$

$$q_2 = \cos\left(\frac{\theta}{2}\right) + \sin\left(\frac{\theta}{2}\right)\mathbf{j}, \quad pitch$$

$$q_3 = \cos\left(\frac{\phi}{2}\right) + \sin\left(\frac{\phi}{2}\right)\mathbf{i}. \quad roll$$

Equation (1) indicates a rotation q_1 followed by rotation q_2 and followed by rotation q_3 as shown in Figure 3.

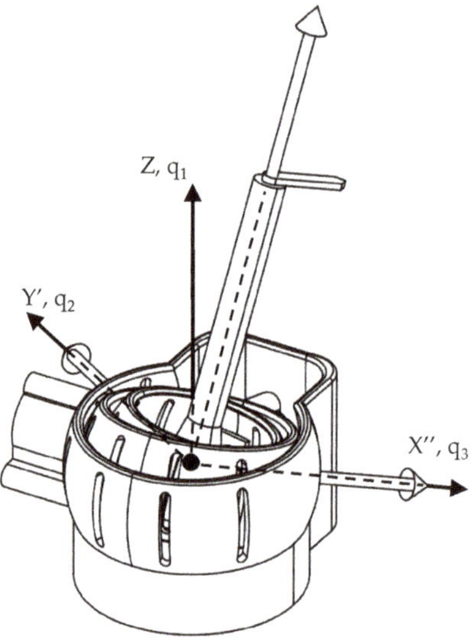

Figure 3. Tait–Bryan angles in the master mechanism.

The slave mechanism drives the patient affected arm by three servo-motors. An inverse kinematics is needed to obtain the motors rotations from the orientation of the joystick q.

With reference to the Figure 4, the equations required to solve the inverse position kinematics are obtained considering the ith. of the 3-RRUR mechanism:

$$\mathbf{a}_i + \mathbf{b}_i + \mathbf{d}_i = \mathbf{s}_i, \tag{2}$$

$$\mathbf{v}_i^T \mathbf{w}_i = 0. \tag{3}$$

The solving procedure is: (*a*) to obtain the orientation λ_i of $\mathbf{d}_i = d\mathbf{w}_i$ by solving Equation (3), (*b*) to obtain the motors rotations by solving Equation (2).

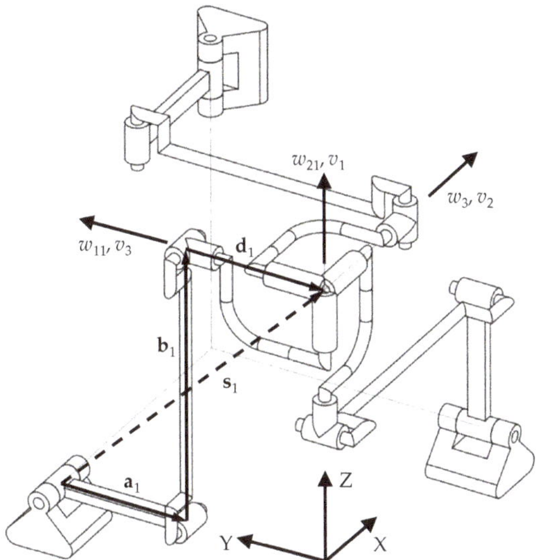

Figure 4. Slave mechanism at home pose.

Under the rotation $q = q_0 + q_1\mathbf{i} + q_2\mathbf{j} + q_3\mathbf{k}$ from the joystick of the master mechanism, $\mathbf{w}_i$ are obtained as (Figure 5):

$$\mathbf{w}_1 = e_1\mathbf{j}e_1^*,$$
$$\mathbf{w}_2 = e_2\mathbf{k}e_2^*,$$
$$\mathbf{w}_3 = e_3\mathbf{i}e_3^*,$$

where e_i^* is the *i*th. conjugate quaternion and

$$e_1 = e_{01} + e_{11}\mathbf{i},$$
$$e_2 = e_{02} + e_{22}\mathbf{j},$$
$$e_3 = e_{03} + e_{33}\mathbf{k}.$$

e_1, e_2, e_3 are the Euler parameter quaternions that represent the unknown rotations λ_i of $\mathbf{w}_i$ around the first three axes of each limb. On the other hand the axes $\mathbf{v}_i$ connected to the joystick are rotated by q.

$$\mathbf{v}_1 = q\mathbf{k}q^*,$$
$$\mathbf{v}_2 = q\mathbf{i}q^*,$$
$$\mathbf{v}_3 = q\mathbf{j}q^*.$$

Equation (3) leads to:

$$(2e_{11}^2 - 1)(2q_0q_1 - 2q_2q_3) - 2e_{01}e_{11}(2q_1^2 + 2q_2^2 - 1) = 0,$$
$$(2e_{22}^2 - 1)(2q_0q_2 - 2q_1q_3) - 2e_{02}e_{22}(2q_2^2 + 2q_3^2 - 1) = 0, \tag{4}$$
$$(2e_{33}^2 - 1)(2q_0q_3 - 2q_1q_2) - 2e_{03}e_{33}(2q_1^2 + 2q_3^2 - 1) = 0.$$

Equation (4) with the Euler parameters normalized equations, namely $e_{0i}^2 + e_{ii}^2 = 1$, form two equations with two unknowns for each limb that can be easily solved. The sought angles λ_i are then obtained as: $\lambda_i = 2atan(e_{ii}/e_{0i})$. It is worth noting that geometrically the solutions are obtained as intersection points of a conics and a circle in the Euler parameters plane $\{e_{0i}, e_{ii}\}$. An example is shown in Figure 6. It may be noted that, because of the symmetry of the solutions, a couple of points provides the same angle leading to only two distinct solutions.

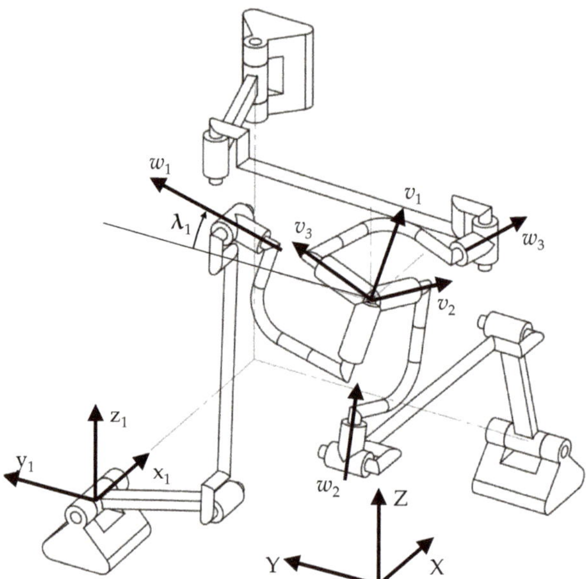

Figure 5. Slave mechanism in an arbitrary pose.

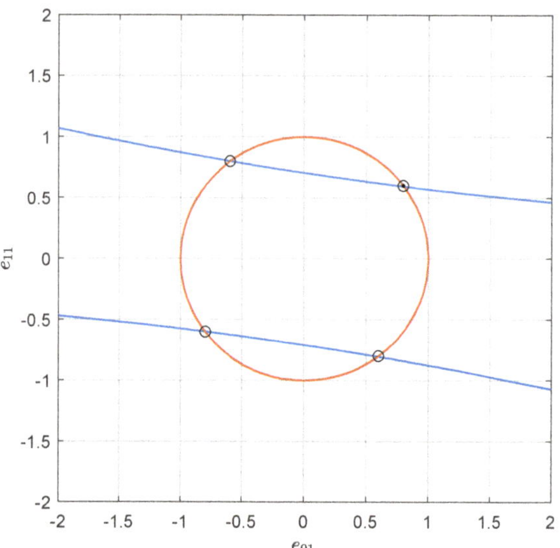

Figure 6. The conics and the circumference in the Euler parameters plane $\{e_{01}, e_{11}\}$.

The rest of the solution procedure consists of solving the inverse kinematics of a three-link planar manipulator, namely Equation (2), with $\lambda_i = \sum_{i=1}^{3} \theta_{1i}$ as known values:

$$ac_{\theta_{1i}} + bc_{(\theta_{1i} + \theta_{2i})} = p_{y_i} - dc_{\lambda_i},$$
$$as_{\theta_{1i}} + bs_{(\theta_{1i} + \theta_{2i})} = p_{z_i} - ds_{\lambda_i}, \tag{5}$$

with (p_{y_i}, p_{z_i}) coordinates of the center of rotation expressed in the reference system of the limb. The procedure is well-known [41] and for this reason it is not reported here in the detail. First, θ_{2i} is obtained by squaring and summing up Equation (5) to have the multiple solutions corresponding to the *elbow-up* or the *elbow-down*, then $s_{\theta_{1i}}$ and $c_{\theta_{1i}}$ can be obtained as the solution of a linear system.

Range of Motion

In the light of the application under study, it is required to know the range of motion of each input link of the slave mechanism.

In doing that the planar three-link model of the ith. limb is considered. According to Figure 7, the maximum Euler angle β and the limiting values of the input angle θ_{1i} can be obtained geometrically. Firstly, the limit of the counterclockwise input rotation θ_{1i_u} is considered. θ_{1i_u} is reached when the first two links of the limb are aligned. This configuration represents an inverse singular configuration for the mechanism (serial singularity) that is avoided in practice:

$$(a+b)c_{\theta_{1i_u}} - ds_\beta = p_{y_i},$$
$$(a+b)c_{\theta_{1i_u}} + dc_\beta = p_{z_i}, \tag{6}$$

Solution of Equation (6) is straightforward. By squaring and summing up the equations an equation of the form $As_\beta + Bc_\beta + C = 0$ is obtained and solved by the half-tan method. Eventually, we obtain $\theta_{1i_u} = atan2(p_{z_i} + ds_\beta, p_{y_i} - dc_\beta)$. Once β is known, the limit of the clockwise input rotation θ_{1i_d} is obtained by solving the Equation (7):

$$ac_{\theta_{1i_d}} + bc_\epsilon + dc_\beta = p_{y_i},$$
$$-as_{\theta_{1i_d}} + bs_\epsilon + ds_\beta = p_{z_i}. \tag{7}$$

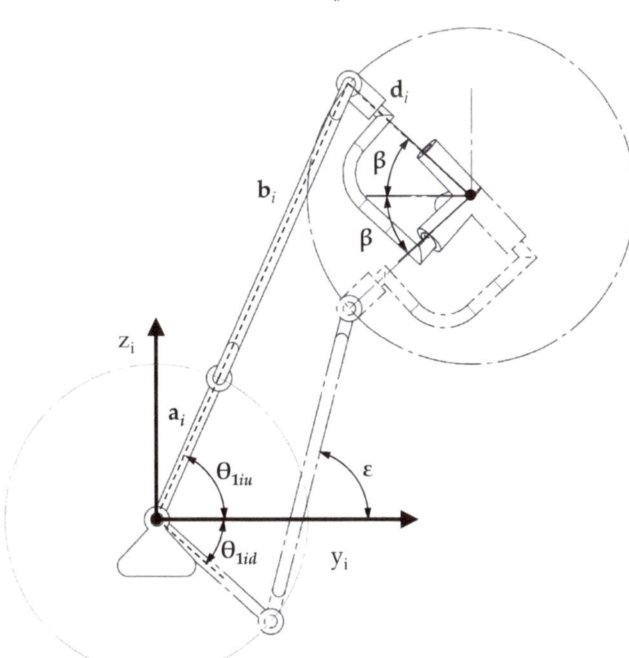

Figure 7. Geometric model for $\theta_{1i_u}, \theta_{1i_d}$ calculation.

A dimensional graphical synthesis was performed in order to avoid singularity configuration during the system's operations. In addition, the Jacobian matrix, which refers to check the inverse singularity is a diagonal matrix whose condition number has been verified [33]. In particular, it is worth noting that the maximum value of the Euler angle attained by the joystick in the simulation of the system was $\tilde{\beta} = 35°$. For this value the condition number of the Jacobian matrix is about 1.7, which is still far away from singular configurations.

4. Simulation of the System

The main goal of the numerical simulations was to produce the motion mapping from the master to the slave mechanism in accordance with the results from the kinematic analysis. Further, the model allowed us to choose the motors to be used in the prototype.

The geometrical dimensions of the slave in the model are: a_i = 72 mm, b_i = 150 mm, d_i = 78 mm.

The simulation was carried out in order to have the same behavior (rotational motion) of the two joysticks. In particular, the motion laws of the three motors of the slave must be related to the angular configuration of the master. It is worth noting that, since the two mechanisms are different, the laws of motion will be different. The outcome of the simulation of the system has to be reproducing the same motion of the joysticks. Figure 8 shows the rotation about z-axis for the first 10 s and an arbitrary spherical motion for the last 10 s. Figure 9 shows the mechanisms configuration at t = 7.5 s when the maximum discrepancy was found between the input and the output curves while performing the basic rotation. The mapping mismatch is shown in the plot, too. For the sake of clarity, the curve is shifted by an opportune offset.

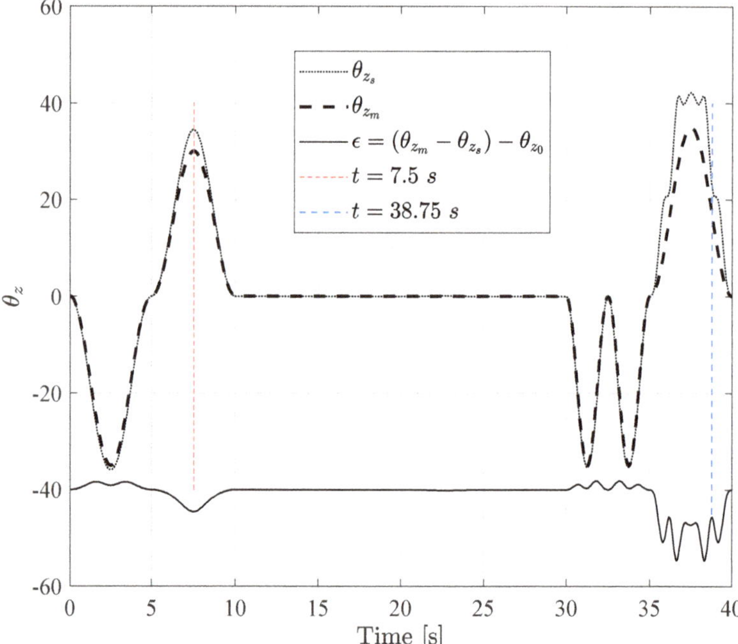

Figure 8. Master to slave R_z mapping.

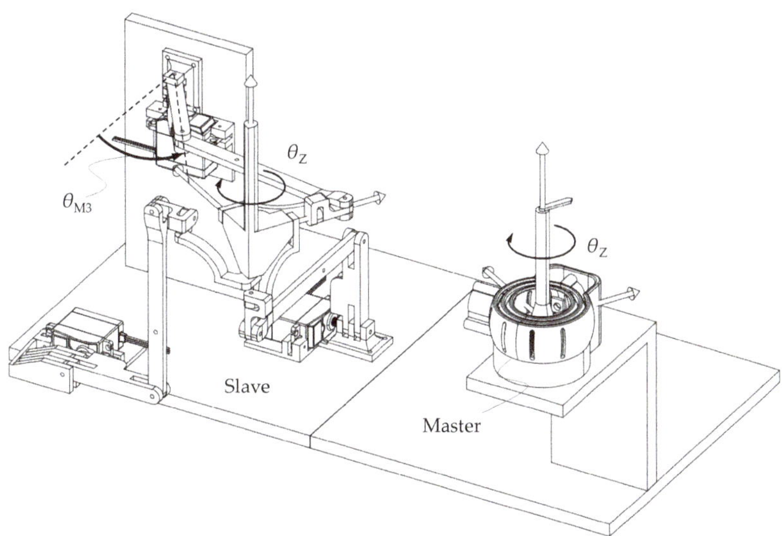

Figure 9. Mechanism configuration at $t = 7.5$ s.

Similarly, Figures 10 and 11 show the rotation about x-axis in the time interval 10–20 s and the rotation about y-axis in the time interval 20–30 s. The corresponding mapping mismatch curves are shown as well. The mechanisms configurations at $t = 17.5$ s and $t = 27.5$ s are shown in Figures 12 and 13.

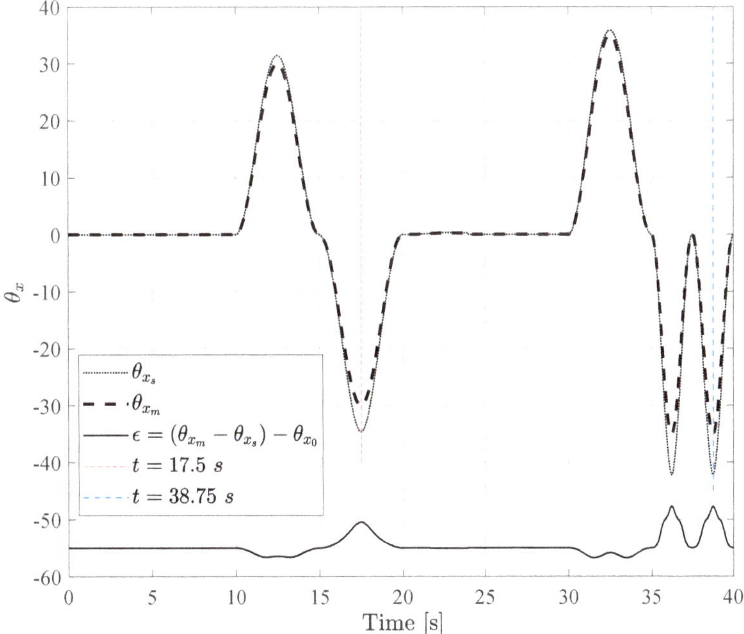

Figure 10. Master to slave R_x mapping.

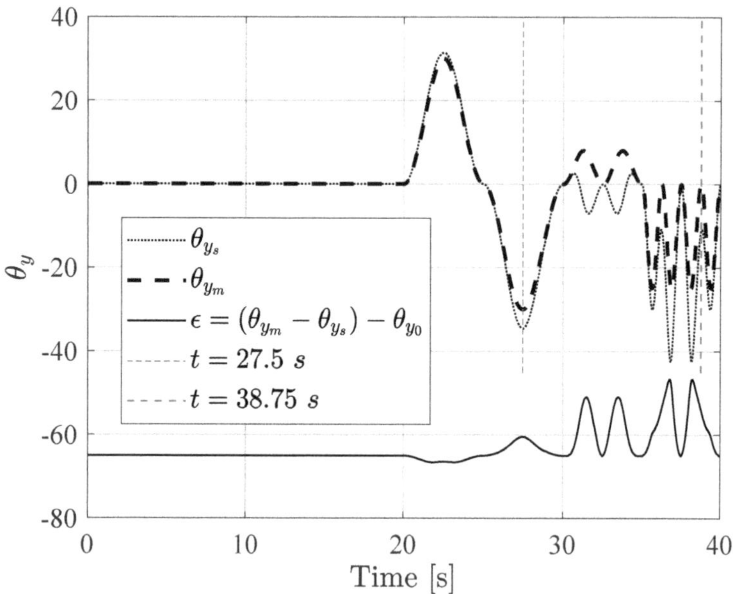

Figure 11. Master to slave R_y mapping.

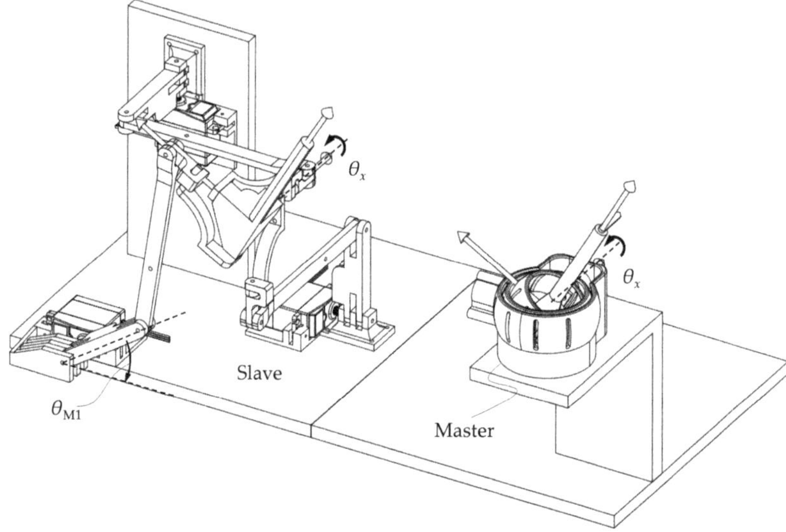

Figure 12. Mechanism configuration at $t = 17.5$ s.

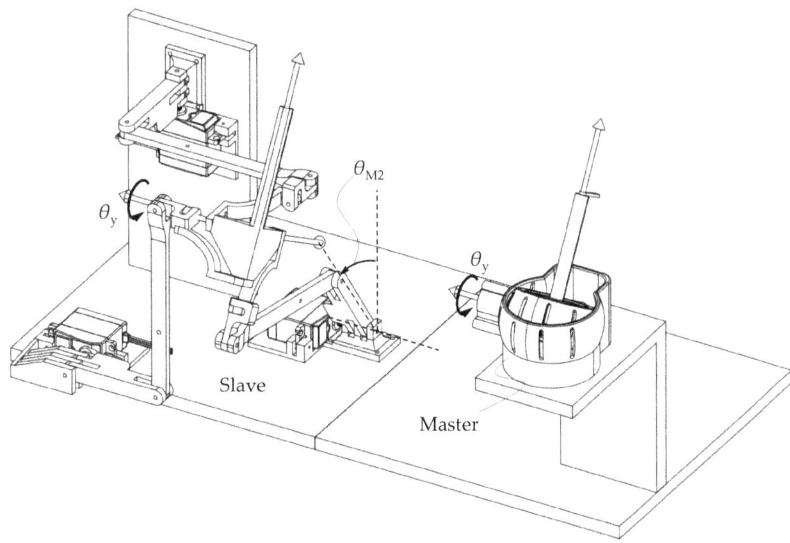

Figure 13. Mechanism configuration at t = 27.5 s.

It may be noted from the simulation plots that, because of the decoupled nature of the slave, the map is close to an identity while doing the basic rotations. A slight mismatch, i.e., 4.5°, was found when the motor rotation approaches its maximum $\theta_{Mi} \rightarrow \tilde{\theta}_{1i_u}$. On the other hand, because of the complex geometry and kinematics of the slave, the motor rotation laws are different from the master counterparts while performing a spherical motion. The limiting values of the motor rotations obtained from the simulation are $\tilde{\theta}_{1i_u} = \tilde{\theta}_{1i_d} = 42°$ and the maximum value of the joystick Euler angle of $\tilde{\beta} = 35°$. As expected, all values are lower than the theoretical counterparts calculated according to Equations (6) and (7), namely $\theta_{1i_u} = 65.7°$, $\theta_{1i_d} = 41°$ and $\beta = 42°$. Figure 14 shows the mechanisms configuration at t = 38.5 s when a spherical rotation is performed.

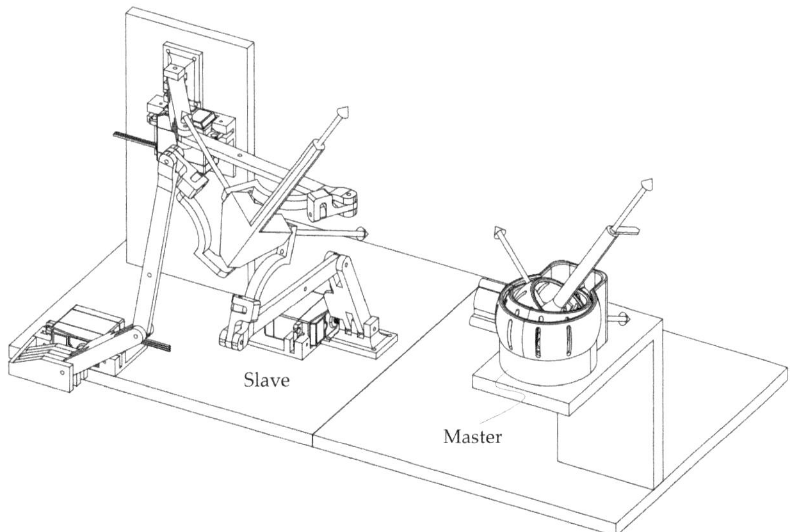

Figure 14. Mechanism configuration at t = 38.5 s.

5. Experimental Set-Up of the System

The layout of the proposed system is shown in Figure 15 following a mechatronic architecture shown in [42,43]. The input signals of the servo-system are the reference signals θ_1, θ_2, θ_3-SET, which correspond to the time position laws of the gimbal. They are sent to control board, which uses a control law G_c for generating the signals θ_{M_1}, θ_{M_2}, θ_{M_3}-SET for the motors. G_c implements the inverse kinematics of the slave mechanism. In the closed-loop regulator, the θ_{M_i}-SET are compared with the feedback signals θ_{M_i}-F/B, which are compensated by the signal conditioning K_θ of the angular transducers connected to the motors shafts. The error ϵ is compensated through the proportional controller P, such that the outputs θ_{M_1}, θ_{M_2}, θ_{M_3}-REF are the command signals of the three small servomotors of the slave system. The θ_{M_i}-OUT signals, which are related to the shaft angular position of the motors, change the configuration of the slave to follow the θ_i-SET signals of the master minimizing the error.

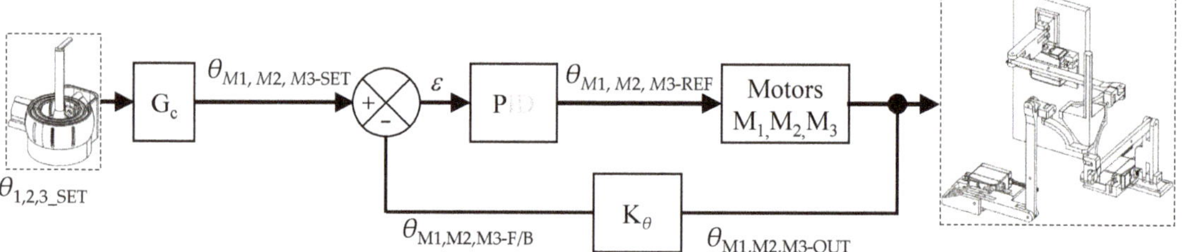

Gc = Control Law from Master to Slave

Figure 15. Block scheme of the control.

Figure 16 shows the overall control scheme, in which a PC, labeled (6), is used for programming and monitoring the operation and finally recording the trial. It is worth noting that after initial programming and calibration, the system is able to work and to interact automatically with the end-user. The end-user grasps the joysticks of the master and slave systems at the same time. The healthy hand plays as the master; therefore, the gimbal (1) is completely passive and follows the movements. The encoders, named as S1, S2 and S3, measure the angular configuration (orientation) of the system and their values, θ_i-SET are sent through a sensors control board (2) to the Arduino control board (3). The resulting signals θ_{M_1}, θ_{M_2}, θ_{M_3}-SET are generated by the G_c control law and sent to the motor control board (4) to operate the motors of the slave system to move the affected hand. It is worth noting that either a mirror movement or the same of the slave can be generated, according to the programming of the system at the beginning of the trial.

The Arduino Electronic Board is used to control the small servomotors to actuate the slave, they have a power supply of 4.8 V, maximum torque 2.45 Nm and rotation speed of $2\ \pi/s$. Figure 16 shows a scheme with the basic components, they are used for the primary task of monitoring and eventually recording the orientation of the healthy and affected hands. Nevertheless, the use of force sensors on the joysticks may be considered in order to increase safety during the operation, i.e., the system stops if the grasping forces in one or both hands increase or decrease rapidly. This option will be further implemented.

The preliminary mechatronic prototype of the system is reported in Figures 17 and 18. As mentioned previously, the system is basically composed by two units, the master is interacting with the full functional upper wrist, while the slave is devoted to the affected upper wrist.

Another important outcome of the proposed system is moving towards the e-healthcare, providing a portable system, relatively at low cost, with mechatronic solutions that allow the remote monitoring and recording of the trials to be supervised even at a distance.

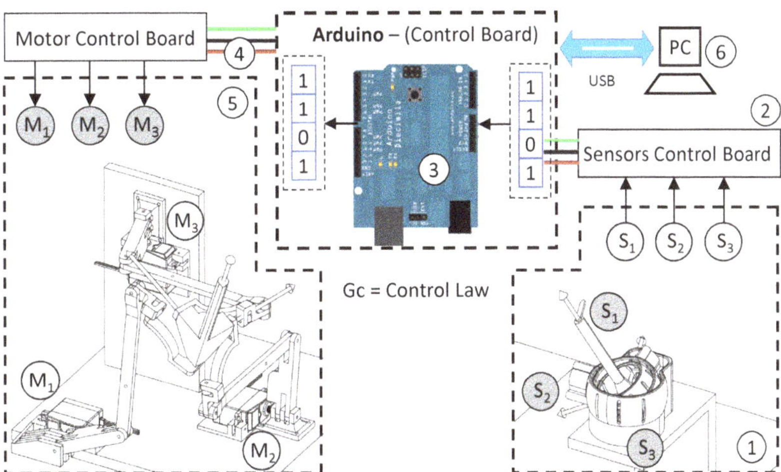

Figure 16. System control scheme.

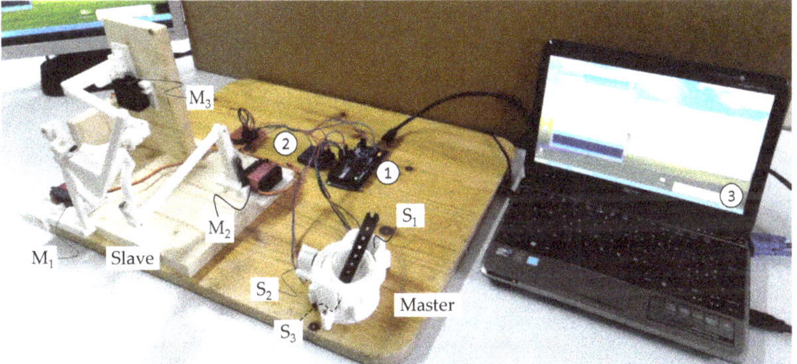

Figure 17. An overview of the proposed test bed of the mechatronic system.

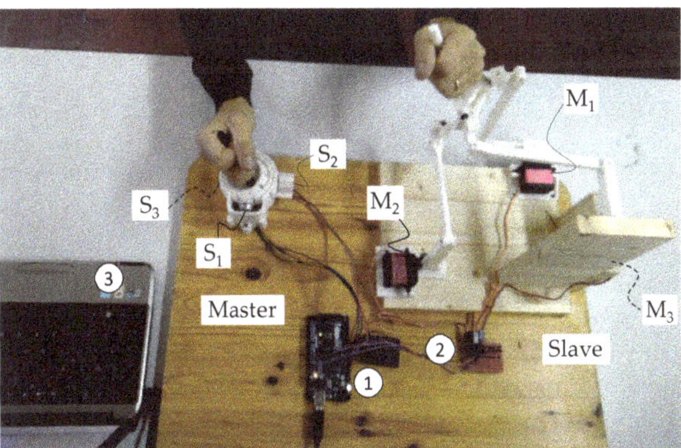

Figure 18. Top view of the proposed test bed of the mechatronic system.

6. Conclusions

We presented the design, analysis and experimental set-up of a master/slave system dedicated to wrist rehabilitation. The basic idea was to convey the motion of the healthy wrist imposed to a gimbal-like mechanism to the affected wrist moved by a mechanism

Article

Redundancy Exploitation of an 8-DoF Robotic Assistant for Doppler Sonography

Elie Gautreau [1], Juan Sandoval [1,*], Aurélien Thomas [1], Jean-Michel Guilhem [2], Giuseppe Carbone [3,*], Saïd Zeghloul [1] and Med Amine Laribi [1]

1   Department of GMSCP, Prime Institute CNRS, ENSMA, University of Poitiers, 86073 Poitiers, France; elie.gautreau@univ-poitiers.fr (E.G.); thomas.aurelien.perso@gmail.com (A.T.); said.zeghloul@univ-poitiers.fr (S.Z.); med.amine.laribi@univ-poitiers.fr (M.A.L.)
2   Independent Researcher, 4 Rue de Coumasaout, 31000 Toulouse, France; jean-michel.guilhem@orange.fr
3   Department of Mechanical, Energy and Management Engineering, University of Calabria, 87036 Rende, Italy
*   Correspondence: juan.sandoval@univ-poitiers.fr (J.S.); giuseppe.carbone@unical.it (G.C.)

Abstract: The design of a teleoperated 8-DoF redundant robot for Doppler sonography is detailed in this paper. The proposed robot is composed of a 7-DoF robotic arm mounted on a 1-DoF linear axis. This solution has been conceived to allow Doppler ultrasound examination of the entire patient's body. This paper details the design of the platform and proposes two alternative control modes to deal with its redundancy at the torque level. The first control mode considers the robot as a full 8-DoF kinematics chain, synchronizing the action of the eight joints and improving the global robot manipulability. The second control mode decouples the 7-DoF arm and the linear axis controllers and proposes a switching strategy to activate the linear axis motion when the robot arm approaches the workspace limits. Moreover, a new adaptive Joint-Limit Avoidance (JLA) strategy is proposed with the aim of exploiting the redundancy of the 7-DoF anthropomorphic arm. Unlike classical JLA approaches, a weighting matrix is actively adapted to prioritize those joints that are approaching the mechanical limits. Simulations and experimental results are presented to verify the effectiveness of the proposed control modes.

Keywords: medical robot; redundancy resolution; human-robot interaction; torque-control; Doppler sonography

Citation: Gautreau, E.; Sandoval, J.; Thomas, A.; Guilhem, J.-M.; Carbone, G.; Zeghloul, S.; Laribi, M.A. Redundancy Exploitation of an 8-DoF Robotic Assistant for Doppler Sonography. *Actuators* **2022**, *11*, 33. https://doi.org/10.3390/act11020033

Academic Editors: Marco Carricato and Edoardo Idà

Received: 29 December 2021
Accepted: 20 January 2022
Published: 24 January 2022

Publisher's Note: MDPI stays neutral with regard to jurisdictional claims in published maps and institutional affiliations.

1. Introduction

Nowadays, the use of medical robot assistants arises as a suitable solution to improve the working conditions of practitioners, cooperating with them to accomplish the medical tasks. Generally, the quality of the executed task is improved through this collaboration, where the medical capabilities of the expert are magnified by the robot, resulting in improved precision in tasks and lower time of execution while guaranteeing the well-being of the practitioner [1]. Thereby, several medical robot assistants are currently used in the operating room, such as the da Vinci Surgical System [2] from Intuitive Surgical, which has been the market leader for years. Similar examples can be found in other surgical specialties such as neurological and spine surgery [3], joint replacement surgery [4] or laparoscopic surgery [5].

Non-invasive applications can significantly benefit from medical robotic assistants. This is the case in Doppler sonography application, where a number of studies have recently been conducted to propose efficient robotic assistants aiding the specialists to improve their working conditions [6–8]. Indeed, the practitioner must adopt uncomfortable postures during the execution of standard ultrasound examinations. This often makes sonographers suffer musculoskeletal disorders early in their careers. To address this issue, several teleoperated robotic solutions have been proposed in recent years, mostly using commercial robotic arms as a probe-holder [9–11]. We have also recently proposed a teleoperated robotic assistant using a 7-DoF anthropomorphic arm as probe-holder [7,8]. The practitioner controls the robot by handling a haptic interface into a comfortable workspace. The main drawback of these proposed systems is the limited robot workspace, excluding the possibility of realizing an exam in the whole patient's body. Mobile solutions, such as the one proposed in [12], overcome this problem but need the aid of a human assistant to hold the mobile robot over the patient. A new version of the system proposed in [8] has recently been introduced in [13], including a motorized linear axis at the base of the robot to enlarge its workspace and allow a complete exam through the entire patient's body without needing manual readjustments of the platform's position. Knowing that the new platform at the patient site has 8-DoF, the degree of redundancy can be exploited in several ways, for instance, to consider a kinematic constraint [14], to avoid collisions [15] or mechanical joint limits [16], or to optimize a performance criterion such as the manipulability index [17]. This paper proposes two control modes to deal with the redundancy resolution of the 8-DoF robot. Knowing that safe human-robot interaction must be guaranteed, these control modes are proposed at the torque level, in order to implement a compliant behavior of the robot.

The first control mode considers the robot as a full 8-DoF kinematics chain. The advantage of this approach is that the eight joints are simultaneously activated, avoiding the presence of certain singularities typically linked to the limits of the workspace when only moving the 7-DoF arm. Therefore, the manipulability of the platform is naturally improved. The second control mode considers the two systems, the 7-DoF arm and the linear axis, separately. The robotic arm is activated in priority whereas the linear axis is only activated once the arm reaches the desired workspace limits. Furthermore, we present a new adaptive Joint-Limit Avoidance (JLA) strategy with the aim of exploiting the redundancy of the 7-DoF anthropomorphic arm. Classical JLA approaches define the diagonal weighting matrix as constant, which lets a continuous generation of null space torques to avoid joint limits, even for the joints that are far from their limits [16]. An improved approach is proposed here, where the weighting matrix is actively adapted to prioritize those joints approaching the mechanical limits. This feature allows an enhanced distribution of the redundant space of the robot. Moreover, when none of the joints is in the vicinity of the limits, zero null space torque is generated, allowing a free motion of the robot's elbow in case the expert wants to manually reconfigure the robot.

The main contributions of the paper are summarized as follows: (a) detailed presentation of an 8-DoF teleoperated platform for Doppler sonography, (b) validation of a fully redundant control mode at the torque level for the 8-DoF robotic system and (c) introduction of a new adaptive JLA strategy for redundant robots controlled by torque by means of an optimal variation of the weighting matrix.

The paper is organized as follows: Section 2 details the medical requirements and the description of the proposed robotic assistant platform. Sections 3 and 4 present the first and the second control modes, respectively. The last section concludes the presented work and opens the perspectives of future works.

2. Robotic-Assistant Platform

This section presents the proposed teleoperated robotic platform for Doppler Sonography. The platform has been designed based on the medical requirements collected through a study of the medical gesture.

2.1. Medical Requirements

Doppler ultrasound is a medical imaging method used by sonographers (i.e., angiologists) to study the cardiovascular system of patients using an ultrasound probe. According to the Society of Diagnostic Medical Sonography, 90% of clinical sonographers have experienced symptoms of work-related musculoskeletal disorders (WRMDs), mainly due to an accumulation of repeated gestures in awkward postures and to the frequent application of downward pressure with the probe [18].

In order to confirm this thesis, a study of the angiologist's gestures has been carried out in [7], and we resume in this section the main obtained results. A Motion Capture (MoCap) system, i.e., Qualisys, was used to record several doppler ultrasound examinations made by a practitioner on real patients and under real conditions. Markers used for MoCap were placed both on the practitioner and on the probe to record the right arm, pelvis and head movements of the practitioner during the performed examinations. Moreover, the probe have been instrumented with a FSR sensor to measure the force applied by the specialist (Figure 1). Each patient has been examined in regions all along the body: carotid, legs, and abdomen. The study highlighted an outing from the expert's joint comfort zone, which can cause WRMSDs with repetitive movements. The maximum measured values of the orientation angles considered correspond to the worst postures. The latter are compared to the comfort zone reference angles defined by ISO 11226, ISO 11228-3 and NF EN 1005-4 norms. An example of head rotation and wrist angles for the carotid exam is given in Figure 2. These values, joint angles for neck twist and wrist joint, prove that the angiologist very often works outside the comfort zone described by the standards.

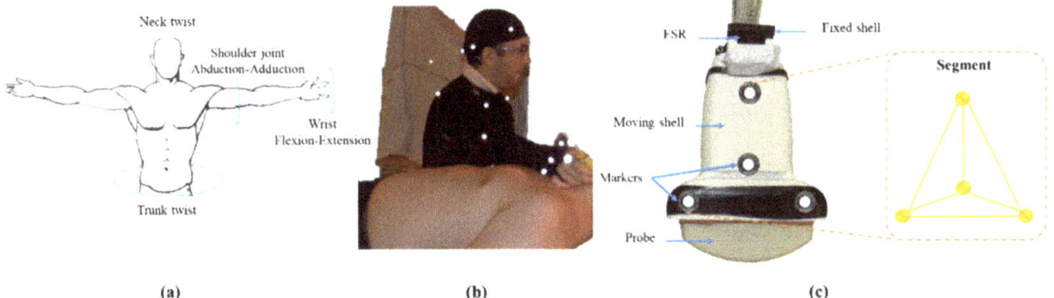

Figure 1. Study of the angiologist's gestures using MoCap: (**a**) Measured angles on the angiologist's body. (**b**) Reflective markers fixed to the angiologist's body (**c**) Reflective markers and FSR sensor fixed to the probe.

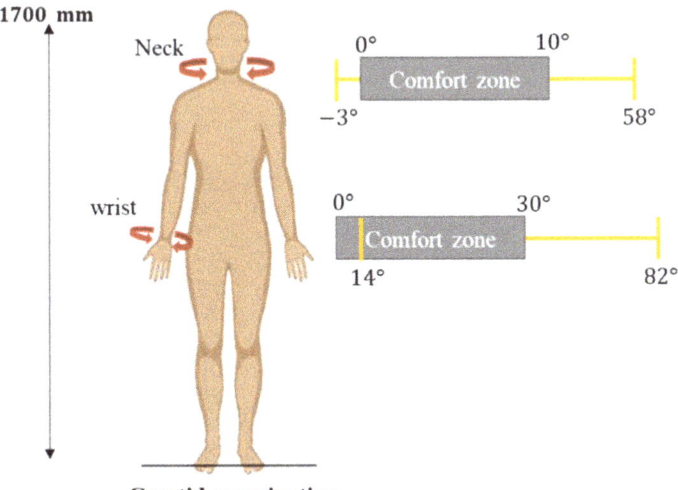

Figure 2. Neck and wrist joint range angles measured on angiologist during carotid exams.

In addition, the efforts measured with the instrumented probe revealed that the intensity of the effort depends on the type of examination and the morphology of the patient and can also lead to the development of MSDs. For instance, the maximum force applied was measured during the examination of an abdomen. A greater effort was required by the angiologist to determine the position of the abdominal aorta, which is particularly recurrent for the category of fat patients.

The interest of this study is twofold. It highlighted (1) the common gestures performed out of the joint comfort zones and (2) the significant efforts to be applied during an examination. The combination of the latter can lead to musculoskeletal disorders during numerous repetitive examinations. In order to overcome these two observations, a robotic teleoperation platform composed of a haptic interface (expert site) and a cobot (patient site) was developed. Thus, the cobot performs the efforts and postures controlled by the doctor through a haptic interface. Since the cobot workspace is not large enough to cover the whole patient's body, it has been mounted on a motorized linear axis, producing an 8-DoF robotic platform.

2.2. Experimental Setup

Figure 3 presents the assembled prototype of the proposed teleoperated robotic platform. The platform at the patient site, i.e., Franka Emika and linear axis, is teleoperated by the practitioner through a 6-DoF haptic interface.

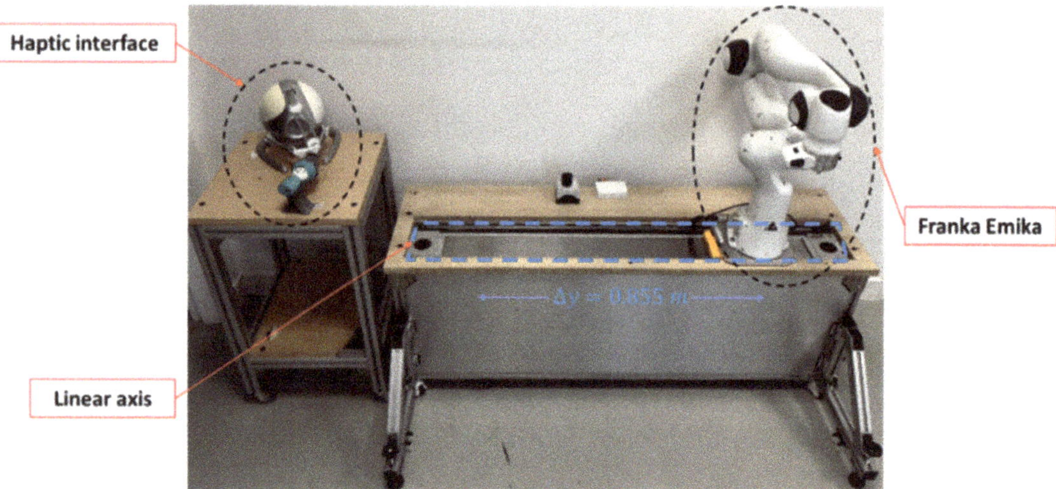

Figure 3. Assembled prototype of the 8-DoF robotic platform for medical applications.

The platform's workspace has been increased by 0.855 m on the *Y-axis* thanks to range of motion of the linear axis. This latter is powered by a Kollmorgen brushless motor via a belt drive. The motor driver is controlled through a Telnet protocol. Moreover, the high-level control of the overall platform is implemented in the Robot Operating System (ROS) middleware.

2.2.1. Expert Site: 6-DoF Haptic Interface

Haptic devices have been widely employed in many areas such as surgery and craniotomy application [19], rehabilitation and manufacturing. In a general framework, haptic devices are based on mechanical interfaces that link tactile information between a human and the device. Various haptic interface architectures have been developed, with either serial or parallel architectures. While serial devices present several drawbacks as inertia, rigidity and positioning issues, parallel interfaces overcome these drawbacks but present limited workspaces and singular configurations issues. Hybrid haptic interface architectures take advantage of the previous ones. Some devices have been developed commercially, such as the Sigma 7 [20] or the 6-DoF interface proposed by Tsumaki et al. [21].

We have developed a 6-DoF hybrid haptic interface, as shown in Figure 4. It is composed of a 3-DoF Novint Falcon interface combined with an inertial measurement unit (IMU) attached with a spherical wrist. Novint Falcon allows translational motions in 3D space, similar to the one proposed by Tsai [22], into a maximum volume of 10.16 cm along each direction, and a maximum force feedback of around 9 N. A LPMS-B2 inertial measurement unit, located inside the artificial probe measures the rotational motions (3-DoF). The artificial probe mounted on the haptic device simulates a real probe for Doppler sonography examinations. Additionally, the system at the expert site includes a foot pedal to prolong the desired motion sent to the robot along the current direction of motion of the haptic interface.

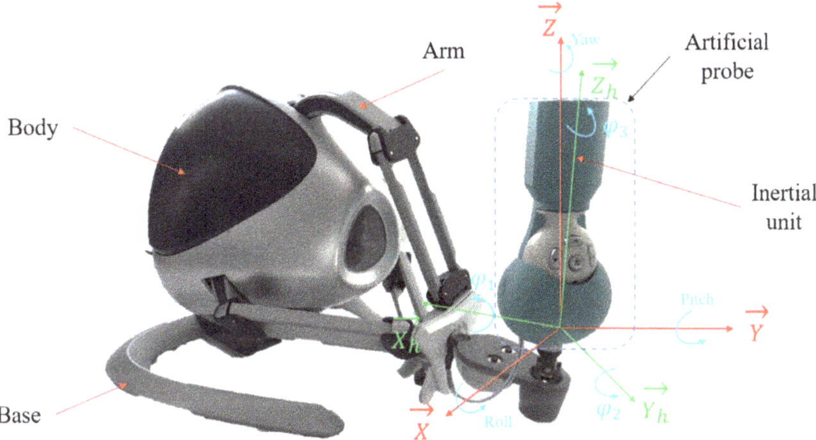

Figure 4. 6-DoF hybrid haptic interface.

2.2.2. Patient Site: 8-DoF Redundant Robot

As mentioned above, we previously proposed the use of a 7-DoF Franka Emika cobot at the patient site for the robotic assistant platform [7]. Nevertheless, the workspace of the 7-DoF cobot is insufficient to perform an examination of the entire patient's body. Therefore, we propose to enlarge the workspace of the robot by mounting it on a belt-driven linear axis (Parker HMRB15) placed parallel to the longitudinal axis of the patient's table (Figure 3). The linear axis is motorized by a 24 V brushless motor coupled with a gear reducer (12.5 mm/lap). More technical details about the linear axis are presented in Table 1.

Table 1. Technical characteristics of the linear axis.

Axis	Parker HMRB15
Motor	Brushless Motor AKM3
Driver	Kollmorgen AKD-P00306
Maximum Imposed Torque	3 Nm
Linear speed imposed	10 mm/s
Motion amplitude	900 mm

When employing the platform, the robotic system is placed next to the patient's table so that the linear axis motion allows the entire table to be covered. It is therefore planned that the patient's waist is aligned with the middle position of the linear axis, facilitating the identification of his positioning by the practitioner.

2.2.3. Communication Framework

Robot Operating System is the middleware used in the platform to guarantee real-time data exchange between the different devices. These applications have been developed in Python and/or C++ and launched in ROS Kinetic version under Ubuntu 16.04 with real-time patched kernel, so that a soft real-time operating system is configured.

The diagram in Figure 5 shows the information transmitted between the different systems. The packages Rosfalcon, Joy and Lmps_IMU are, respectively, those of the Novint Falcon, the foot pedal and the inertial measurement unit composing the hybrid haptic interface. Concerning the operator site, the package Franka_ros allows to read/write commands to the cobot's controller and the package Axis communicates with the controller of the linear axis. Finally, the motion of the axis is determined by the Pos_criteria package, according to the control mode implemented (see Sections 3 and 4).

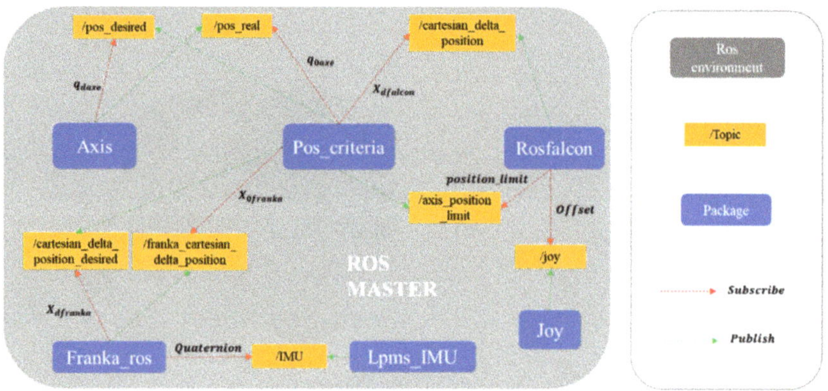

Figure 5. Data exchange of the platform in ROS environment.

Communication between the different systems is achieved through Bluetooth (inertial unit), USB (haptic interface), Ethernet (Franka Emika) and Telnet (linear axis) protocols. In the case of the linear axis, telnet protocol has been chosen for its ease of use, as it uses a Telnet library, i.e., Telnetlib in python.

As the linear axis has been implemented as an upgrade, the package Axis has been created and linked to the existing devices. This package enables the control of the linear axis in either position, speed or torque (compliance). The latter gives us the possibility to achieve compliance control. The ROS node to control the linear axis in torque is depicted in Figure 6.

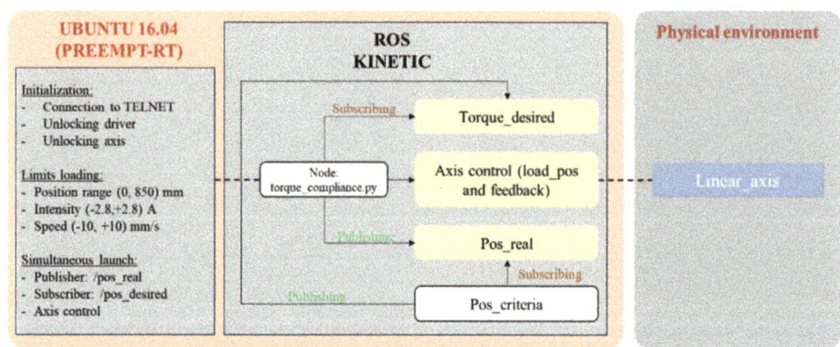

Figure 6. Block diagram of the Axis torque node.

The layout of the three nodes corresponding to the position, speed and torque control remains the same. The speed limit stands low as this application aims to deal with patients, thus security is an important issue. Moreover, in the case of high speed, where robot joints can be damaged, the robot will detect this and freeze, preventing the robot from being used.

Two alternative control modes for the platform at the patient site are proposed in the following sections. The first control mode, presented in Section 3, considers the kinematics of the robot as a fully 8-DoF system. Therefore, the motions of the 8 axes are synchronized to fulfill the desired task and the two degrees of redundancy guarantee the avoidance of singularities of the robot arm. This control mode has for now been validated in simulation. The second control mode, presented in Section 4 and validated experimentally, decouples the motion of the linear axis and the robot arm. The linear axis is only activated if the robot arm approaches its workspace limits. Furthermore, the remaining degree of redundancy of the robot arm is employed to avoid the joint limits through a new adaptive JLA strategy.

3. Fully Redundant Control Mode

3.1. Robot Modelling

As mentioned above, the first control mode considers the robot model of an 8-DoF system. Here below we detail the kinematic and dynamic model of the proposed platform. First, the Denavit–Hartenberg parameters' table [23] of the platform is presented in Figure 7, where joint $i = 0$ corresponds to the linear axis and joints $i = 1, \ldots, 7$ are related to the cobot arm.

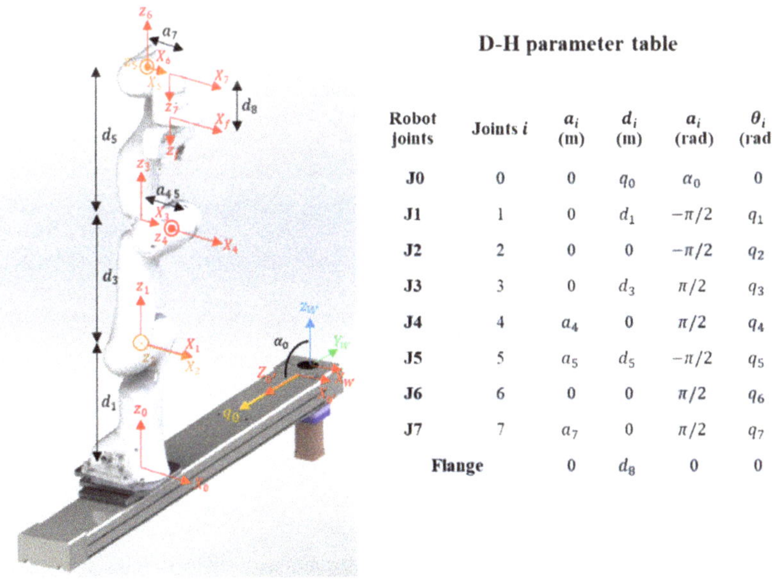

D-H parameter table

Robot joints	Joints i	a_i (m)	d_i (m)	α_i (rad)	θ_i (rad)
J0	0	0	q_0	α_0	0
J1	1	0	d_1	$-\pi/2$	q_1
J2	2	0	0	$-\pi/2$	q_2
J3	3	0	d_3	$\pi/2$	q_3
J4	4	a_4	0	$\pi/2$	q_4
J5	5	a_5	d_5	$-\pi/2$	q_5
J6	6	0	0	$\pi/2$	q_6
J7	7	a_7	0	$\pi/2$	q_7
Flange		0	d_8	0	0

Figure 7. D-H parameters of the 8-DoF robotic assistant platform.

Considering that a Doppler ultrasound exam requires that the probe fully moves in the space, a task-space dimension of $m = 6$ is defined. Then, let define the non-squared Jacobian matrix $\mathbf{J}(\mathbf{q}) \in \Re^{m \times n}$ with $n = 8$ denoting the number of robot joints. Therefore, the relation between the joint-space velocity $\dot{\mathbf{q}} \in \Re^n$ and the task-space velocity $\dot{\mathbf{x}} \in \Re^m$ yields,

$$\dot{\mathbf{x}} = \mathbf{J}(\mathbf{q}) \cdot \dot{\mathbf{q}} \tag{1}$$

Since the robotic platform is torque-controlled, let us now define its dynamic equation of motion in joint-space as,

$$\mathbf{M}(\mathbf{q})\ddot{\mathbf{q}} + \mathbf{C}(\mathbf{q},\dot{\mathbf{q}})\dot{\mathbf{q}} + \mathbf{g}(\mathbf{q}) = \boldsymbol{\tau}_c + \boldsymbol{\tau}_{ext} + \boldsymbol{\tau}_f \tag{2}$$

This model depends on the inertial matrix $\mathbf{M}(\mathbf{q}) \in \Re^{n \times n}$, the centrifugal and Coriolis matrix $\mathbf{C}(\mathbf{q},\dot{\mathbf{q}}) \in \Re^{n \times n}$ and the vector of gravitational torques $\mathbf{g}(\mathbf{q}) \in \Re^n$. Moreover, vectors $\boldsymbol{\tau}_c \in \Re^n$, $\boldsymbol{\tau}_f \in \Re^n$ and $\boldsymbol{\tau}_{ext} \in \Re^n$ represents the output, friction and external torques, respectively.

This section may be divided by subheadings. In the following, we explain the details of the control law defining $\boldsymbol{\tau}_c$, as well as results of the robot's behavior performed in a customized simulator.

3.2. Control Law

A compliant behavior of the robotic platform is suitable to reduce the effects of undesired collisions between the robot and the patient. The following control law allows to reproduce the effects of a mechanical damper-spring system at the cartesian space during the execution of a desired trajectory $\mathbf{x}_d \in \Re^m$,

$$\mathbf{F}_{task} = \mathbf{K}_p(\mathbf{x}_d - \mathbf{x}) - \mathbf{K}_d\dot{\mathbf{x}} \tag{3}$$

where the stiffness and damping effects can be adjusted with the diagonal constant matrices $\mathbf{K}_p \in \Re^{m \times m}$ and $\mathbf{K}_d \in \Re^{m \times m}$, respectively [24].

A joint-torque control law implementing the compliant behavior of Equation (3) can be defined as follows,

$$\boldsymbol{\tau}_c = \mathbf{J}^{\mathsf{T}} \cdot \mathbf{F}_{task} + \boldsymbol{\tau}_{\aleph} + \boldsymbol{\tau}_{comp} \tag{4}$$

This control law also includes the torque vector $\boldsymbol{\tau}_{comp} \in \Re^n$, compensating the gravitational and dynamic effects, and the torque vector $\boldsymbol{\tau}_{\aleph} \in \Re^n$, exploiting the redundancy of the robot. As mentioned above, redundancy can be used in several ways according to the specific needs of the application. For instance, a suitable way to exploit the redundancy is to stabilize the internal motion, yielding,

$$\boldsymbol{\tau}_{\aleph} = \aleph(\mathbf{q})\left[K_{p_{null}}(\mathbf{q}_{\mathbf{init}} - \mathbf{q}) - K_{d_{null}}\dot{\mathbf{q}}\right] \tag{5}$$

Joint torques produced by Equation (5) attempt to keep the joint positions as best as possible at the initial joint configuration $\mathbf{q}_{\mathbf{init}} \in \Re^n$. The weight of this law for each joint can be tuned with the values of diagonal constant matrices $K_{p_{null}} \in \Re^{n \times n}$ and $K_{d_{null}} \in \Re^{n \times n}$. In order to guarantee that this control law is only performed in the null-space of the robot, avoiding undesired perturbations in the cartesian-space, the torque vector is premultiplied by a null-space projector $\aleph(\mathbf{q}) = (\mathbf{I}_{n \times n} - \mathbf{J}^{\mathsf{T}}\mathbf{J}^{+})$, defined in terms of the Moore–Penrose pseudoinverse of $\mathbf{J}(\mathbf{q})$, i.e., $\mathbf{J}^{+} = \mathbf{J}^{\mathsf{T}}(\mathbf{J} \cdot \mathbf{J}^{\mathsf{T}})^{-1}$.

3.3. Performance Analysis

To achieve a compliance control mode of the robotic platform, a dynamic simulator was developed in Matlab-Simulink, using the Simscape toolbox. For purpose of realistic simulations, the CAD of the real robotic platform was added to the simulator. The dynamic model was calculated by Simulink based on the geometric and inertial parameters of the assembled multi-body system. This new tool allowed us to verify the performance of the robot when controlling it by torque, based in [25]. Figure 8 shows the flowchart for the simulator operating principle.

The control law is an association of three laws:

- Recursive Newton–Euler algorithm: This algorithm has been implemented to calculate the gravitational, centrifugal and Coriolis compensation torques.

- Null-space control law: The use of a null torque vector enables the robot's internal motions stabilization for a given task. Adjusting the values of damping $K_{d_{null}}$ and stiffness $K_{p_{null}}$ allows the choice of priority motion of either the robot or the linear axis. In order to let the robot and the linear axis moving together, the constant matrices have been set to $K_{p_{null}} = diag(100,7,4,4,5,4,3,4)$ and each value of $K_{d_{null}}$ has been set to $K_{d_{null\ 1,1}} = 26\sqrt{K_{p_{null\ 0,0}}}$ for linear axis and $K_{d_{null\ j,j}} = 0.9\sqrt{K_{p_{null\ j,j}}}$ for the cobot ($j = 1,\ldots,7$). It is worth mentioning that these values were determined empirically and are highly dependent of the model uncertainties, e.g., friction forces.

- Cartesian compliance law: The required torque is computed to achieve the imposed task at the cartesian space. Stiffness $\mathbf{K}_p$ and damping $\mathbf{K}_d$ matrices have been set to $\mathbf{K}_p = diag(500,200,500,40,40,40)$ and $K_{d_{l,l}} = 2.2\sqrt{K_{p_{l,l}}}$. Empirical values of stiffness and damping are consistent with real values allowed by the Franka cobot.

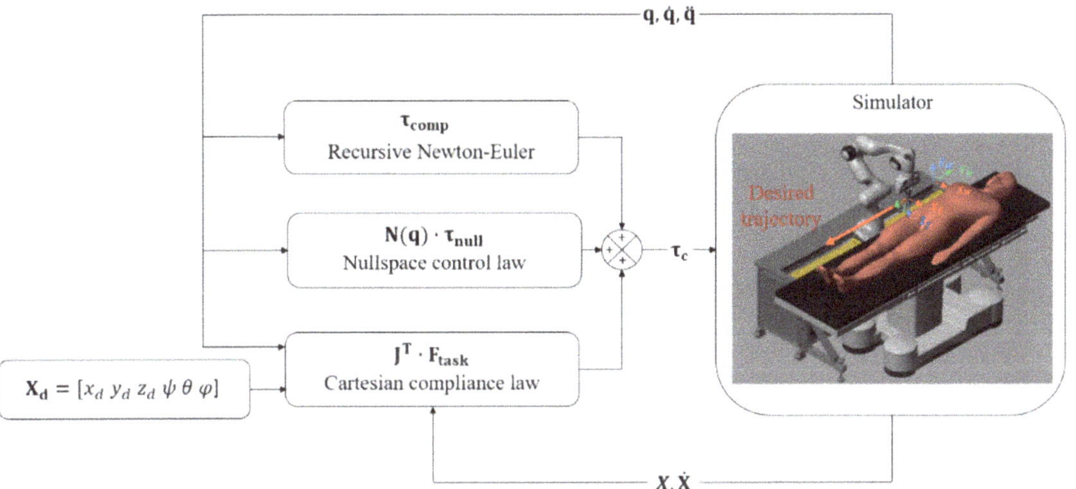

Figure 8. Control schema implemented in the dynamic simulator.

A linear trajectory with a total length of 0.8 m along the *Y-axis* has been imposed to validate the correct functioning of the robotic platform when controlled as an 8-DoF redundant robot. The desired linear trajectory has been chosen in such a way that it exceeds the Franka cobot workspace.

In the initial configuration, the robot end-effector cartesian position is set to 0 m along the *Y-axis*. To perform the desired trajectory, a simultaneous motion is conducted by the cobot and the linear axis, as shown in Figures 9 and 10. Indeed, the values of $K_{p_{null}}$ have a high influence on the portion of motion performed by the axis and by the cobot along the *Y-axis*. In this case, we put a larger value to the first diagonal value, related to the action of the linear axis, so that its motion is launched easily. However, in the case that the motion of the linear axis is not a priority, this constant value could be reduced to prioritize the motion of the cobot. A comparison between the desired and the performed trajectories are also presented in Figure 11. It can be seen that the desired trajectory is entirely executed by the platform. Errors between the real and desired motions along the three cartesian axis remain low: $x_{error} = 1.6 \cdot 10^{-5}$, $y_{error} = 1.39 \cdot 10^{-4}$ and $z_{error} = 2.9 \cdot 10^{-5}$, considering that a natural error is induced by the compliance law implemented in Equation (3).

A suitable performance criterion to be analyzed in this new platform is the manipulability index [13], i.e., $\mathbf{M} = \sqrt{\mathbf{J}^T\mathbf{J}}$. It is well-known that the manipulability index decreases when the manipulator approaches a singularity configuration, i.e., $det(\mathbf{J}) = 0$. Naturally, if the linear trajectory is only performed by the cobot arm, the manipulability index would decrease to zero since the workspace limits would be reached. In contrast, the use of the linear axis and the simultaneous activation of the cobot and the axis allows it to preserve higher manipulability values, such as shown in Figure 12. The manipulability of the robotic platform (8-DoF) is always increasing because of the linear axis's motion. It is also calculated the manipulability of the Franka cobot arm separately, proving that it does not fall, since it is not reaching the limit of its working space.

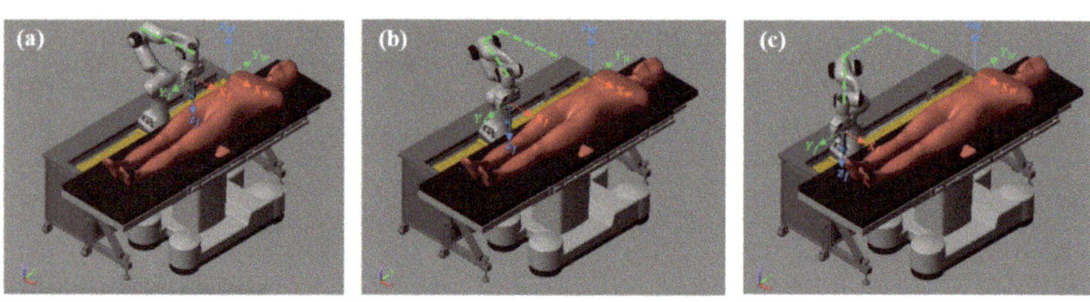

Figure 9. Sequential evolution of the robot configuration during the execution of a linear trajectory of 0.8 m along the *Y-axis*. (**a**) Initial, (**b**) intermediate and (**c**) final robot configuration.

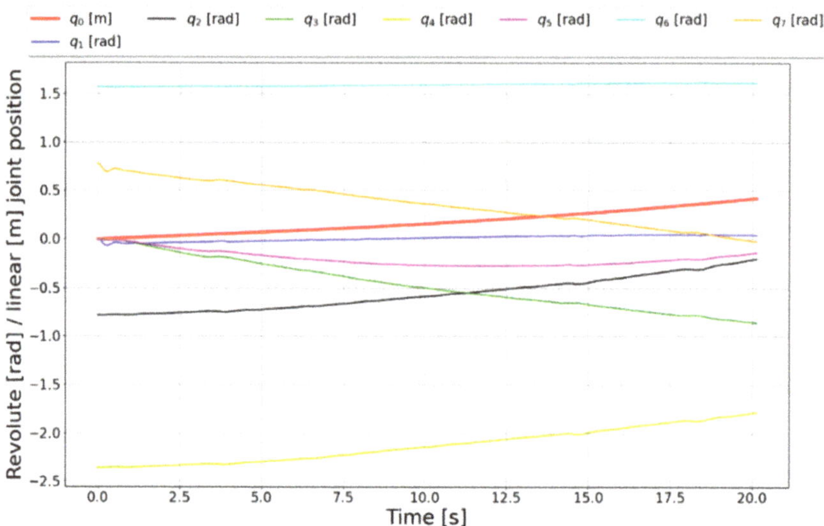

Figure 10. Joint position trajectories of the 8-DoF robot assistant during the execution of a linear trajectory along the *Y-axis*.

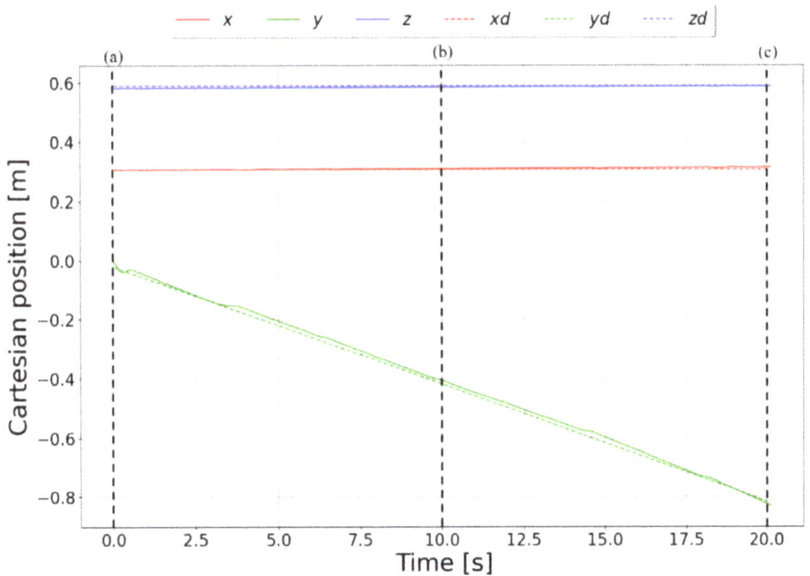

Figure 11. Cartesian position trajectory of the 8-DoF robot assistant during the execution of a linear trajectory along the *Y-axis*.

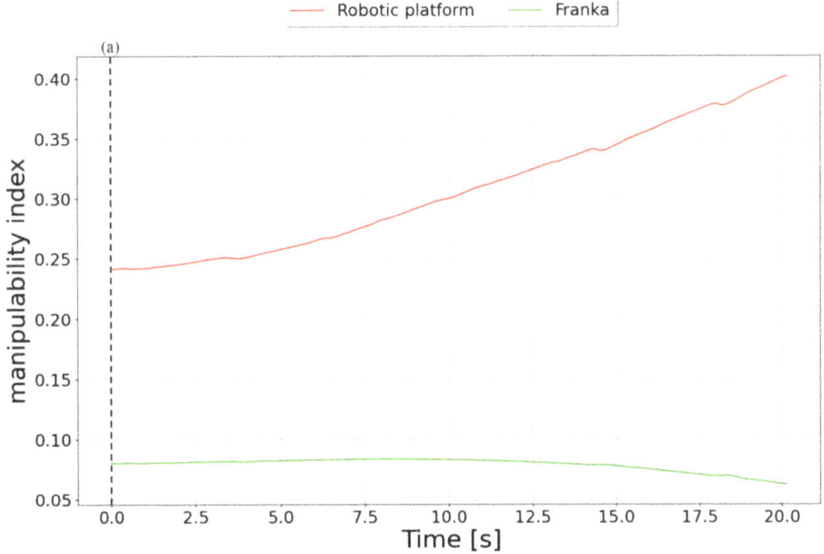

Figure 12. Manipulability Index of the 8-DoF robotic platform and the one of the Franka cobot (7-DoF) during the execution of a linear trajectory along the *Y-axis*.

4. Decoupled Redundant Control Mode

In this section we propose a decoupled control mode for the proposed 8-DoF redundant robotic platform. This means that both the 7-DoF cobot and the linear axis are controlled independently as two separate systems. The details of this control mode are depicted here below.

4.1. Manipulator/Axis Switching Strategy

The control strategy adopted to associate the linear axis and the 7-DoF robot is achieved by decoupling the two devices. The motion distribution between the cobot and the linear axis is achieved through a central ROS node called pos_criteria (Figure 6). A desired cartesian position limit *Xlim* is defined to establish a switching condition determining whether the robot or the linear axis moves along the direction of motion of the linear axis, i.e., *Y-axis*. Upper and lower limits are imposed to (-0.45, 0.45) m. These limits are largely lower than the maximum robot workspace (0.855 m). Two scenarios depending on the position condition are shown in Figure 13.

a. While the robot's end-effector does not reach the limits, the linear axis remains fixed, and the robot moves along the *Y-axis* to fulfil the desired trajectory.

b. If the limits are exceeded, the linear axis is launched, and the robot's end-effector remains fixed along the *Y-axis*.

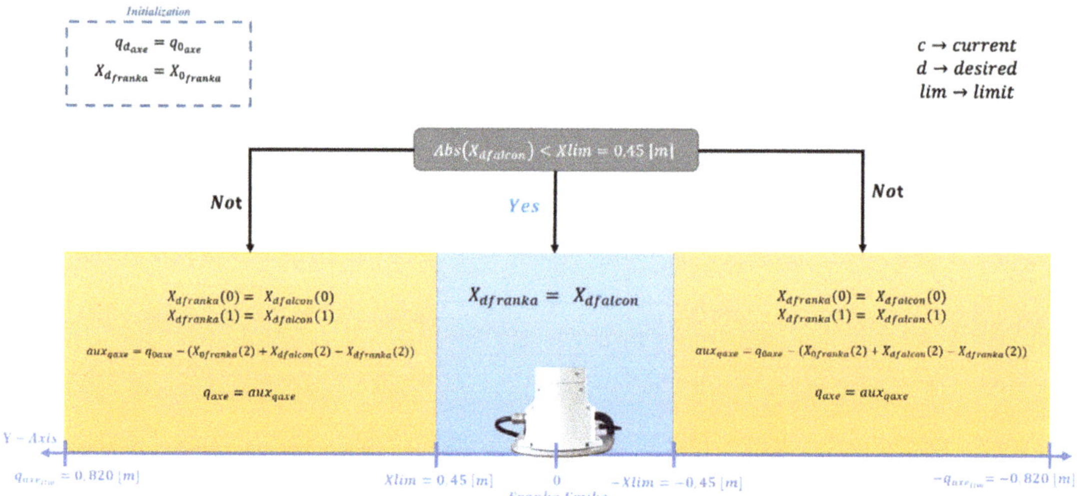

Figure 13. Movement distribution between the Franka Emika cobot and the linear axis in the *pos_criteria* node.

Evidently, the cobot always executes the movements along the remaining directions since the linear axis only can be actuated along the *Y-axis*.

We note that ∓0.820 m corresponds to the length of the linear axis, where 0 m is the linear axis origin (middle point). The desired position transmitted by the haptic device is $q_{d_{axe}}$. Figure 14 depicts the proposed switching strategy.

4.2. Adaptive Joint-Limit Avoidance Strategy

A suitable way to exploit the redundancy of the $n = 7$-DoF cobot in the application of Doppler sonography is to enlarge its cartesian workspace, mainly rotational, allowing the ultrasound probe to execute large rotational movements. To achieve this, a suitable strategy consists of moving away the joint positions from the mechanical limits $\left[q_{min_i}, q_{max_i}\right]$. Thereby, the null-space control approach can be defined as follows,

$$\boldsymbol{\tau_{\aleph}} = \aleph(\mathbf{q})\boldsymbol{\tau}_{JLA} = \aleph(\mathbf{q})\left[K_{JLA}\left(\frac{\partial \mathbf{w}_{JLA}(\mathbf{q})}{\partial \mathbf{q}}\right)^T - \mathbf{D}_{JLA}\dot{\mathbf{q}}\right] \qquad (6)$$

The torque vector $\boldsymbol{\tau}_{JLA} \in \Re^n$ maximizes the objective function $\mathbf{w}_{JLA}(\mathbf{q}) \in \Re$ and is projected to the null-space of $J(q)$ through $\aleph(\mathbf{q})$ to guarantee a compatibility with the cartesian task performed by the cobot. The second term of $\boldsymbol{\tau}_{JLA}$ allows to stabilize the internal motion of the robot. Let us define the objective function as,

$$\mathbf{w}_{JLA}(\mathbf{q}) = -\frac{1}{2n}\sum_{i=1}^{n}\left(\frac{\mathbf{q}_i - \mathbf{q}_{c_i}}{\mathbf{q}_{max_i} - \mathbf{q}_{min_i}}\right)^2 \tag{7}$$

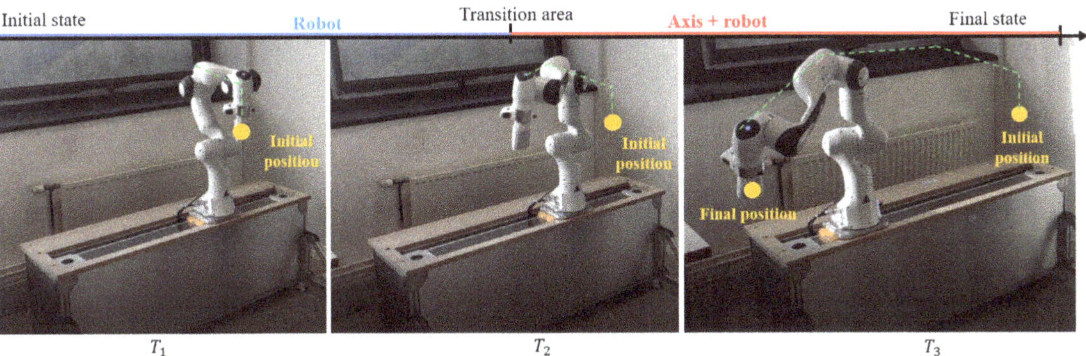

Figure 14. Switching robot/axis motion strategy. During a linear trajectory, the robot is initially activated to execute the desired motion. Once *Xlim* is reached (transition area), the axis is activated instead of the robot to execute the motion along the *Y-axis* (the robot keeps moving along the remaining axes).

The diagonal weighting matrix $\boldsymbol{K}_{JLA} \in \Re^{n \times n}$ enhances to ponderate each joint according to a prediction of those joints having higher risks to reach the mechanical limits during the execution of the cartesian task. Classical JLA strategies set $\boldsymbol{K}_{JLA}$ as constant, which is not an optimal way to exploit the redundant motion of the robot. First, if $\boldsymbol{K}_{JLA}$ is full-rank, it causes a permanent generation of null-space torques for all the joints that are not exactly at the middle point of the joint range $\mathbf{q}_{c_i}$, even if these joints are far of the mechanical limits. Moreover, the simultaneous avoidance of mechanical limits for different combination of joints can be incompatible with respect to the cartesian task and, therefore, insolvable by the robot.

In order to mitigate the limits of the classical approach, let us propose a new way to ponderate the weight of each joint when applying the JLA strategy of Equation (6). The goal is to adapt the weighting value of a joint i according to its proximity to the mechanical limits. Let's denote $\left(\mathbf{q}_{lim_{min_i}}, \mathbf{q}_{lim_{max_i}}\right)$ as the joint position thresholds indicating that the joint is approaching the mechanical limits. Between these two values the weighting value $\boldsymbol{K}_{JLA_i}$ is set to zero since it is estimated that the joint is far enough from the mechanical limits. Once one of the thresholds is reached, $\boldsymbol{K}_{JLA_i}$ linearly increases its value from the threshold and until the mechanical limit. Figure 15a compares the evolution of $\boldsymbol{K}_{JLA_i}$ for both, the classical JLA strategy and the proposed adaptive JLA strategy. Moreover, the consequences for the generation of the null-space torque term $\boldsymbol{K}_{JLA}\left(\frac{\partial \mathbf{w}_{JLA}(\mathbf{q})}{\partial \mathbf{q}}\right)^{T}$ are shown. When the proposed strategy is applied to the 7 joints, the proposed strategy generates zero null-space torques if all the joints move between the threshold limits (Figure 15b), which is an interesting advantage since it allows a manual reconfiguration of the elbow robot if needed. Moreover, the proposed strategy avoids the calculation of unneeded null-space torques and limits the case of incompatible torques acting in the nulls-space.

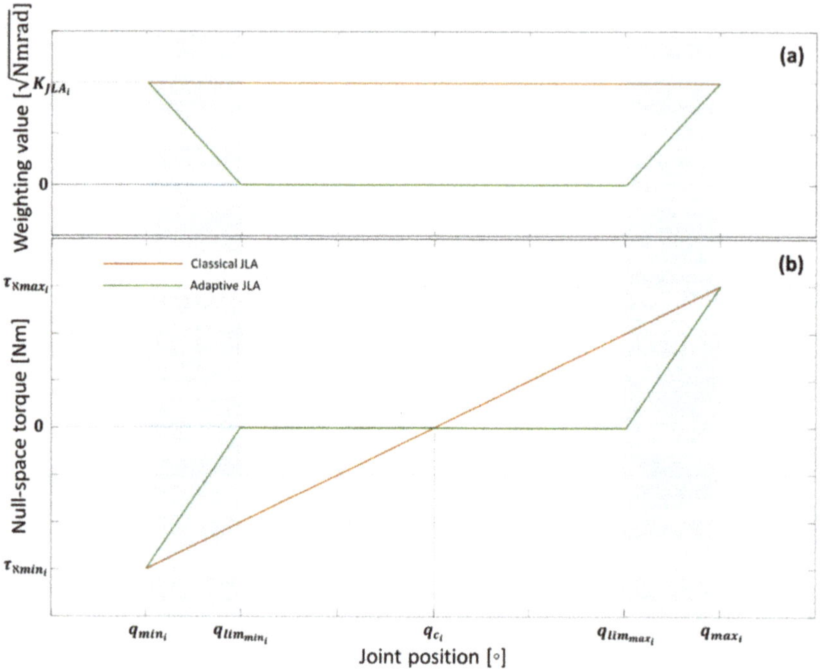

Figure 15. Evolution of (**a**) the weighting value and (**b**) the null-space torque for classical and adaptive JLA strategies along the position range of joint *i*.

4.3. Study Case

In order to validate the usefulness of the proposed adaptive joint-limit avoidance strategy, a situation typically encountered in a Doppler Ultrasound exam is investigated. Since the self-rotation axis of the ultrasound probe is coincident with the last robot joint, i.e., 7th joint, self-rotation movement only requests motions in the last robot joint. This assumption is correct if no null-space torques are applied to the control input. Therefore, the range of the self-rotation is limited to the one of the 7th joints, i.e., $(-166°, 166°)$ for the Franka Emika robot. This is frequently not sufficient to carry out common exams. The degree of redundancy of the anthropomorphic robot can then be used to enlarge the rotational workspace of the robot. Figure 16 shows the setup of the study case, where the robot holds the probe over the patient's body and a self-rotation motion is executed according to the orders of the expert site. For the sake of repeatability, the same desired self-rotation trajectory is applied for all the performed tests presented below.

First, a desired self-rotation motion is performed by the robot without a joint-limit avoidance strategy. Figure 17 compares the desired (ϕ_d) and real (ϕ_r) orientation trajectory, respectively. It is worth mentioning that the time delay evidenced between the two curves is due to the compliant behavior imposed by the control approach. It can be shown that the robot is unable to execute the desired trajectory since it reaches the mechanical limit of the 7th joint, as confirmed by the joint position trajectories (Figure 18). These results also confirm that only the 7th joint is requested to move in the absence of a null-space torque.

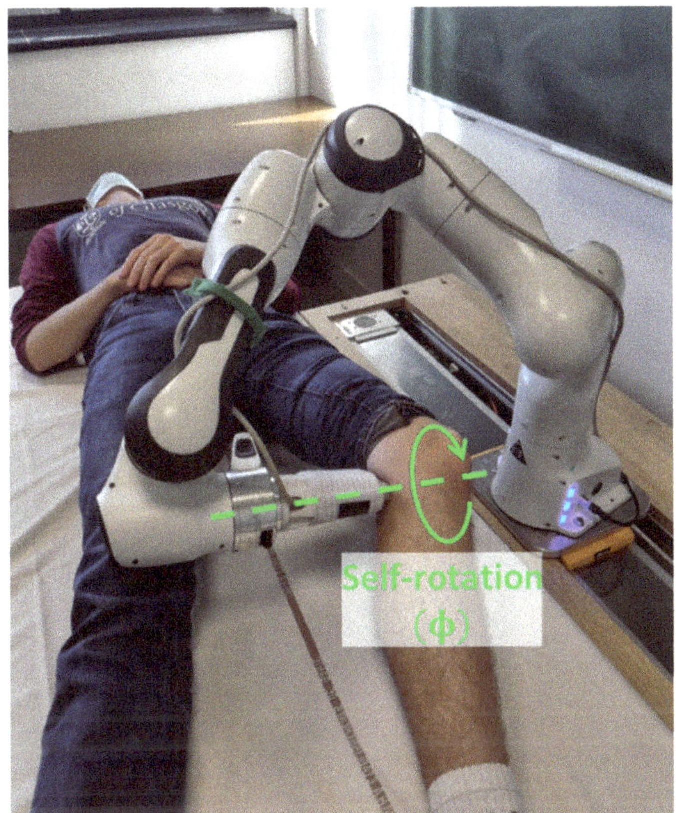

Figure 16. Exemplary image of the study case (no human subjects participated to the experiments). A self-rotation motion is imposed to the haptic probe during a Doppler ultrasound exam (see the accompanying Video S1).

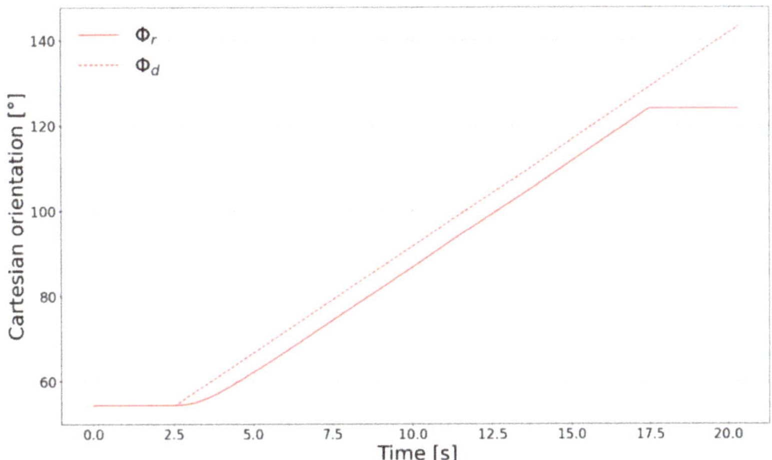

Figure 17. Comparison between the desired probe self-rotation ϕ_d and the executed self-rotation ϕ_r in the absence of a JLA strategy.

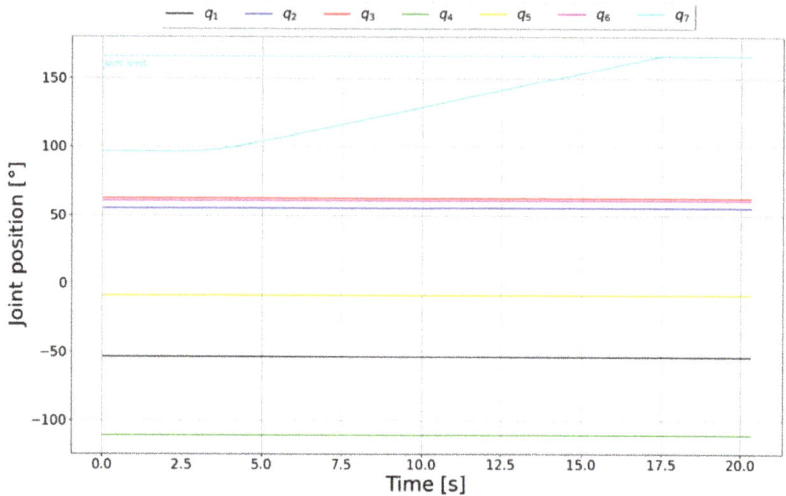

Figure 18. Joint position trajectories in the absence of a JLA strategy.

The same case is investigated when applying the joint-limit avoidance strategy of Equation (6). Initially, the classical JLA approach with constant values at the diagonal weighting matrix is evaluated. The performed experiences are resumed in Table 2 and the results are presented in Figure 19. Since we seek to avoid that the 7th joint reaches the mechanical limits, the first tests have been performed setting a non-zero constant value only to the 7th joint. Experiments have been performed with three different values ($K_{JLA_7} = 6$, 8 and 10 $\sqrt{Nmrad}$) and results in Figure 19 confirm that these values effectively enlarge the rotational robot's workspace moving away from the 7th joint from the mechanical limits. To do this, the robot's elbow motion, related to the degree of redundancy of the cobot, is reoriented accordingly. However, none of these tests allows completely following the desired trajectory since the elbow motion causes the 2nd joint to reach the mechanical limits. Therefore, three supplementary tests have been performed setting non-zero values to joints two and seven (see cases 4–6 in Table 2). Although the obtained results show that the rotational workspace can be enlarged depending on the combination of the constant values. This classical strategy is still limited because of the linear evolution of the null-space torques along with the joints' range, such as were explained in the previous section.

Table 2. Different weighting matrix choices for classical joint-limit avoidance strategy.

Cases	$K_{JLA}\left(\sqrt{Nmrad}\right)$
1	$diag([0, 0, 0, 0, 0, 0, 6])$
2	$diag([0, 0, 0, 0, 0, 0, 8])$
3	$diag([0, 0, 0, 0, 0, 0, 10])$
4	$diag([0, 6, 0, 0, 0, 0, 6])$
5	$diag([0, 8, 0, 0, 0, 0, 8])$
6	$diag([0, 10, 0, 0, 0, 0, 10])$

Finally, we tested the proposed adaptive JLA strategy presented in Section 4.2. The control parameters of the strategy were tuned according to Table 3, where a margin of security of 30° has been set up between the joint limit thresholds and the mechanical limits provided by the constructor.

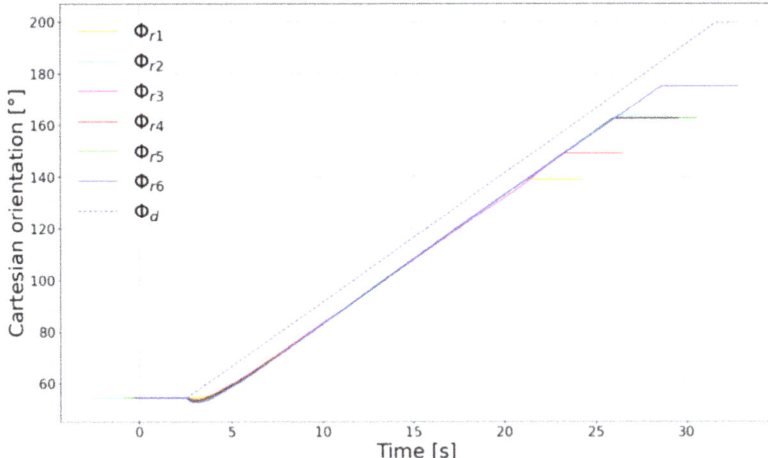

Figure 19. Comparison between the desired probe self-rotation ϕ_d and the executed self-rotation ϕ_r for the different constant weighting matrix choices presented in Table 1.

Table 3. Joint parameters applied for the adaptive JLA strategy. The mechanical limits $\mathbf{q}_{min}$ and $\mathbf{q}_{max}$ are provided by the constructor.

	Mechanical Limits		Joint Limit Thresholds		Weighting Limit
Joint i	q_{min} (°)	q_{max} (°)	$q_{lim_{min}}$ (°)	$q_{lim_{max}}$ (°)	$K_{JLA_i}\left(\sqrt{Nmrad}\right)$
1	−166	166	−136	136	20
2	−101	101	−71	71	20
3	−166	166	−136	136	20
4	−176	−4	−146	−34	20
5	−166	166	−136	136	20
6	−1	215	29	185	20
7	−166	166	−136	136	20

Figures 20–22 show the obtained results for the execution of the desired self-rotation trajectory. It can be seen in Figure 20 that the robot completely executes the desired trajectory due to the adaptive JLA strategy. It is worth mentioning that the offset observed at the end of the trajectory is caused by the tuning of low stiffness values in $\mathbf{K}_p$. Figures 21 and 22 also show that only joints two and seven reach the joint limit thresholds and, therefore, the other weighting joints rest at zero, avoiding the generation of useless null-space torques. Furthermore, unlike the classical JLA strategy, in this case, if none of the joints reaches the corresponding limit thresholds, no null-space torques are generated and the elbow robot can freely be moved by hand.

However, in order to endorse the correct functioning of the entire platform, a test phase in real conditions on multiple exams must be undertaken. This step will validate the conducted work and will initiate the improvement of the haptic interface to meet the angiologist's requirements. This first work opens the door to study multiple control strategies exploiting the degrees of redundancy of the platform.

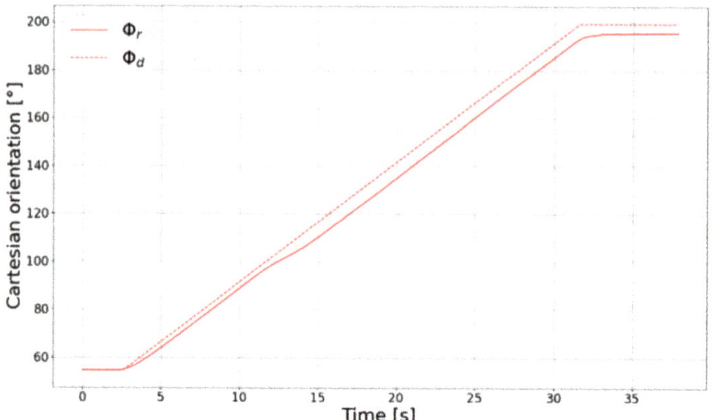

Figure 20. Comparison between the desired probe self-rotation ϕ_d and the executed self-rotation ϕ_r when applying the adaptive JLA strategy.

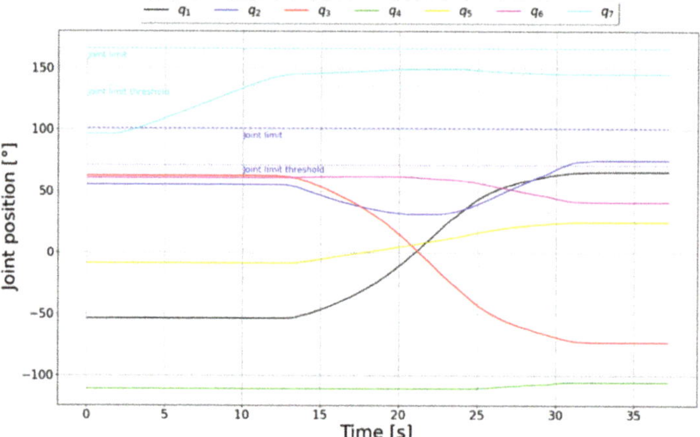

Figure 21. Joint position trajectories when applying the adaptive JLA strategy.

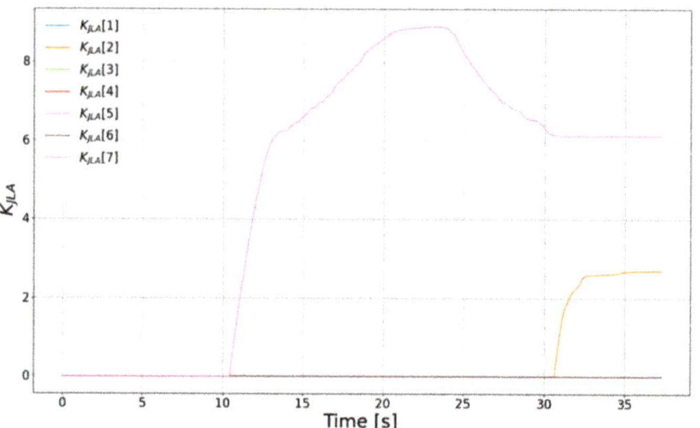

Figure 22. Evolution of the diagonal weighting values related to each robot joint for the adaptive JLA strategy.

5. Conclusions and Future Work

This paper presents an innovative redundant robotic assistant that combines a 7-DoF collaborative Franka robot and a linear axis to achieve a 7 + 1 DoF platform. This robot assistant is conceived to assist sonographers in Doppler Sonography exams and, unlike existing systems, is suitable for full body scanning without requiring manual repositioning of the robot base. The details of the kinematic design have been presented here, including the operator site (6-DoF hybrid haptic interface) and the patient site (7-DoF cobot + 1-DoF linear axis). Moreover, two alternative torque-based control laws have been proposed to deal with the degrees of redundancy of the platform, whereas a compliant behavior has been implemented for the execution of the main medical task, which is a valuable safety feature in case of undesired contacts with the patient. The first mode, called "fully redundant control mode", comprises the 8-DoF in the kinematic model so that the motion of the 8 axes is synchronized. Moreover, the prioritization of the linear axis motion is tuned by a null-space torque added to the torques performing the main task. The second control mode, "decoupled control mode", considers the two systems, the cobot and the linear axis, separately, and a switching strategy is proposed to launch the motion of the linear axis when the cobot is approaching the workspace limits. In addition, an adaptive JLA strategy has been proposed in this paper to exploit the redundancy of the cobot. In fact, unlike the classical strategies tuning a constant weighting matrix for the JLA strategy and generating unsuitable remaining torques, the new strategy proposes a variable weighting matrix, whose values are adapted according to the proximity of each robot joint to its mechanical limit. Therefore, the priority is judiciously given by order to the joints closest to their limits. In the opposite case, if none of the joints is near the mechanical limit, no additional null-space torques are generated. The application of this strategy naturally enlarges the workspace of the robot, as it has been shown for a study case presented for the examination of the inner side of a knee.

Future work will be dedicated to the medical validation of the experimental platform. It is planned to validate the usefulness of the proposed platform during Doppler sonography exams compared to the classical gesture, mostly in terms of time of execution and ergonomic conditions. Moreover, technical adjustments such as the velocity of motion of the linear axis and the choice of the optimal type of control will be studied. Finally, this work opens the door to the study of the proposed adaptive JLA strategy for several applications and robotic systems.

Supplementary Materials: The following supporting information can be downloaded at: https://www.mdpi.com/article/10.3390/act11020033/s1; Video S1: Caption of Figure 16 refers the accompanying video.

Author Contributions: E.G. and J.S. have designed the experiments and co-wrote the paper; A.T. has participated to the development of the platform; J.-M.G. is the Doppler sonographer who has validated the platform; the research work has been supervised by M.A.L., G.C. and S.Z. All authors have read and agreed to the published version of the manuscript.

Funding: This research was funded by the University of Poitiers and PPRIME Institute.

Institutional Review Board Statement: Not applicable.

Informed Consent Statement: Not applicable.

Conflicts of Interest: The authors declare no conflict of interest.

References

1. Hoeckelmann, M.; Rudas, I.J.; Fiorini, P.; Kirchner, F.; Haidegger, T. Current Capabilities and Development Potential in Surgical Robotics. *Int. J. Adv. Robot. Syst.* **2015**, *12*, 61. [CrossRef]
2. Intuitive Surgical Inc. History—Da Vinci Surgery Is Born. August 2012. Available online: http://www.intuitivesurgical.com/company/history/is_born.html (accessed on 10 October 2014).
3. Lefranc, M.; Peltier, J. Evaluation of the ROSA™ Spine robot for minimally invasive surgical procedures. *Expert Rev. Med. Devices* **2016**, *13*, 899–906. [CrossRef] [PubMed]

4. Pearle, A.D.; Kendoff, D.; Stueber, V.; Musahl, V.; Repicci, J.A. Perioperative management of unicompartmental knee arthroplasty using the MAKO robotic arm system (MAKOplasty). *Am. J. Orthop.* **2009**, *38* (Suppl. S2), 16–19. [PubMed]
5. Voros, S.; Haber, G.P.; Menudet, J.F.; Long, J.A.; Cinquin, P. ViKY Robotic scope holder: Initial clinical experience and preliminary results using instrument tracking. *IEEE/ASME Trans. Mechatron.* **2010**, *15*, 879–886. [CrossRef]
6. Carbone, G.; Nakadate, R.; Solis, J.; Ceccarelli, M.; Takanishi, A.; Minagawa, E.; Sugawara, M.; Niki, K. Workspace Analysis and Design Improvement of a Carotid Flow Measurement System. *Proc. Inst. Mech. Eng. Part H J. Eng. Med.* **2010**, *224*, 1311–1323. [CrossRef] [PubMed]
7. Sandoval, J.; Laribi, M.A.; Zeghloul, S.; Arsicault, M.; Guilhem, J.-M. Cobot with Prismatic Compliant Joint Intended for Doppler Sonography. *Robotics* **2020**, *9*, 14. [CrossRef]
8. Sandoval, J.; Laribi, M.A.; Zeghloul, S.; Arsicault, M. Towards a Safe Physical Human-Robot Interaction for Tele-Operated System: Application to Doppler Sonography. In *Mechanism Design for Robotics, Proceedings of the IFToMM Symposium on Mechanism Design for Robotics, Udine, Italy, 11–13 September 2018*; Springer: Singapore, 2018; pp. 335–343.
9. Mathiassen, K.; Fjellin, J.E.; Glette, K.; Hol, P.K.; Elle, O.J. An Ultrasound Robotic System Using the Commercial Robot UR5. *Front. Robot. AI* **2016**, *3*, 1. [CrossRef]
10. Virga, S.; Zettinig, O.; Esposito, M.; Pfister, K.; Frisch, B.; Neff, T.; Navab, N.; Hennersperger, C. Automatic force-compliant robotic ultrasound screening of abdominal aortic aneurysms. In Proceedings of the 2016 IEEE/RSJ International Conference on Intelligent Robots and Systems (IROS), Daejeon, Korea, 9–14 October 2016; pp. 508–513. [CrossRef]
11. Chatelain, P.; Krupa, A.; Navab, N. Confidence-driven control of an ultrasound probe: Target-specific acoustic window optimization. In Proceedings of the 2016 IEEE International Conference on Robotics and Automation (ICRA), Stockholm, Sweden, 16 May 2016; pp. 3441–3446. [CrossRef]
12. Vieyres, P.; Poisson, G.; Courrèges, F.; Smith-Guerin, N.; Novales, C.; Arbeille, P. A Tele-Operated Robotic System for Mobile Tele-Echography: The Otelo Project. In *M-Health*; Istepanian, R.S.H., Laxminarayan, S., Pattichis, C.S., Eds.; Springer: Boston, MA, USA, 2006; pp. 461–473. [CrossRef]
13. Gautreau, E.; Thomas, A.; Sandoval, J.; Laribi, M.A.; Zeghloul, S. Design of an 8-DoF Redundant Robotic Platform for Medical Applications. In *Mechanism Design for Robotics*; Zeghloul, S., Laribi, M.A., Arsicault, M., Eds.; Springer: Cham, Switzerland, 2021; Volume 103, pp. 297–304. [CrossRef]
14. Sandoval, J.; Poisson, G.; Vieyres, P. A new kinematic formulation of the RCM constraint for redundant torque-controlled robots. In Proceedings of the 2017 IEEE/RSJ International Conference on Intelligent Robots and Systems (IROS), Vancouver, BC, Canada, 24–28 September 2017; pp. 4576–4581.
15. Yang, C.; Amarjyoti, S.; Wang, X.; Li, Z.; Ma, H.; Su, C.Y. Visual Servoing Control of Baxter Robot Arms with Obstacle Avoidance Using Kinematic Redundancy. In *Intelligent Robotics and Applications. ICIRA 2015*; Lecture Notes in Computer Science; Liu, H., Kubota, N., Zhu, X., Dillmann, R., Zhou, D., Eds.; Springer: Cham, Switzerland, 2015; Volume 9244. [CrossRef]
16. Atawnih, A.; Papageorgiou, D.; Doulgeri, Z. Kinematic control of redundant robots with guaranteed joint limit avoidance. *Robot. Auton. Syst.* **2016**, *79*, 122–131. [CrossRef]
17. Yoshikawa, T. Manipulability of Robotic Mechanisms. *Int. J. Robot. Res.* **1985**, *4*, 3–9. [CrossRef]
18. Evans, K.; Roll, S.; Baker, J. Work-Related Musculoskeletal Disorders (WRMSD) Among Registered Diagnostic Medical Sonographers and Vascular Technologists: A Representative Sample. *J. Diagn. Med. Sonogr.* **2009**, *25*, 287–299. [CrossRef]
19. Essomba, T.; Sandoval, J.; Laribi, M.A.; Wu, C.-T.; Breque, C.; Zeghloul, S.; Richer, J.-P. Kinematic and Force Experiments on Cadavers for the Specification of a Tele-Operated Craniotomy Robot. In *Mechanism Design for Robotics, Proceedings of the 29th International Conference on Robotics in Alpe Adria Danube Region (RAAD2019), Kaiserslautern, Germany, 19–21 June 2019*; Springer Science and Business Media LLC: Berlin/Heidelberg, Germany, 2020; pp. 447–454.
20. Tobergte, A.; Helmer, P.; Hagn, U.; Rouiller, P.; Thielmann, S.; Grange, S.; Albu-Schaffer, A.; Conti, F.; Hirzinger, G. The sigma.7 haptic interface for MiroSurge: A new bi-manual surgical console. In Proceedings of the 2011 IEEE/RSJ International Conference on Intelligent Robots and Systems, San Francisco, CA, USA, 25–30 September 2011; pp. 3023–3030. [CrossRef]
21. Tsumaki, Y.; Naruse, H.; Nenchev, D.; Uchiyama, M. Design of a compact 6-DOF haptic interface. In Proceedings of the IEEE International Conference on Robotics and Automation, Leuven, Belgium, 23–27 May 1998; Volume 3, pp. 2580–2585.
22. Tsai, L.-W. Multi-Degree-of-Freedom Mechanisms for Machine Tools and the Like. U.S. Patent 5,656,905, 12 August 1997.
23. Siciliano, B.; Khatib, O. *Springer Handbook of Robotics*; Springer International Publishing: Berlin/Heidelberg, Germany, 2016.
24. Dietrich, A.; Bussmann, K.; Petit, F.; Kotyczka, P.; Ott, C.; Lohmann, B.; Albu-Schäffer, A. Whole-body impedance control of wheeled mobile manipulators. *Auton. Robot.* **2015**, *40*, 505–517. [CrossRef]
25. Trabelsi, A.; Sandoval, J.; Ghiss, M.; Laribi, M.A. Development of a Franka Emika Cobot Simulator Platform (CSP) Dedicated to Medical Applications. In *Advances in Mechanism and Machine Science*; Springer: Singapore, 2021; Volume 102, pp. 95–103. [CrossRef]

Article

Application of the Half-Order Derivative to Impedance Control of the 3-PUU Parallel Robot

Luca Bruzzone *, Pietro Fanghella and Davide Basso

DIME Department, University of Genoa, 16145 Genoa, Italy; pietro.fanghella@unige.it (P.F.); dvd.genova@gmail.com (D.B.)
* Correspondence: luca.bruzzone@unige.it; Tel.: +39-010-3352967

Abstract: This paper presents an extension of impedance control of robots based on fractional calculus. In classical impedance control, the end-effector reactions are proportional to the end-effector position errors through the stiffness matrix K, while damping is proportional to the first-order time-derivative of the end-effector coordinate errors through the damping matrix D. In the proposed approach, a half-derivative damping is added, proportional to the half-order time-derivative of the end-effector coordinate errors through the half-derivative damping matrix HD. The discrete-time digital implementation of the half-order derivative alters the steady-state behavior, in which only the stiffness term should be present. Consequently, a compensation method is proposed, and its effectiveness is validated by multibody simulation on a 3-PUU parallel robot. The proposed approach can be considered the extension to MIMO robotic systems of the $PDD^{1/2}$ control scheme for SISO mechatronic systems, with potential benefits in the transient response performance.

Keywords: impedance control; fractional calculus; half-order derivative; parallel kinematics machine

Citation: Bruzzone, L.; Fanghella, P.; Basso, D. Application of the Half-Order Derivative to Impedance Control of the 3-PUU Parallel Robot. *Actuators* **2022**, *11*, 45. https://doi.org/10.3390/act11020045

Academic Editors: Marco Carricato and Edoardo Idà

Received: 20 December 2021
Accepted: 30 January 2022
Published: 1 February 2022

Publisher's Note: MDPI stays neutral with regard to jurisdictional claims in published maps and institutional affiliations.

1. Introduction

In a wide range of robotic applications, for example, assembly of electronic boards or handling of objects to be placed on horizontal pallets, the full mobility (6-DOF) of the end-effector is not necessary, since the tasks can be proficiently performed by means of a 3-DOF translational motion or by a 4-DOF motion with three translations and one rotation around a vertical axis (Schoenflies motion [1]). The rotational degree of freedom of the Schoenflies motion is often obtained by adding a 1-DOF wrist to a translational mechanism.

Considering serial robots, translational and Schoenflies motions are realized in most cases by Cartesian robots or by SCARA robots [2–5]. Considering parallel kinematics machines (PKMs), translational motion can be obtained by parallel Cartesian robots [6–8], characterized by three legs that are planar serial mechanisms moved by three orthogonal linear actuators perpendicularly to their planes, or by other closed-loop schemes which are not purely translational in general, but become purely translational in case of specific orientations of the joint axes [1,9–11]. If necessary, translational PKMs can be upgraded to Schoenflies motions by a 1-DOF wrist, but there are also some designs of PKMs which perform native Schoenflies motions [12,13].

No matter if it is serial or parallel, a robot can be controlled in position if the task can be correctly performed regulating only the end-effector trajectory, or by hybrid position/force control when the proper execution of the task requires accurate force regulation in some phases and in some directions [14]. An intermediate approach, which is widely adopted since it does not require force sensors, is represented by impedance control [15,16].

The basic concept of impedance control is that the robot end-effector, subject to external forces, follows a trajectory with a predetermined spatial compliance. The relationship between the force/moment exerted by the end-effector on the environment and the end-effector position/orientation error is defined by means of the stiffness and damping matrices. This approach allows to obtain a compliant behavior in the directions which must be force-controlled, and a stiff behavior in the directions which must be position-controlled.

There are many possible ways to the define the orientation of a rigid body in space [17], and this results in different possible approaches for the definition of the rotational stiffness/damping in impedance control [18–21]. On the contrary, for translational robots, the approach to impedance control is much less diversified and simpler, since a point position in space is always represented using an orthogonal reference frame. Additionally, in case of a robot with Schoenflies motion, impedance control is simplified, since the rotation is mono-dimensional, and the rotational behavior is decoupled from the translational behavior. In this work, we will consider only translational impedance control, considering that it can be easily extended to robots with Schoenflies motion.

For translational robots, the impedance behavior of the end-effector is usually defined by means of the stiffness and damping matrices, which respectively represent the zero-order and the first-order terms of a three-dimensional PD control in the external coordinates, expressed in the principal reference frame. These matrices are non-diagonal in the world frame when the principal stiffness and damping directions are not parallel to the axes of the fixed reference frame [11].

Some authors have proposed nonlinear impedance algorithms, in which nonlinear stiffness and damping are imposed to the end-effector, in particular for a better execution of cooperative human–robot tasks, or to maintain position/orientation within a specified region even in case of excessive forces/moments [22–24].

An alternative method to define the impedance behavior is based on fractional calculus, which introduces derivatives and integrals of non-integer order [25]. Accordingly, the damping term can be defined proportional not to the first-order derivative of the end-effector error, but to a derivative with generic non-integer order μ. In the scientific literature, there are some examples of fractional-order impedance controls of robots [26,27]. This kind of impedance control generalizes to a three-dimensional system the fractional-order PD^{μ} control scheme for SISO systems [28]. Fractional-order impedance can be used, for example, to perform contact force tracking control more accurately than traditional impedance control [29].

In the proposed work, the stiffness/damping behavior imposed by the impedance control is linear, but a half-order term, based on the fractional derivative of order 1/2 of the position error, is added to the zero-order and first-order terms of classical impedance control. This impedance control generalizes to a three-dimensional system the fractional-order $\text{PDD}^{1/2}$ control scheme developed, so far, for single axes [30].

As a matter of fact, it is evident that traditional linear impedance control of a n-DOF mechanical system (KD) is the n-dimensional version of the PD control of a 1-DOF mechanical system, with the n-dimensional stiffness term K, proportional to the position error, corresponding to the proportional term, and the n-dimensional damping term D, proportional to the first-order derivative of the position error, corresponding to the derivative term. The proposed KDHD impedance control is obtained from the KD scheme with the addition of an n-dimensional half-order damping term HD, proportional to the derivative of order 1/2 (half-derivative) of the position error; therefore, the aim of this work is to investigate if the addition of the half-derivative term in the implementation of the impedance algorithm can bring the same benefits that have been shown theoretically and experimentally for the $\text{PDD}^{1/2}$ control of a 1-DOF inertial system with respect to the classical PD scheme [31].

As a case study, the KDHD impedance control is applied to a 3-PUU parallel robot and compared to the classical KD impedance control.

In the remainder of the paper, Section 2 discusses the theoretical definition and the digital implementation of the half-derivative of a function of time, Section 3 proposes the KDHD impedance control, highlighting the differences with respect to the classical KD impedance control, and Section 4 presents the 3-PUU architecture and its kinematics. Section 5 compares the KDHD and the KD impedance controls applied to the 3-PUU robot by multibody simulation, and Section 6 outlines conclusions and future developments.

2. Half-Order Derivative: Definition and Digital Implementation Issues

Fractional calculus generalizes the concept of derivative and integral to non-integer order [25]. According to the Grünwald–Letnikov definition, suitable for robust discrete-time implementation [32], the fractional differential operator for a continuous function of time, $x(t)$, is defined as:

$$\frac{d^\alpha}{dt^\alpha}x(t) = \lim_{h \to 0}\frac{1}{h^\alpha}\sum_{k=0}^{\left[\frac{t-a}{h}\right]}(-1)^k\frac{\Gamma(\alpha+1)}{\Gamma(k+1)\Gamma(\alpha-k+1)}x(t-kh),\tag{1}$$

where $\alpha \in \mathbb{R}^+$ is the differentiation order, a and t are the fixed and variable limits, Γ is the Gamma function, h is the time increment, and $[y]$ is the integer part of y. For real-time digital implementation with sampling time T_s, Equation (1) can be rewritten in the following form [31]:

$$x(t)^{(\alpha)} \cong x(kT_s)^{(\alpha)} \cong \frac{1}{T_s^\alpha}\left(\sum_{j=0}^{k}w_j^\alpha x((k-j)T_s)\right), k = [(t-a)/T_s],\tag{2}$$

where:

$$w_0^\alpha = 1, \; w_j^\alpha = \left(1-\frac{\alpha+1}{j}\right)w_{j-1}^\alpha, \, j = 1, \, 2, \, \ldots .\tag{3}$$

The calculation of Equation (2) requires considering all the $k + 1$ sampled values of the time history of x; therefore, the incessant increase of the number of addends is an issue for real-time implementation, and consequently, only a fixed number of n previous steps is used in Equation (2), realizing a nth order digital filter, with fixed memory length $L = nT_s$. This is acceptable according to the so-called short-memory principle [33], which states that considering only the recent past of the function in the interval $[t–L, t]$ does not yield relevant approximations in the evaluation of the fractional derivative.

Nevertheless, the truncation of the summation of Equation (2) to $n + 1$ addends is an issue if the fractional derivative is applied to impedance control. As a matter of fact:

$$\lim_{k \to \infty}\sum_{j=0}^{k}w_j^\alpha = 0,\tag{4}$$

but with finite n, this summation is non-null:

$$W_\alpha(n) = \sum_{j=0}^{n}w_j^\alpha > 0.\tag{5}$$

Moreover, W_α tends to zero quite slowly, as shown in Figure 1, as an example, for $\alpha = 1/2$ (half-derivative).

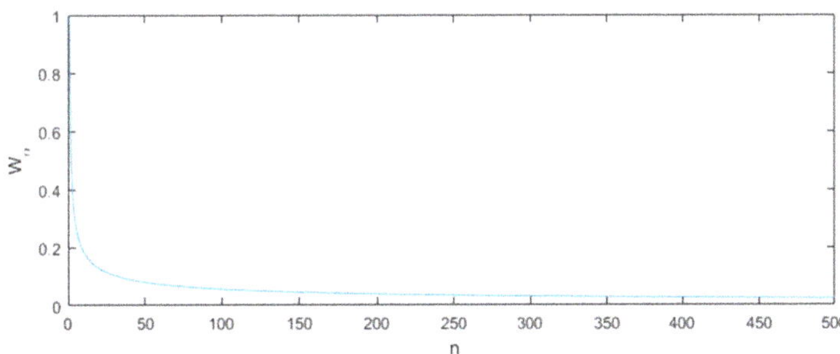

Figure 1. Summation of the approximated fractional-derivative filter terms (W_α) as a function of the filter order, for $\alpha = 1/2$ (half-derivative).

As a consequence, the numerically evaluated fractional derivative of a constant c is non-null, but equal to $cW_\alpha(n)/T_s^\alpha$, tending to zero as n tends to infinite. This influences the behavior of impedance control in the steady state, giving rise to a constant position error. In this condition, only the stiffness term should be non-null, but if fractional-order damping is numerically evaluated with a finite-order digital filter, the half-derivative damping term is also present, and should be properly compensated.

3. KDHD Impedance Control

The classical formulation of impedance control with gravity compensation of a non-redundant parallel robot is expressed by the following control law [11]:

$$\boldsymbol{\tau} = \left(\mathbf{J}_p^T\right)^{-1}\left[\mathbf{K}_{KD}(\mathbf{x}_d - \mathbf{x}(\mathbf{q})) + \mathbf{D}_{KD}(\mathbf{x}_d - \mathbf{x}(\mathbf{q}))^{(1)}\right] + \boldsymbol{\tau}_g(\mathbf{q}), \tag{6}$$

where $\mathbf{J}_p$ is the Jacobian matrix for a parallel robot, which defines the relationship between the time-derivative of the external coordinates $\mathbf{x}$ and the time-derivative of the internal coordinates $\mathbf{q}$:

$$\dot{\mathbf{q}} = \mathbf{J}_p\dot{\mathbf{x}}, \tag{7}$$

In Equation (6), $\mathbf{K}_{KD}$ and $\mathbf{D}_{KD}$ are the stiffness and damping matrices, $\mathbf{x}_d$ is the reference trajectory expressed in external coordinates, and $\boldsymbol{\tau}_g$ is the gravity compensation vector.

Let us note that the matrices $\mathbf{K}_{KD}$ and $\mathbf{D}_{KD}$, on the basis of the robot mobility, define the translational impedance, the rotational impedance, or both [21]. In case of robots with Schoenflies motion, their size is 4×4, but the rotational behavior is evidently decoupled from the translational behavior; therefore, the stiffness and damping matrices are block-diagonal, with a 3×3 submatrix representing the translational behavior and a fourth diagonal element representing the rotational behavior. In the following, for the sake of simplicity, we will consider translational robots with 3×3 stiffness and damping matrices, bearing in mind that the proposed approach can be easily extended to robots with Schoenflies motion.

The stiffness and damping matrices are in general non-diagonal in the fixed reference frame, W. On the basis of the task requirements, it is possible to choose a principal reference frame, P, in which it is convenient to define decoupled stiffness and damping; therefore, the stiffness and damping matrices are defined diagonal in P and are transformed into the frame W by means of the rotation matrix between W and P [21]:

$$\mathbf{K}_{KD} = \left(\mathbf{R}_W^P\right)^T \mathbf{K}_{KD}^P \, \mathbf{R}_W^P, \tag{8}$$

$$\mathbf{D}_{KD} = \left(\mathbf{R}_W^P\right)^T \mathbf{D}_{KD}^P \, \mathbf{R}_W^P. \tag{9}$$

In the proposed KDHD impedance control, the damping term is not proportional only to the first-order derivative of the position error in external coordinates, but also to its half-derivative (derivative of order 1/2), in order to implement the extension from PD to $PDD^{1/2}$ in the three-dimensional space:

$$\boldsymbol{\tau} = \left(\mathbf{J}_p^T\right)^{-1}\left[\mathbf{K}_{KDHD}(\mathbf{x}_d - \mathbf{x}(\mathbf{q})) + \mathbf{D}_{KDHD}(\mathbf{x}_d - \mathbf{x}(\mathbf{q}))^{(1)} + \mathbf{HD}_{KDHD}(\mathbf{x}_d - \mathbf{x}(\mathbf{q}))^{(1/2)}\right] + \boldsymbol{\tau}_g(\mathbf{q}) \tag{10}$$

Obviously, the half-derivative damping matrix $\mathbf{HD}_{KDHD}$ can also be defined in the principal reference frame, P, and then transformed to the frame W:

$$\mathbf{HD}_{KDHD} = \left(\mathbf{R}_W^P\right)^T \mathbf{HD}_{KDHD}^P \, \mathbf{R}_W^P. \tag{11}$$

Let us note that in the definition of the KDHD impedance control law, a non-redundant parallel robot has been considered. In case of non-redundant serial robots, the approach is similar, but it is more convenient to adopt the Jacobian matrix, which transfers from internal coordinates' derivatives to external coordinates' derivatives [2]:

$$\dot{\mathbf{x}} = \mathbf{J}_s \dot{\mathbf{q}}.$$ (12)

Consequently, for application to non-redundant serial robots, the impedance control laws (6) and (10) can be rewritten in the following form:

$$\boldsymbol{\tau} = \mathbf{J}_s^T \left[\mathbf{K}_{KD}(\mathbf{x}_d - \mathbf{x}(\mathbf{q})) + \mathbf{D}_{KD}(\mathbf{x}_d - \mathbf{x}(\mathbf{q}))^{(1)} \right] + \boldsymbol{\tau}_g(\mathbf{q}),$$ (13)

$$\boldsymbol{\tau} = \mathbf{J}_s^T \left[\mathbf{K}_{KDHD}(\mathbf{x}_d - \mathbf{x}(\mathbf{q})) + \mathbf{D}_{KDHD}(\mathbf{x}_d - \mathbf{x}(\mathbf{q}))^{(1)} + \mathbf{HD}_{KDHD}(\mathbf{x}_d - \mathbf{x}(\mathbf{q}))^{(1/2)} \right] + \boldsymbol{\tau}_g(\mathbf{q})$$ (14)

4. Kinematic and Dynamic Model of the 3-PUU Parallel Robot

The proposed KDHD impedance control has been tested by multibody simulation on the 3-PUU parallel robot, whose scheme is shown in Figure 2.

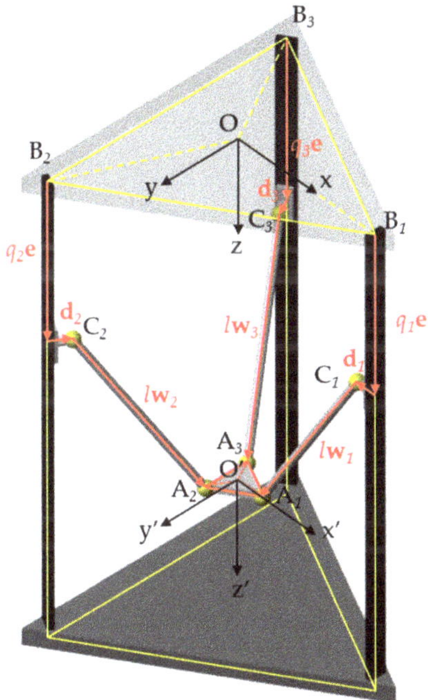

Figure 2. Geometrical model of the 3-PUU parallel robot.

The reference frame O(x,y,z) of the base platform is located at the center of the equilateral triangle $B_1B_2B_3$, whose vertices lie on a circumference with center O and radius R, while the reference frame O'(x',y',z') of the moving platform is located at the center of the equilateral triangle $A_1A_2A_3$, whose vertices lie on a circumference with center O' and radius r. The unit vector **e**, perpendicular to the plane of the triangle $B_1B_2B_3$, defines the direction of the three prismatic joints, and the vector of the internal coordinates $\mathbf{q} = [q_1, q_2, q_3]^T$ is composed by the distances of the centers of the three sliders from the points B_1, B_2, and B_3; consequently, the three vectors, $q_1\mathbf{e}$, $q_2\mathbf{e}$, and $q_3\mathbf{e}$, represent the displacements of the sliders' centers with respect to B_1, B_2, and B_3. For constructive reasons, the centers of the universal joints mounted on the sliders do not lie on the prismatic joints' axes, but are shifted by the three vectors $\mathbf{d}_i$, $i = 1 \ldots 3$, with equal module d and the direction of the vector (O–B_i).

Six universal joints are located at the points C_1, C_2, C_3, A_1, A_2, and A_3. In order to obtain purely translational motion of the moving platform, in each PUU kinematic chain, the first revolute axis of the upper U joint must be parallel to the first revolute axis of the lower U joint, and the second revolute axis of the upper U joint must be parallel to the second revolute axis of the lower U joint [34].

5. Comparison of the KD and KDHD Impedance Control Laws by Multibody Simulation

5.1. Modeling and Simulation Overview

The model of the 3-PUU parallel robot has been implemented in the multibody simulation environment Simscape Multibody™ by MathWorks (Figure 3). The robot geometrical and inertial parameters considered in the simulations are collected in Table 1.

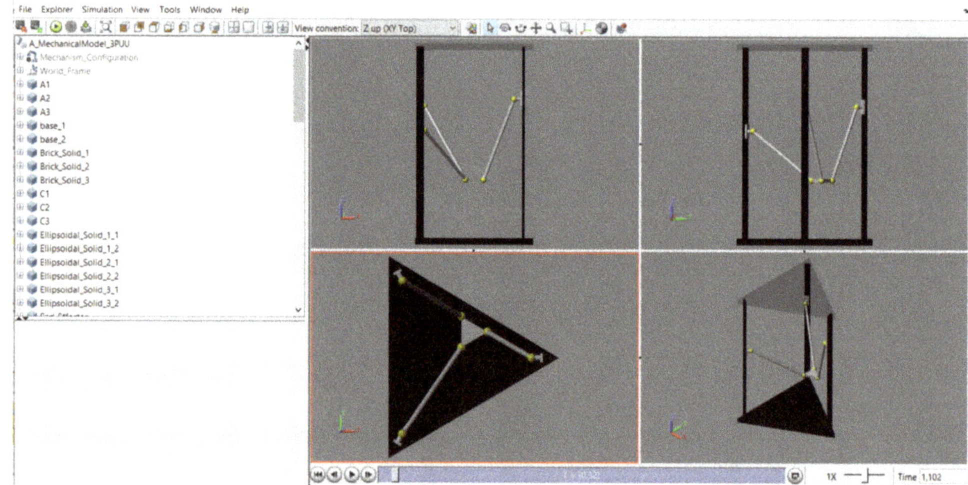

Figure 3. Multibody model of the 3-PUU parallel robot.

Table 1. Geometrical and inertial parameters of the 3-PUU parallel robot.

Symbol	Parameter	Value	Unit
r	moving platform radius	0.0693	m
R	base platform radius	0.3864	m
d	distance from slider center and universal joint center	0.04	m
l	leg length	0.5	m
m_{mp}	moving platform mass (including three universal joints)	5	kg
m_l	mass of one leg	1	kg
m_s	mass of one slider (including one universal joint)	1.5	kg

The proposed KDHD impedance control law has been compared to KD impedance control in the following case studies.

(A) End-effector reference trajectory, $\mathbf{x}_d$, characterized by displacements along the three directions with trapezoidal speed laws, no external force on the end-effector, and the end-effector behavior is isotropic: the stiffness and damping matrices are diagonal with three equal elements. This case represents an approach/depart motion without contact with the environment.

(B) Constant end-effector reference position and external force applied to the end-effector. The stiffness and damping matrices are diagonal in the world frame (the fixed reference frame, W, coincides with the principal stiffness/damping reference frame, P), but the end-effector behavior is not isotropic, since the three diagonal elements are not equal. This case study evidences how with the KDHD scheme, the approximation of the half-derivative calculation by means of a digital filter with a fixed memory length alters the stiffness imposed by impedance control, as discussed in Section 2; consequently, the stiffness matrix must be properly compensated. A compensation method (KDHDc) is proposed, and its effectiveness is validated by simulation.

(C) Constant end-effector reference position and external force applied to the end-effector, as in case B, but with non-diagonal stiffness and damping matrices, since the fixed reference frame, W, does not coincide with the principal stiffness/damping reference frame, P. Additionally, in this case, the effectiveness of the proposed stiffness-compensated KDHDc impedance control is validated by simulation.

In all the case studies, gravity force is neglected, since its effect is exactly compensated by the gravity compensation term, which is equal for all the considered impedance controls schemes. This allows to better highlight the differences between the KD, KDHD, and KDHDc schemes, eliminating the equal contribution to the actuation forces due to gravity.

Impedance control is based on the measurement of the internal coordinates of the robot without measurement of the force exerted by the end-effector on the environment, as in the case of hybrid position/force control. The internal coordinates of the robot, both for rotational and for linear actuators, are measured by digital encoders, which are not affected by noise; therefore, we have decided not to take into account noise on the measured position of the three sliders.

On the other hand, some noise is certainly present on the actuation forces, due to the electrical effects of the current loop of the motor drivers; nevertheless, the sensitivity of the closed-loop transfer function to disturbances on the direct path, which comprises the motor driver, is much lower than the sensitivity to disturbances on the feedback path. Consequently, noise has been neglected in the first stage of the research.

5.2. Case Study A

In this case study, the stiffness and damping matrices are diagonal and isotropic both for the KD and KDHD impedance controls. For the KD scheme:

$$\mathbf{K}_{KD} = k_{KD}\mathbf{I}, \tag{15}$$

$$\mathbf{D}_{KD} = d_{KD}\mathbf{I}, \tag{16}$$

where $\mathbf{I}$ is the identity matrix. The diagonal value, d_{KD}, of the damping matrix can be selected starting from the nondimensional damping coefficient, ζ_{KD}, according to the following expression [35]:

$$d_{KD} = 2\zeta_{KD}\sqrt{k_{KD}m_{mp}}, \tag{17}$$

in which the mass, m_{mp}, of the moving platform (including the three lower universal joints) is considered.

For the KDHD scheme, the stiffness matrix and the first-order and half-order damping matrices are:

$$\mathbf{K}_{KDHD} = k_{KDHD}\mathbf{I}, \tag{18}$$

$$\mathbf{D}_{KDHD} = d_{KDHD}\mathbf{I}, \tag{19}$$

$$\mathbf{HD}_{KDHD} = hd_{KDHD}\mathbf{I}. \tag{20}$$

The diagonal values d_{KDHD} and hd_{KDHD} of the damping matrices can be selected starting from the nondimensional coefficients ζ_{KDHD} and ψ_{KDHD}, according to the following expressions:

$$d_{KDHD} = 2\zeta_{KDHD}\sqrt{k_{KDHD}m_{mp}}, \tag{21}$$

$$hd_{KDHD} = \psi_{KDHD}^{3/4}m_{mp}^{1/4}. \tag{22}$$

The coefficients ζ_{KDHD} and ψ_{KDHD} non-dimensionally represent the derivative and half-derivative damping terms [35]. In [36], three couples of PD and PDD$^{1/2}$ tunings are compared in the control of a linear axis (Table 2).

Table 2. Nondimensional parameters of the compared PD and PDD$^{1/2}$ tunings.

KD/KDHD Comparison	PD Control/ KD Impedance Control	PDD$^{1/2}$ Control/ KDHD Impedance Control	
	ζ_{KD}	ζ_{KDHD}	ψ_{KDHD}
I	0.8	0.46	0.7990
II	1	0.45	1.4266
III	1.2	0.48	2.1510

These three couples of PD and PDD$^{1/2}$ tunings have been selected as starting points of the research using the nondimensional approach discussed in [36] to derive the coefficients ζ and ψ of a PDD$^{1/2}$ controller from the coefficient ζ of a reference PD controller, with application to a nondimensional second-order purely inertial linear system, with transfer function $G(s) = 1/s^2$. This method can be summarized as follows:

- A PD closed-loop control with a given ζ is applied to the position control of $G(s)$, considering a step input.
- The settling energy of the step response is calculated.
- There are infinite combinations of ζ and ψ for a PDD$^{1/2}$ controller with the same proportional gain which are characterized by the same settling energy of the PD. The ζ–ψ combination which minimizes the settling time is selected.

Table 2 collects the PDD$^{1/2}$ ζ–ψ combinations associated to three reference PD controllers, with ζ = 0.8, 1, and 1.2. The basic idea of this approach is to obtain a PDD$^{1/2}$ tuning with a similar control effort as the corresponding PD tuning, but with improved readiness. Since this approach is nondimensional, the association between the coefficient ζ of the PD and the ζ–ψ combination of the PDD$^{1/2}$ controller is not influenced by the system mass/inertia nor by the proportional gain; moreover, even if this approach is based on the step input, the benefits in terms of control readiness of the PDD$^{1/2}$ controller are also demonstrated with different reference inputs [36].

The stiffness and damping matrices for the KD and KDHD impedance controls have been obtained starting from the nondimensional values of Table 2, imposing $k_{KD} = k_{KDHD} = 1 \cdot 10^3$ N/m and using Equations (15) to (22); then, the two control laws have been compared in the presence of a trapezoidal reference trajectory, $\mathbf{x}_d$, characterized by four phases:

1. Constant velocity from $(0, 0, z_{d,in})$ to $(0.1, 0.1, 0.1 + z_{d,in})$ [m], where $z_{d,in}$ is the initial z-coordinate of the end-effector, which does not influence the simulation results due to the architecture of the PKM, with three sliders aligned along the fixed frame z-axis. The duration of this phase is t_{ramp}.
2. $\mathbf{x}_d$ remains constant in $(0.1, 0.1, 0.1 + z_{d,in})$ [m] for t_{stop}.
3. $\mathbf{x}_d$ returns to $(0, 0, z_{d,in})$ [m] with constant velocity in t_{ramp}.
4. $\mathbf{x}_d$ remains constant in $(0, 0, z_{d,in})$ [m] for t_{stop}.

Some simulation results are shown in Figures 4 and 5, with reference to the KD-KDHD comparison number I (ζ_{KD} = 0.8, Table 2), ramp time t_{ramp} = 0.5 s, and stop time = 3 s.

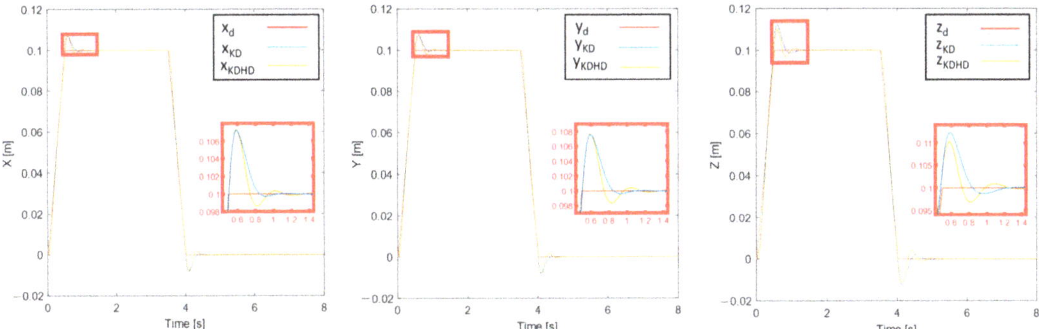

Figure 4. Case study A, $k_{KD} = k_{KDHD} = 1 \cdot 10^3$ N/m, KD-KDHD comparison number I, ramp time $t_{ramp} = 0.5$ s, stop time $t_{stop} = 3$ s, external coordinates.

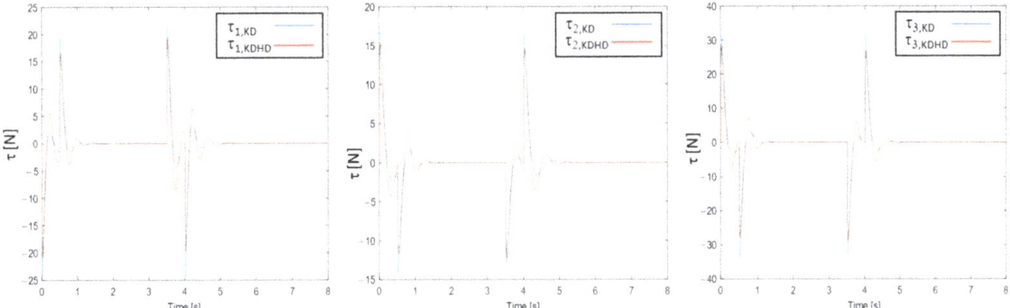

Figure 5. Case study A, $k_{KD} = k_{KDHD} = 1 \cdot 10^3$ N/m, KD-KDHD comparison number I, ramp time $t_{ramp} = 0.5$ s, stop time $t_{stop} = 3$ s, actuation forces.

Table 3 summarizes the results of the case study A with the three considered KD-KDHD tuning couples of Table 2, and with two different ramp times (0.5 and 1 s), while the stop time has been kept constant at 3 s, sufficient to reach the settling time to within 2% after phases 1 and 3. The half-derivative is calculated using Equation (2) with sampling time $T_s = 0.005$ s and filter order $n = 10$, and these values are suitable for a real-time digital implementation on a commercial controller. The total control effort related to the three motors, reported in Table 3, is defined as:

$$E_{tot} = \sum_{i=1}^{3} \int_{0}^{T_{sim}} \tau_i^2 dt, \tag{23}$$

where T_{sim} is the simulation time.

Comparing the performances of the two impedance controllers in the six cases of Table 3, it is possible to observe the following outcomes:

-   The KDHD control has lower integral absolute errors of all the external coordinates in all the cases (average variation with respect to KD: −25.0%), and has lower maximum actuation forces for all three motors in all the cases (average variation with respect to KD: −17.6%). This is the most interesting aspect, because it means that the KDHD control can reduce the tracking error with same size of actuators; moreover, this confirms that the extension from KD to KDHD has benefits similar to the extension from PD to PDD$^{1/2}$ [31], supposed as the starting point of the investigation.

- Considering the maximum error, in most cases, the KDHD is better than the KD, but not for all the external coordinates and for all the cases, such as for the integral absolute error; however, the average reduction of the maximum error is -3.4%.
- The settling time to within 2% of the commanded displacement is in general better for the KDHD, with an average reduction over the six cases and over the three external coordinates of -25.8%.
- The total control effort is higher for the KDHD with respect to KD, with an average increase over the six cases and over the three external coordinates of $+24.7\%$.

Table 3. KD/KDHD impedance control comparison, case study A.

Comparison		t_{ramp} (s)	Maximum Motor Force (N)			Total Control Effort (N²s)	Maximum Error (mm)			Integral Absolute Error (IAE) (mm·s)			Settling Time to Within 2% (s)		
			$\tau_{1,max}$	$\tau_{2,max}$	$\tau_{3,max}$	E_{tot}	$e_{x,max}$	$e_{y,max}$	$e_{z,max}$	IAE_x	IAE_y	IAE_z	$t_{s,x}$	$t_{s,y}$	$t_{s,z}$
I	KD	0.5	25.48	17.38	34.74	422.78	8.6	9.1	12.8	5.9	6.3	11.7	0.2415	0.2555	0.2760
	KDHD		19.48	15.43	28.52	470.35	8.3	8.8	11.9	5.4	5.7	9.9	0.1980	0.2100	0.3925
II	KD	0.5	31.36	21.19	42.29	470.86	7.6	8.0	11.2	5.5	5.7	10.0	0.2450	0.2590	0.2950
	KDHD		23.06	18.25	33.77	557.19	7.3	7.7	10.4	4.2	4.5	7.2	0.1700	0.1800	0.2960
III	KD	0.5	37.36	25.15	49.89	533.46	6.8	7.2	10.1	5.3	5.5	8.9	0.2465	0.2615	0.3145
	KDHD		27.47	21.67	40.23	640.80	6.3	6.7	9.0	3.1	3.3	5.3	0.1425	0.1520	0.1520
I	KD	1	12.35	8.34	16.28	95.94	4.1	4.4	6.1	3.1	3.3	5.8	0.2050	0.2200	0.2380
	KDHD		9.71	7.72	14.39	122.15	4.2	4.4	6.0	2.9	3.1	5.2	0.1640	0.1755	0.1825
II	KD	1	15.26	10.25	20.12	105.05	3.6	3.8	5.5	2.9	3.0	5.1	0.1915	0.2080	0.2330
	KDHD		11.49	9.12	16.86	140.73	3.7	3.9	5.2	2.2	2.3	3.8	0.1365	0.1475	0.1515
III	KD	1	18.18	12.23	23.99	117.16	3.2	3.4	4.9	2.8	2.9	4.8	0.1755	0.1925	0.2285
	KDHD		13.69	10.84	20.09	160.68	3.2	3.4	4.5	1.6	1.7	2.8	0.1110	0.1205	0.1270

In [31], a detailed discussion of the $PDD^{1/2}$ tuning of a second-order, purely inertial system is outlined, and these tuning criteria can be applied to the KDHD impedance control tuning, bearing in mind that the nonlinearity of a MIMO system as an impedance-controlled PKM introduces some alterations in the three-dimensional extension of the $PDD^{1/2}$ concept.

Let us note that the considered trapezoidal reference position law is characterized by velocity steps and impulsive accelerations in 0, t_{ramp}, $(t_{ramp} + t_{stop})$, and $(2t_{ramp} + t_{stop})$; consequently, the position error cannot be completely eliminated, not even adopting a model-based control scheme, since infinite actuation forces would be required. Accordingly, some overshoot is unavoidable in the transitory states after the discontinuities in t_{ramp} and $(2t_{ramp} + t_{stop})$. A smoother reference law for position, for example, with cycloidal or trapezoidal speed, would decrease the position error; nevertheless, here, a trapezoidal position law was adopted to highlight the differences between the KD and the KDHD impedance controls in trajectory tracking. The position error can be reduced by using higher stiffness, with increasing force peaks; for example, Figures 6 and 7 represent the external coordinates and the actuation forces for the same KD-KDHD comparison of Figures 4 and 5 (comparison number I, ramp time $t_{ramp} = 0.5$ s, stop time $t_{stop} = 3$ s), but with $k_{KD} = k_{KDHD} = 1 \cdot 10^4$ N/m. The maximum error of the external coordinates is lower, but the actuation forces are higher.

5.3. Case Study B

In this case study, the stiffness and damping matrices are diagonal in the world frame, but the desired end-effector behavior is not isotropic; therefore, the three diagonal elements of each matrix are not equal as in case study A. We impose higher compliance on the z-axis through the following values: $k_{KD,x} = k_{KDHD,x} = k_{KD,y} = k_{KDHD,y} = 2 \cdot 10^4$ N/m, $k_{KD,z} = k_{KDHD,z} = 1 \cdot 10^3$ N/m. The diagonal values of the damping matrices are obtained as in case A, starting from the nondimensional parameters of Table 2 and using Equations (17), (21), and (22) separately for each axis. A force $F = [100,100,100]^T$ N is applied to the end-effector at $t = 0$ s. The half-derivative is calculated adopting the same discrete-time implementation as in case A ($T_s = 0.005$ s, $n = 10$).

Figures 8 and 9 show the simulation results with reference to the KD-KDHD comparison number II ($\zeta_{KD} = 1$, Table 2), in terms of external coordinates and actuation forces. Observing Figure 8, it is possible to note that the steady-state displacements of the external coordinates using the impedance control KD (blue) and KDHD (violet) are different. This is due to the fact, already discussed in Section 2, that the approximation of the half-derivative calculation by means of a digital filter with a fixed memory length alters the stiffness imposed by impedance control. As a matter of fact, considering that the half-derivative of a constant c is non-null, but equal to $cW_{1/2}(n)/(T_s)^{1/2}$, as discussed at the end of Section 2, the following stiffness-compensated KDHD impedance control (KDHDc) is proposed:

$$\boldsymbol{\tau} = \left(\mathbf{J}^T\right)^{-1}\left[\left(\mathbf{K}_{KDHD} - \frac{W_{1/2}(n)}{T_s^{1/2}}\mathbf{HD}_{KDHD}\right)(\mathbf{x}_d - \mathbf{x}(\mathbf{q})) + \right.$$
$$\left. + \mathbf{D}_{KDHD}(\mathbf{x}_d - \mathbf{x}(\mathbf{q}))^{(1)} + \mathbf{HD}_{KDHD}(\mathbf{x}_d - \mathbf{x}(\mathbf{q}))^{(1/2)}\right] + \boldsymbol{\tau}_g(\mathbf{q}) \tag{24}$$

The effectiveness of this stiffness compensation is validated by the fact that applying the KDHDc control law, the external coordinates (Figure 8, yellow) tend to the same values obtained by applying the KD control law, which is not affected by the stiffness alteration due to the numerical evaluation of the half-derivative; as expected, these steady-state values correspond to the force/stiffness ratios $F_x/k_{KD,x}$, $F_y/k_{KD,y}$, and $F_z/k_{KD,z}$ for the three directions. Let us note that, due to the definition of W_α, this compensation is correct only in the steady state with a constant position reference, which are the conditions for which it has been introduced to obtain a correspondence between steady-state force and displacement.

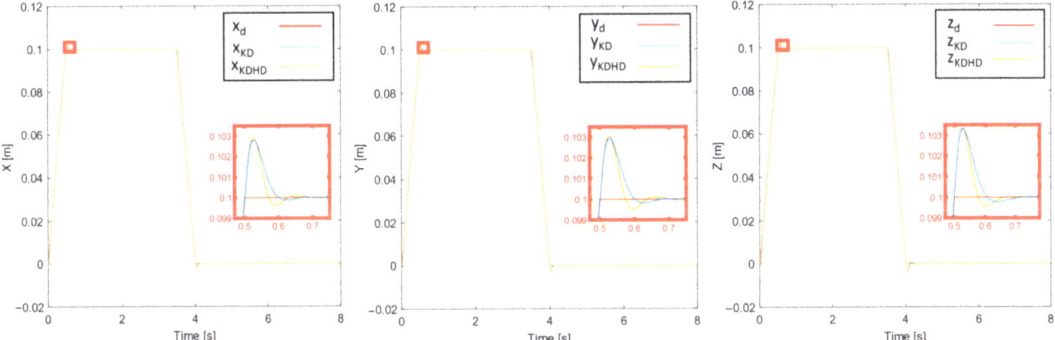

Figure 6. Case study A, $k_{KD} = k_{KDHD} = 1\cdot10^4$ N/m, KD-KDHD comparison number I, ramp time $t_{ramp} = 0.5$ s, stop time $t_{stop} = 3$ s, external coordinates.

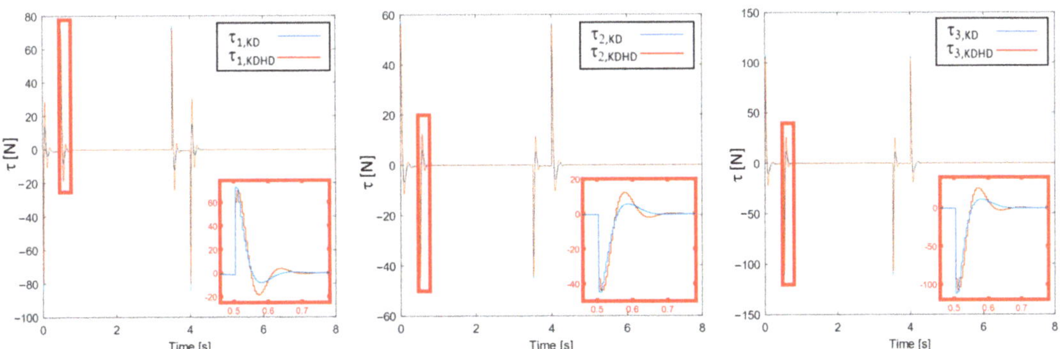

Figure 7. Case study A, $k_{KD} = k_{KDHD} = 1\cdot10^4$ N/m, KD-KDHD comparison number I, ramp time $t_{ramp} = 0.5$ s, stop time $t_{stop} = 3$ s, actuation forces.

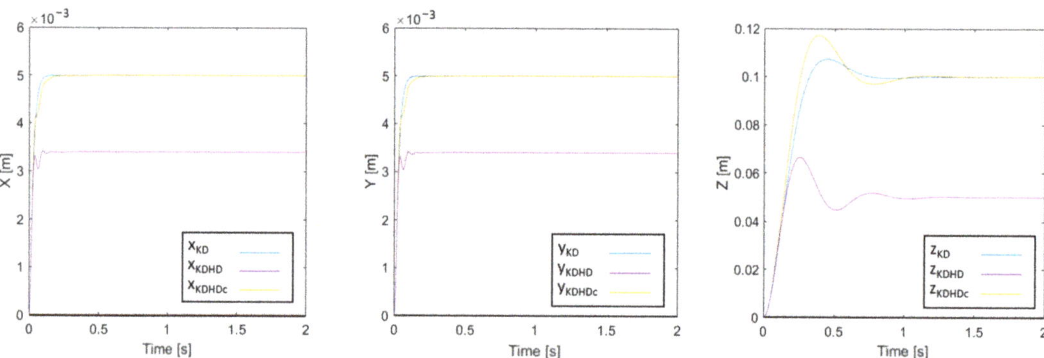

Figure 8. Case study B, KD-KDHD comparison number II, external coordinates.

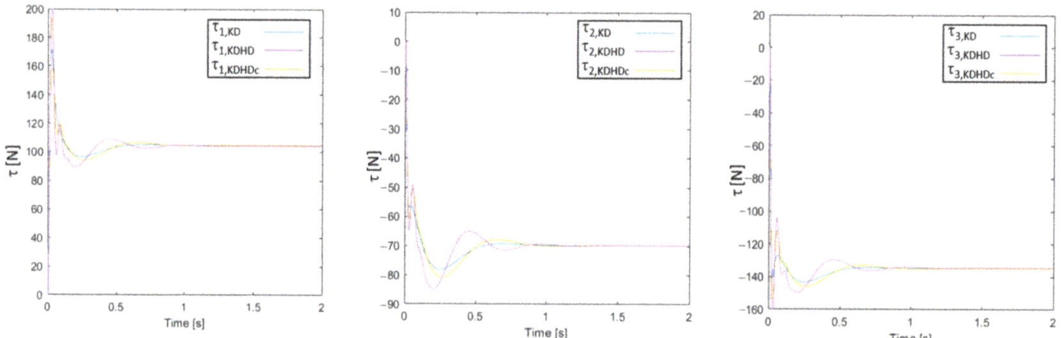

Figure 9. Case study B, KD-KDHD comparison number II, actuation forces.

5.4. Case Study C

In this case study, the stiffness compensation of the impedance control law KDHDc (Equation (24)) is applied when the fixed reference frame, W, does not coincide with the principal stiffness/damping reference frame, P. This occurs when impedance control must impose compliance in a specific direction, not coincident with one axis of the reference frame W, and higher stiffness in the remaining ones. In this case, a rotation of 45° around the z-axis is considered; therefore, the rotation matrix between W and P is:

$$\mathbf{R}_W^P = \begin{bmatrix} \sqrt{2}/2 & \sqrt{2}/2 & 0 \\ -\sqrt{2}/2 & \sqrt{2}/2 & 0 \\ 0 & 0 & 1 \end{bmatrix} \tag{25}$$

The stiffness matrix expressed in the principal reference frame, P, is characterized by the following diagonal values: $k_{KDp,x} = k_{KDHDp,x} = 1 \times 10^3$ N/m, and $k_{KDp,y} = k_{KDHDp,y} = k_{KDp,z} = k_{KDHDp,z} = 2 \times 10^4$ N/m, with higher compliance along the x-axis of the reference P. A force of $F = [0,100,100]^T$ N in the frame W is applied to the end-effector at $t = 0$ s. The half-derivative is calculated adopting the same discrete-time implementation as in cases A and B ($T_s = 0.005$ s, $n = 10$). Figure 10 shows the simulation results with reference to the comparison number II ($\zeta_{KD} = 1$, Table 2), in terms of external coordinates for the KD and KDHDc control laws.

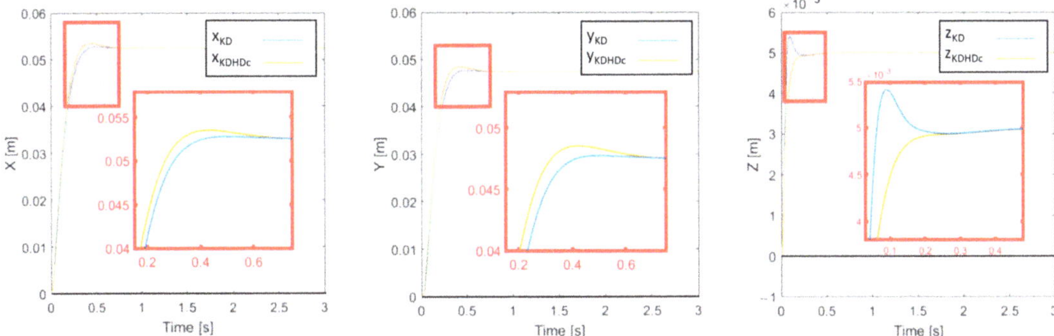

Figure 10. Case study C, KD-KDHD comparison number II, external coordinates.

The effectiveness of the KDHDc stiffness compensation is also validated in this case by the fact that the external coordinates tend to the same steady-state values of the KD control law. The steady-state displacement along x_W is slightly higher than the steady-state displacement y_W, while the steady-state displacement along z_W is much lower. This is coherent with the fact that the principal direction with lower stiffness is x_P, which is rotated by 45° around the z-axis; therefore, considering only the x-y plane (Figure 11), a force along x_W causes a larger displacement along x_P with a positive direction, and a smaller displacement along y_P with a negative direction. This results, in the frame W, in a slightly higher displacement along x_W with respect to y_W. In this case study, the ratio between the stiffness along x_P and the ones along y_P and z_P is 1/20; imposing a lower ratio, the displacement would more precisely follow the desired compliance direction.

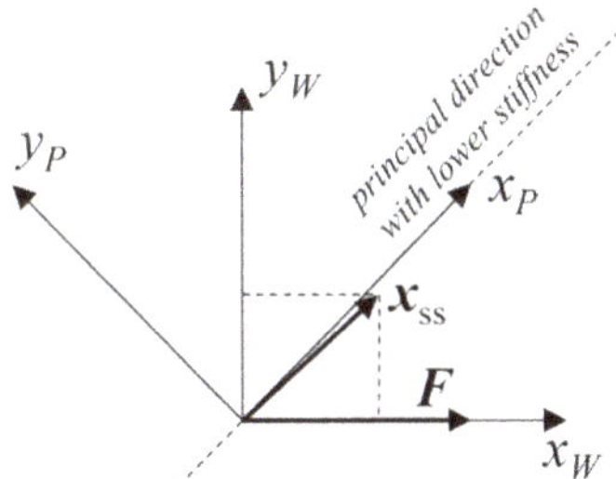

Figure 11. Case study C, **F**: external force; x_{ss}: steady-state displacement (x-y plane).

6. Conclusions

In this paper, an extension of classical impedance control (KD), with compliance defined by the stiffness (**K**) and damping (**D**) matrices, has been proposed and tested by multibody simulation on a 3-PUU parallel robot. This extension is based on fractional calculus, and in particular on the half-order derivative (derivative of order 1/2), adding a half-derivative damping defined by the matrix **HD**.

This work was inspired by the research about the PDD$^{1/2}$ control scheme for SISO systems, which is a PD with the addition of the half-derivative term, and extended it to a particular class of MIMO systems, impedance-controlled robots. In this work, a PKM was considered, but the approach can also be applied to serial robots by using Equations (12) to (14).

This work is only a starting point, and the effects of the introduction of the half-derivative damping must be further investigated. However, it is possible to outline the following conclusions:

- Even if applied to a nonlinear MIMO system, the introduction of the half-derivative term allows to tune the system behavior differently from a classical KD impedance control. The gains' set used in the case study A attained a reduction of maximum actuation force and total control effort, but the system can be tuned differently in order to improve other performance indexes, exploiting the previous works on $PDD^{1/2}$ control.

- The addition of the half-derivative damping necessarily introduced, with the discrete-time implementation by a finite-order digital filter, an alteration to the stiffness of the impedance control in steady state, as shown by Equation (5) and Figure 1. This hinders the capability of impedance control to regulate the contact force between the end-effector and the environment, which is the main scope of impedance control. To solve this issue, a stiffness-compensated KDHD impedance control algorithm (KDHDc) has been proposed (Equation (24)), and its effectiveness has been verified by simulation.

The saturation of the currents, and consequently of the actuation forces on the sliders, has not been taken into account to avoid the influence of an additional parameter, introducing a highly nonlinear effect and making the KD-KDHD comparison more complex. In any case, the simulations showed that the maximum actuation forces, with the considered control gains, were lower with the KDHD impedance control, so saturation cannot undermine the benefits of the proposed scheme.

In the future research, a systematic analysis of the influence of the control parameters of the KDHDc impedance control algorithm will be carried out in order to highlight the benefits of the proposed approach, also considering serial chains and robots with rotational mobility.

Author Contributions: L.B. conceived the control algorithm and designed the simulation campaign; D.B. developed the multibody model and performed simulations; P.F. supervised the scientific methodology; L.B. and P.F. prepared the manuscript. All authors have read and agreed to the published version of the manuscript.

Funding: This research received no external funding.

Institutional Review Board Statement: Not applicable.

Informed Consent Statement: Not applicable.

Data Availability Statement: Data is contained within the article. The simulation data presented in this study are available in figures and tables.

Conflicts of Interest: The authors declare no conflict of interest.

References

1. Hervé, J.M. The Lie group of rigid body displacements, a fundamental tool for mechanism design. *Mech. Mach. Theory* **1999**, *34*, 719–730. [CrossRef]
2. Craig, J. *Introduction to Robotics. Mechanics and Control*; Addison-Wesley: Boston, MA, USA, 1989.
3. Makino, H.; Furuya, N. Selective compliance assembly robot arm. In Proceedings of the First International Conference on Assembly Automation (ICAA), Brighton, UK, 25–27 March 1980; pp. 77–86.
4. Bruzzone, L.; Bozzini, G. A statically balanced SCARA-like industrial manipulator with high energetic efficiency. *Meccanica* **2011**, *46*, 771–784. [CrossRef]
5. Arawade, S. State of Art Review on SCARA Robotic Arm. *Int. J. Adv. Res. Sci. Commun. Technol.* **2021**, *5*, 145–152. [CrossRef]
6. Kong, X.; Gosselin, C.M. Kinematics and singularity analysis of a novel type of 3-CRR 3-DOF translational parallel manipulator. *Int. J. Robot. Res.* **2002**, *21*, 791–798. [CrossRef]
7. Gosselin, C.M.; Kong, X.; Foucault, S.; Bonev, I.A. A fully-decoupled 3-DOF translational parallel mechanism. In Proceedings of the 4th Chemnitz Parallel Kinematics Seminar (PKS 2004), Chemnitz, Germany, 20–21 April 2004; pp. 595–610.
8. Bruzzone, L.; Molfino, R.M. A novel parallel robot for current microassembly applications. *Assem. Autom.* **2006**, *26*, 299–306. [CrossRef]
9. Clavel, R. Delta, a fast robot with parallel geometry. In Proceedings of the 18th International Symposium on Industrial Robots, Lausanne, Switzerland, 26–28 April 1988; pp. 91–100.
10. Frisoli, A.; Checcacci, D.; Salsedo, F.; Bergamasco, M. Synthesis by screw algebra of translating in-parallel actuated mechanisms. In *Advances in Robot Kinematics*; Lenarcic, J., Stanisic, M.M., Eds.; Kluwer Academic: Dordrecht, The Netherlands, 2000; pp. 433–440.
11. Bruzzone, L.; Molfino, R.M.; Zoppi, M. An impedance-controlled parallel robot for high-speed assembly of white goods. *Ind. Robot.* **2005**, *32*, 226–233. [CrossRef]

12. Fang, Y.; Tsai, L.-W. Structure synthesis of a class of 4-DoF and 5-DoF parallel manipulators with identical limb structures. *Int. J. Robot. Res.* **2002**, *21*, 799–810. [CrossRef]
13. Company, O.; Marquet, F.; Pierrot, F. A new high-speed 4-DoF parallel robot synthesis and modelling issues. *IEEE Trans. Robot. Autom.* **2003**, *19*, 411–420. [CrossRef]
14. Raibert, M.H.; Craig, J.J. Hybrid Position/Force Control of Manipulators. *ASME J. Dyn. Sys. Meas. Control* **1981**, *103*, 126–133. [CrossRef]
15. Caccavale, F.; Siciliano, B.; Villani, L. Robot Impedance Control with Nondiagonal Stiffness. *IEEE Trans. Autom. Control* **1999**, *44*, 1943–1946. [CrossRef]
16. Valency, T.; Zacksenhouse, M. Accuracy/Robustness Dilemma in Impedance Control. *J. Dyn. Syst. Meas. Control* **2003**, *125*, 310–319. [CrossRef]
17. Angeles, J. *Rational Kinematics*; Springer: New York, NJ, USA, 1988.
18. Bonev, I.A.; Ryu, J. A new approach to orientation workspace analysis of 6-DOF parallel manipulators. *Mech. Mach. Theory* **2001**, *36*, 15–28. [CrossRef]
19. Bruzzone, L.; Callegari, M. Application of the rotation matrix natural invariants to impedance control of rotational parallel robots. *Adv. Mech. Eng.* **2010**, *2010*, 284976. [CrossRef]
20. Caccavale, F.; Siciliano, B.; Villani, L. The role of Euler parameters in robot control. *Asian J. Control* **1999**, *1*, 25–34. [CrossRef]
21. Bruzzone, L.; Molfino, R.M. A geometric definition of rotational stiffness and damping applied to impedance control of parallel robots. *Int. J. Robot. Autom.* **2006**, *21*, 197–205. [CrossRef]
22. Ikeura, R.; Inooka, H. Variable impedance control of a robot for cooperation with a human. In Proceedings of the 1995 IEEE International Conference on Robotics and Automation, Nagoya, Japan, 21–27 May 1995; Volume 3, pp. 3097–3102.
23. Tsumugiwa, T.; Yokogawa, R.; Hara, K. Variable impedance control with virtual stiffness for human-robot cooperative peg-in-hole task. In Proceedings of the IEEE/RSJ International Conference on Intelligent Robots and Systems, Lausanne, Switzerland, 30 September–4 October 2002; Volume 2, pp. 1075–1081.
24. Shimizu, M. Nonlinear impedance control to maintain robot position within specified ranges. In Proceedings of the 2012 SICE Annual Conference (SICE), Akita, Japan, 20–23 August 2012; pp. 1287–1292.
25. Miller, K.S.; Ross, B. *An Introduction to the Fractional Calculus and Fractional Differential Equations*; John Wiley & Sons: New York, NY, USA, 1993.
26. Kizir, S.; Elşavi, A. Position-Based Fractional-Order Impedance Control of a 2 DOF Serial Manipulator. *Robotica* **2021**, *39*, 1560–1574. [CrossRef]
27. Liu, X.; Wang, S.; Luo, Y. Fractional-order impedance control design for robot manipulator. In Proceedings of the ASME 2021 International Design Engineering Technical Conferences and Computers and Information in Engineering Conference, virtual, online, 17–19 August 2021; Volume 7, p. V007T07A028.
28. Podlubny, I. Fractional-order systems and $PI^\lambda D^\mu$ controllers. *IEEE Trans. Autom. Control* **1999**, *44*, 208–213. [CrossRef]
29. Fotuhi, M.J.; Bingul, Z. Novel fractional hybrid impedance control of series elastic muscle-tendon actuator. *Ind. Robot.* **2021**, *48*, 532–543. [CrossRef]
30. Bruzzone, L.; Fanghella, P. Comparison of PDD1/2 and PDμ position controls of a second order linear system. In Proceedings of the IASTED International Conference on Modelling, Identification and Control, Innsbruck, Austria, 17–19 February 2014; pp. 182–188.
31. Bruzzone, L.; Fanghella, P.; Baggetta, M. Experimental assessment of fractional-order $PDD^{1/2}$ control of a brushless DC motor with inertial load. *Actuators* **2020**, *9*, 13. [CrossRef]
32. Machado, J.T. Fractional-order derivative approximations in discrete-time control systems. *J. Syst. Anal. Model. Simul.* **1999**, *34*, 419–434.
33. Das, S. *Functional Fractional Calculus*; Springer: Berlin/Heidelberg, Germany, 2011.
34. Lu, S.; Ding, B.; Li, Y. Minimum-jerk trajectory planning pertaining to a translational 3-degree-of-freedoom parallel manipulator through piecewise quintic polynomials interpolation. *Adv. Mech. Eng.* **2020**, *12*, 1–18. [CrossRef]
35. Bruzzone, L.; Bozzini, G. PDD1/2 control of purely inertial systems: Nondimensional analysis of the ramp response. In Proceedings of the IASTED International Conference on Modelling, Identification and Control, Innsbruck, Austria, 14–16 February 2011; pp. 308–315.
36. Bruzzone, L.; Fanghella, P. Fractional-order control of a micrometric linear axis. *J. Control Sci. Eng.* **2013**, *2013*, 947428. [CrossRef]

 *actuators*

Article

Influence of the Dynamic Effects and Grasping Location on the Performance of an Adaptive Vacuum Gripper

Matteo Maggi [†], Giacomo Mantriota [†] and Giulio Reina [*,†]

Department of Mechanics, Mathematics and Management, Polytechnic of Bari, Via Orabona 4, 70126 Bari, Italy; matteo.maggi@poliba.it (M.M.); giacomo.mantriota@poliba.it (G.M.)
* Correspondence: giulio.reina@poliba.it
† These authors contributed equally to this work.

Abstract: A rigid in-plane matrix of suction cups is widely used in robotic end-effectors to grasp objects with flat surfaces. However, this grasping strategy fails with objects having different geometry e.g., spherical and cylindrical. Articulated rigid grippers equipped with suction cups are an underinvestigated solution to extend the ability of vacuum grippers to grasp heavy objects with various shapes. This paper extends previous work by the authors in the development of a novel underactuated vacuum gripper named Polypus by analyzing the impact of dynamic effects and grasping location on the vacuum force required during a manipulation cycle. An articulated gripper with suction cups, such as Polypus, can grasp objects by adhering to two adjacent faces, resulting in a decrease of the required suction action. Moreover, in the case of irregular objects, many possible grasping locations exist. The model explained in this work contributes to the choice of the most convenient grasping location that ensures the minimum vacuum force required to manipulate the object. Results obtained from an extensive set of simulations are included to support the validity of the proposed analytical approach.

Keywords: underactuation; vacuum grasping; suction cups; grasping configurations; inertial effects

check for
updates

Citation: Maggi, M.; Mantriota, G.; Reina, G. Influence of the Dynamic Effects and Grasping Location on the Performance of an Adaptive Vacuum Gripper. *Actuators* **2022**, *11*, 55. https://doi.org/10.3390/act11020055

Academic Editor: Giorgio Grioli

Received: 20 December 2021
Accepted: 10 February 2022
Published: 12 February 2022

Publisher's Note: MDPI stays neutral with regard to jurisdictional claims in published maps and institutional affiliations.

1. Introduction

Robots are widely employed in many fields, from industrial to medical, and regardless of the specific application, all robots have an end-effector to interact with the environment. In factories or warehouses, common tasks such as grasping, holding, and manipulating objects are often performed by simple grippers with two fingers [1] able only to pinch with a small stroke. These grippers are simple and cheap, but they can grasp a narrow range of objects. Dexterous manipulators to handle generic objects is an old [2] but open problem; indeed, there are several challenges for researchers to design novel grippers to accommodate some tasks [3]. Several designs have been proposed for manipulators; a brief classification has been proposed in [4] where grippers are divided by possible configuration, actuation, application, and size. In [5], the authors presented a versatile gripper with three re-configurable fingers to switch between power and precision grasping to emulate human hand ability.

An interesting strategy to grasp an object is to control the adhesion; this can be achieved mainly with vacuum [6,7], electroadhesion [8], or a gecko-inspired structure [9,10]. Vacuum grippers have great potential; indeed, in the 2015 Amazon Picking Challenge, nine out of 26 groups used vacuum, including the winner [11,12]. In addition, a world leader robotic company (Boston Dynamics) is developing a vacuum gripper called Stretch able to move in a warehouse and equipped with a matrix of suction cups for pick-and-place. Other examples of vacuum grippers are the octopus-inspired grippers [13–18] that fall in the soft robotics, which allows high flexibility and adaptability at the expense of limited payload.

In previous work by the authors [19], a novel underactuated vacuum gripper was introduced and named as Polypus, after the contraction of POLYthecnic of Bari's octoPUS.

The system, which is equipped with suction cups, features high adaptability to objects thanks to different grasping strategies, e.g., unilateral and power, and a potential high payload for its modular design. Moreover, Polypus is outfitted with a locking device in the hinges to freeze relative angles between phalanxes once it assumed the shape of the object, increasing considerably the payload and making the actuation unnecessary during the hold. It is actuated by wires that pass through phalanxes and are fixed to the tip of the finger. The main novel contributions of this research lay in:

- The extension of the Polypus grasping model from quasi-static to dynamic behavior, i.e., by taking into account the inertial effects connected with the manipulation task of an object under grasp.
- The assessment of the minimum vacuum force (MVF) required to manipulate a given object thorough different types of grasping configurations, in terms of contact points, e.g., placing the gripper in different positions of the object even with bending fingers to enhance the potentiality of Polypus compared to a fix matrix of suction cups.

The validity of the proposed analytical approach is verified via an extensive set of simulations of pick and place tasks with varying execution time and grasping location, and objects of different shapes and properties. Section 2 briefly describe the proposed gripper, Sections 3 and 4, respectively, explain the theoretical model to predict the vacuum force needed to manipulate an object and provide simulation results. This work ends with a section that draws the main relevant conclusions.

2. Polypus Design

This section briefly recalls the main features of Polypus that is shown in Figure 1 as a CAD rendering. The interested reader is referred to [19] for an in-depth description of the system. The proposed gripper has a modular rigid framework made up of a palm and N fingers divided in M phalanxes; each phalanx is equipped with a suction cup. Polypus presents $N \times (M - 1)$ degrees of underactuation having only N actuators, one for each finger. As a typical solution to determine the sequential closure of phalanxes, torsional springs are foreseen in the hinges. Figure 2 illustrates the predetermined sequential closure on a sphere. The first step is of course the approach to the object (Figure 2, section 1). Once Polypus is close enough, actuators start pulling the wires and, being the elastic constant in the hinges closest to the palm the lowest, the first hinge rotates before all others (Figure 2, section 2). Once the first suction cup touches the object, the first hinge stops rotating, and the hinge with the second lowest elastic constant starts its approaching motion (Figure 2, section 3). This sequence continues until all the suction cups are in contact with the object (Figure 2, section 4). Another key feature of Polypus design is the existence of a locking device located in the hinges. Once the shape of the object has been assumed, this locking device makes Polypus a rigid body, significantly reducing the required vacuum force. An example of a manipulator with a locking device but without suction cups has been presented in [20].

Figure 1. Rendering of the proposed vacuum gripper Polypus.

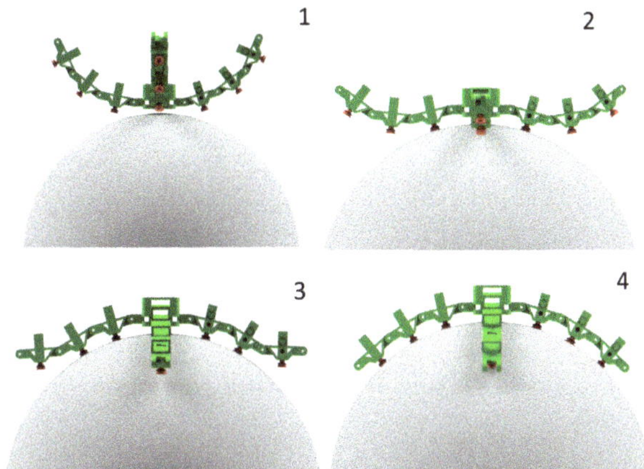

Figure 2. Example of sequential closure with a sphere having a radius four times the phalanx length.

3. Grasping Model

A model to evaluate the minimum vacuum force (MVF) is presented that explicitly takes into account the inertial effects connected to the manipulation task. The MVF is defined as the required force generated by each suction cup to avoid slipping and detachment during the manipulation. Since Polypus is modular, this model could be useful for example to assess the number of phalanxes to manipulate an object. Starting with an arbitrary number of phalanxes, the model provides the MVF; this value defines the dimension of suction cups. If the required suction cups are too big, the number of phalanxes should be increased. Moreover, different grasping locations could be compared to choose the most convenient. The following three subsections explain, respectively, the considered forces with relative assumptions, the trajectory of the manipulation, and the optimization framework.

3.1. Model Description

Each phalanx is identified by a couple of indexes (i, j) where the first indicates the finger and the second indicates the phalanx number starting from the closest to the palm. For the hypothesis, the contact between a suction cup and the object is modeled as a point neglecting pressure distribution because the object is much bigger than the suction cup [21]. Another assumption is that the module of vacuum forces is the same for all the phalanxes i.e., the same suction cups and degree of vacuum.

For a generic phalanx, the forces and relevant points are shown in Figure 3a and defined as follows:

- $F_{v_{i,j}}^{i,j+1} = \left[0, F_v, 0\right]^T$: The vacuum force is perpendicular to the suction cup pointing toward Polypus.

- $F_{n_{i,j}}^{i,j+1} = \left[0, -F_{n_{i,j}}, 0\right]^T$: The normal contact force has the same direction of the vacuum force but the opposite verse.

- $F_{t_{i,j}}^{i,j+1} = \left[-F_{tx_{i,j}}, 0, F_{tz_{i,j}}\right]^T$: The friction force has a generic direction in the contact plane with an upper limit fixed by friction constraint.

- $P_{i,j}^{i,j+1} = \left[-\frac{L}{2}, -h, 0\right]^T$: The contact point is in the middle of the suction cup. L is the length of the phalanx i.e., distance between two hinges, and h is the height of the phalanx i.e., distance between the line connecting two hinges and the contact point.

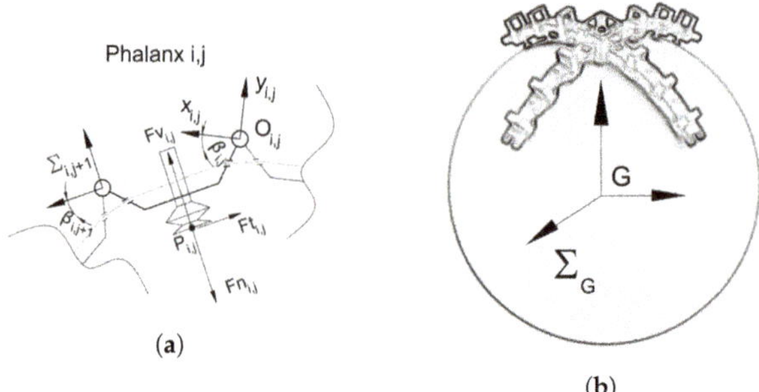

Figure 3. Schematic representation of reference frames used to develop the theoretical model with reference to the single phalanx (**a**) and object under grasp (**b**).

In the notation used above, the superscript indicates from which reference frame the quantity is seen, while the subscript indicates where the quantity is referred (e.g., P_b^a is the contact point of phalanx b seen from frame a). Knowing the pose of Polypus that is defined by relative angles between phalanxes (i.e., $\beta_{i,j}$ in Figure 3a), all the forces can be written in the object frame (Σ_G in Figure 3b) using homogeneous transformations. These forces and the relative location are useful to evaluate the equilibrium of the object, as will be formalized in the Section 3.3.

3.2. Trajectory Generation

As a running example throughout the paper, a common 2D trajectory is considered where the object under grasp is rotated by Polypus following a pure rotation, θ_z, about the fixed point, C, connecting two target positions via a piecewise constant acceleration law. Figure 4 shows an example of this trajectory where the initial and final orientation of Polypus is respectively $\theta_I = 0$ and $\theta_F = 90$ deg. Given the total simulation time T and setting the ratio for the time sub intervals, e.g., $T_0 = 0$, $T_1 = \frac{1}{3}T$, and $T_2 = \frac{2}{3}T$, Polypus starts with a null velocity and accelerates for a time defined by $(T_1 - T_0)$ with a constant acceleration; then, it continues with a constant cruise velocity for $(T_2 - T_1)$. The movement ends with a deceleration section $(T - T_2)$ to reach zero angular velocity. Defining the acceleration as:

$$\alpha_a = \frac{2(\theta_F - \theta_I)}{(T_1 - T_0)(T + T_2 - T_1 - T_0)}, \tag{1}$$

the whole trajectory is the following:

$$\alpha_z(t) = \begin{cases} \alpha_a & \text{for} \quad T_0 \leq t < T_1 \\ 0 & \text{for} \quad T_1 \leq t < T_2 \\ -\alpha_a \frac{T_1 - T_0}{T - T_2} & \text{for} \quad T_2 \leq t \leq T \end{cases} \tag{2}$$

$$\omega_z(t) = \begin{cases} \alpha_a(t - T_0) & \text{for} \quad T_0 \leq t < T_1 \\ \alpha_a(T_1 - T_0) & \text{for} \quad T_1 \leq t < T_2 \\ \alpha_a \frac{(T_1 - T_0)(T - t)}{T - T_2} & \text{for} \quad T_2 \leq t \leq T \end{cases} \tag{3}$$

$$\theta_z(t) = \begin{cases} \theta_I + \alpha_a \frac{(t - T_0)^2}{2} & \text{for} \quad T_0 \leq t < T_1 \\ \theta_I + \alpha_a(T_1 - T_0)(t - \frac{T_1 + T_0}{2}) & \text{for} \quad T_1 \leq t < T_2 \\ \theta_I + \alpha_a(T_1 - T_0)(t - \frac{T_1 + T_0}{2} - \frac{(t - T_2)^2}{2(T - T_2)}) & \text{for} \quad T_2 \leq t \leq T \end{cases} \tag{4}$$

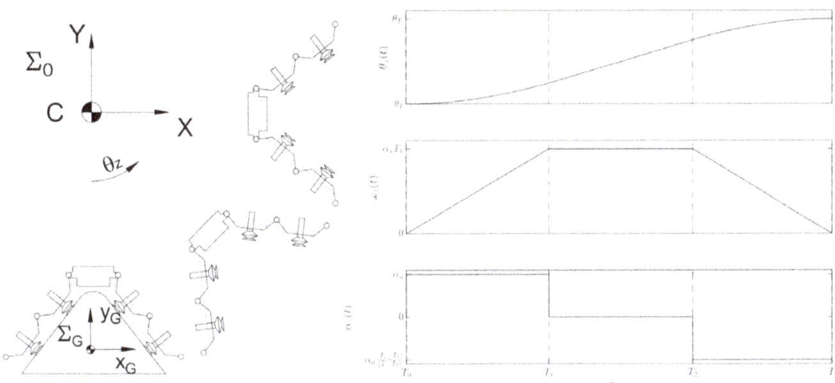

Figure 4. Example of trajectory for the manipulation.

Lastly, the acceleration vector pertaining to the object center of mass, G, and gravity vector, g, can be expressed in the object frame (Σ_G) as follows

$$\ddot{G}(t) = \begin{bmatrix} \alpha_z(t)\|G - C\|, & \omega_z(t)^2\|G - C\|, & 0 \end{bmatrix}^T \tag{5}$$

$$g(t) = 9.81 \begin{bmatrix} -\sin\theta_z(t), & -\cos\theta_z(t), & 0 \end{bmatrix}^T \text{m/s}^2. \tag{6}$$

Moreover, vectors with angular velocity and acceleration expressed in the object reference frame have to be set up, e.g., $\dot{\theta}(t) = \begin{bmatrix} 0 & 0 & \omega_z(t) \end{bmatrix}^T$.

Although common 2D trajectory examples have been used in most of the simulations, Section 5 presents a use case where Polypus follows a 3D trajectory that mimics the movement of an object taken from a shelf and moved behind the robot similarly to the trajectory required by the Amazon picking challenge [11]. For the general case, i.e., $\dot{\theta}(t) = \begin{bmatrix} \omega_x(t) & \omega_y(t) & \omega_z(t) \end{bmatrix}^T$ and $\ddot{\theta}(t) = \begin{bmatrix} \alpha_x(t) & \alpha_y(t) & \alpha_z(t) \end{bmatrix}^T$, previous Equations (5) and (6) can be expressed as follows

$$\ddot{G}(t) = \ddot{\theta} \times (G - C) + \dot{\theta} \times [\dot{\theta} \times (G - C)] \tag{7}$$

$$g(t) = R_0^G \begin{bmatrix} 0 \\ -9.81 \\ 0 \end{bmatrix} \text{m/s}^2 \tag{8}$$

where R_0^G is the rotation matrix from the fixed reference frame to Σ_G. The interested reader is referred to previous works, e.g., [22,23] for more details on the generation of operational space trajectories.

3.3. Optimization

The aim of the optimization is to find the MVF while performing the manipulation; indeed, the following optimization will be performed for each time $t \in T$ with increment dt, and the time dependency will be omitted to simplify the notation. The cost function to be minimized is the module of the vacuum force F_v; recall that as mentioned in Section 3.1, the module is the same for all the phalanxes. The optimization variables are vacuum forces as well as contact forces, while the constraints are:

- Equilibrium of the object forces and torques expressed in the frame attached to the object (i.e., Σ_G in Figure 3b).

$$\sum_{i=1}^{N}\sum_{j=1}^{M}[(\boldsymbol{F}_{v_{i,j}} + \boldsymbol{F}_{n_{i,j}} + \boldsymbol{F}_{t_{i,j}})Q(i,j)] + m\boldsymbol{g} - m\ddot{\boldsymbol{G}} = 0 \tag{9}$$

$$\sum_{i=1}^{N}\sum_{j=1}^{M}[(\boldsymbol{r}_{G,P_{i,j}} \times \boldsymbol{F}_{v_{i,j}} + \boldsymbol{r}_{G,P_{i,j}} \times \boldsymbol{F}_{n_{i,j}} + \boldsymbol{r}_{G,P_{i,j}} \times \boldsymbol{F}_{t_{i,j}})Q(i,j)] - I_G\ddot{\boldsymbol{\theta}} - \dot{\boldsymbol{\theta}} \times I_G\dot{\boldsymbol{\theta}} = 0 \tag{10}$$

where $\ddot{\boldsymbol{G}}$ and $\ddot{\boldsymbol{\theta}}$ are, respectively, the linear and angular acceleration vector of the CoM, m is the mass of the object, I_G is the inertia tensor relative to the CoM when expressed in Σ_G, $\boldsymbol{r}_{G,P}$ is the vector pointing in G with origin in P, and Q is a matrix to activate the suction cups, the element (i,j) is 1 if the suction cup (i,j) is in contact with the object; otherwise, it is 0.

- Positiveness of $F_{v_{i,j}}$ and $F_{n_{i,j}}$.

$$F_v \geq 0 \tag{11}$$

$$F_{n_{i,j}} \geq 0 \tag{12}$$

- Coulomb's law of friction.

$$F_{tx_{i,j}}^2 + F_{tz_{i,j}}^2 \leq (\mu F_{n_{i,j}})^2 \tag{13}$$

At the end of each optimization, the value of the MVF (F_v) is saved in a vector and plotted as a function of time or angular position of Polypus in the next section.

4. Results

Simulations were performed using the *optimproblem* environment available under Matlab R2020b. The convex optimization problem was solved using the Interior point method where the initial guess was set equal to the solution found at the previous time. The manipulation of four different objects has been considered. In all simulations; it is assumed that Polypus fingers feature three phalanxes. Each phalanx has a length of 85 mm, a height of 45 mm, and a width of 40 mm. The palm is a square with a 100 mm side; therefore, the total tip to tip length of Polypus is 610 mm. For each object type, multiple simulations have been performed varying the speed of the manipulation (i.e., varying the inertial effect), the friction coefficient, and the grasping configuration of Polypus. One of the objectives of the simulations is to determine the extent of the vacuum force required to manipulate an object. In turn, this value can be compared with the maximum force a suction cup can generate. Considering as a reference the VASB Festo bellows and a degree of vacuum of -70 kPa, the maximum holding force results in 34 N for the 30 mm suction cup and 56 N for the 40 mm suction cup.

4.1. Case I: Flat Object—2 Fingers

For this case study, two fully extended fingers are considered to grasp a box with dimensions $61 \times 50 \times 10$ cm and homogeneous mass distribution for a total weight of 5 kg. A schematic representation of the manipulation (i.e., θ_z varies from 0 to 90 deg with a piecewise constant acceleration) is reported in the inset at the bottom of Figure 5; note that object proportions have been distorted for graphical purposes. The center of rotation (C) is 15 cm above the center of the palm. Figure 5 shows the MVF as a function of the angle θ_z for different mean angular velocities; the friction coefficient is set equal to 0.6. The dotted line is a really slow manipulation considering a total rotation time (T) of 30 s. Here, dynamic effects are negligible, resulting in a peak of MVF of 15.89 N for $\theta_z = 59$ deg. On the other hand, the continuous line is a fast rotation (i.e., 0.6 s to span 90 deg) that requires almost doubled vacuum force (28.82 N). Jumps in the MVF come from the piecewise constant acceleration; indeed, the most critical part of the rotation is the beginning where inertia and weight forces are synchronized. In the intermediate section, where angular velocity is constant, there is only the centripetal contribution, which slightly increases the

MVF. In the decelerating part, from 68 to 90 deg, inertia force is opposite to weight force, causing a drop in MVF. It should be noticed that at the end of the manipulation, 13.6 N of vacuum force are needed for a static hold (i.e., as soon as the acceleration becomes zero). Another set of simulations with the same object has been performed, varying the friction coefficient from 0.2 to 1. The manipulation time is 1.2 s, which includes 1 s for the rotation (equally divided in three sectors with constant acceleration) and 0.1 s of static holds at the beginning and the end. The MVF here is plotted in Figure 6 as a function of time in the upper plot, while the bottom section of Figure 6 plots the manipulation angle θ_z, and the two graphs in the same figure share the x-axis. Seeing the MVF as a function of time allows the analysis of zero velocity sectors; thus, the increase of the vacuum force at the end of the manipulation becomes noticeable. As expected, lower friction coefficients require greater suction action, the maximum value of MVF is 42.8 N when μ is 0.2, 18.5 N for $\mu = 0.6$, and 14.8 N considering $\mu = 1$. Therefore, Fest VASB section cups of 30 mm are enough in all scenarios except when friction is 0.2.

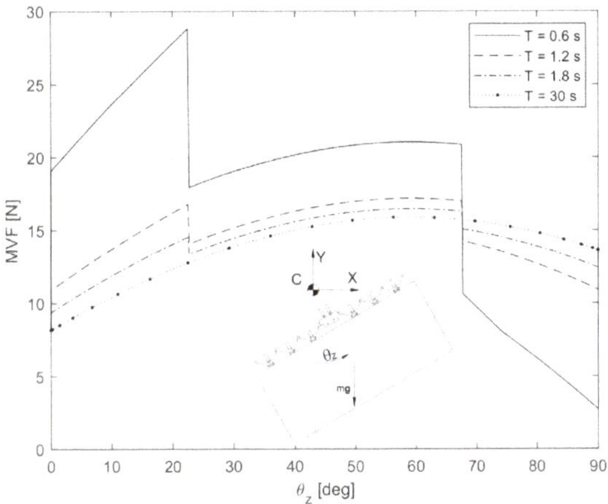

Figure 5. Minimum vacuum force required to rotate a box expressed as a function of the inclination angle θ_z that is shown in the bottom inset. Different curves refer to increasing time T, i.e., the total time used for the rotation as defined in Equations (2)–(4).

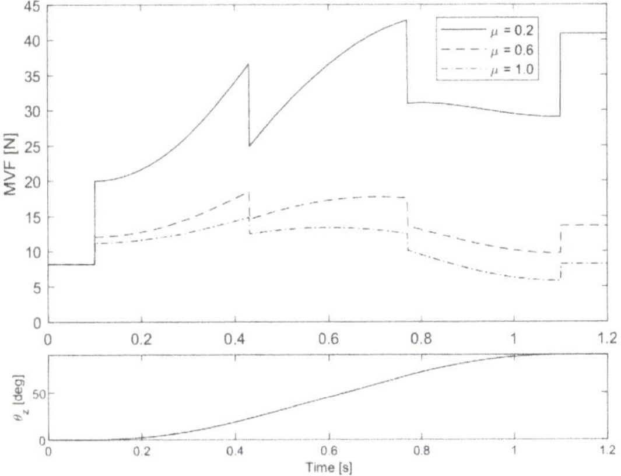

Figure 6. (**Upper plot**): Minimum vacuum force required during the rotation of a box as a function of time. Results are presented for different friction coefficients. (**Bottom plot**): Inclination angle θ_z. Note that the two plots shares the x-axis.

4.2. Case II: Trapezoid Object—2 Fingers

The second object is a trapezoid with sides 45 degrees bent, the bigger base is 81 cm in length, and the height is 38 cm. These dimensions lead to the same center of mass location of the previous case (i.e., 25 cm away from the upper side). In addition, in this case, two fingers are considered, and the mass of the trapezoid is the same as the box (5 kg). A schematic representation of the object and manipulation angle (θ_z) is shown in the inset of Figure 7; the center of the rotation is 15 cm above the center of the palm. The same figure presents the value of the MVF considering a friction coefficient of 0.6 and different total rotation time (T), which implies diverse average rotation velocities. In the fastest case, the 90 deg rotation is achieved in 0.6 s, resulting in a maximum value of MVF of 25.3 N; if Polypus completes the rotation in 1.2 s, an MVF of 15.4 N is required, and 12.2 N is the peak of MVF for a very slow rotation i.e., 30 s. As in the previous case, the dynamic effects increase the required vacuum force mostly in the first part of the manipulation where Polypus is accelerating, while at the end of the rotation, the value of MVF is close to zero, being tangential acceleration and weight force opposite in direction. As soon as Polypus comes to a rest, an MVF of 4 N will be needed, as for the dotted line. The upper graph of Figure 8 plots results for different friction coefficients considering a rotation time of 1 s and static holds of 0.1 s. While the lower graph of Figure 8 illustrates the rotation angle θ_z as a function of time, the two plots share the x-axis. Here, even when friction is 0.2, suction cups with a diameter of 30 mm would be enough. This consideration highlights the positive effect of grasping objects with low friction from two adjacent faces.

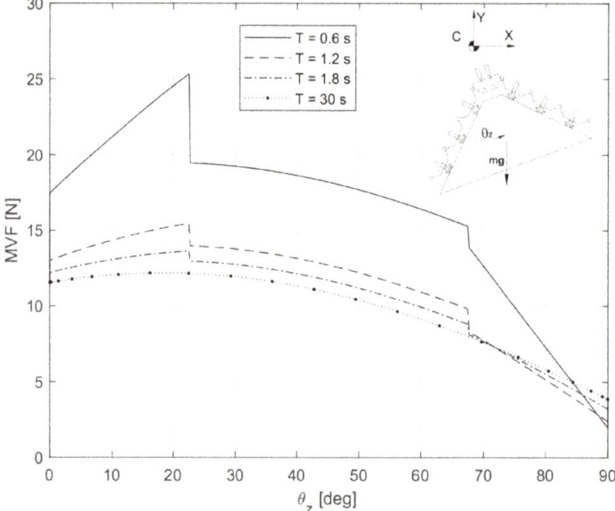

Figure 7. Minimum vacuum force required during the rotation of a trapezoid object as a function of the angle θ_z that is shown in the inset. Different curves refer to increasing time T, i.e., the total time used for the rotation, as defined in Equations (2)–(4).

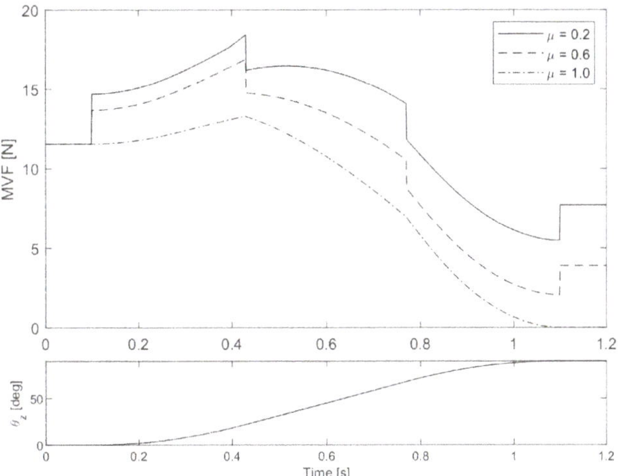

Figure 8. In the **upper plot**, minimum vacuum force required during the rotation of a trapezoid object as a function of time; results are presented for different friction coefficients. In the **bottom plot** rotation angle θ_z, the two plots shares the x-axis.

4.3. Case III: Sphere—Four Fingers

In this case, Polypus manipulates spheres of different sizes with four fingers. However, the mass of the spheres is always two times the mass of previous objects (10 kg). The CoM is in the middle of the sphere. Figure 9 shows from left to right rendered images of the sphere with radius 3, 4, and 6 times the phalanx length; in all the cases, the sphere is too big for power grasping, and thus, vacuum is always required. The friction coefficient is assumed to be 0.6. The upper graph of Figure 10 reports a schematic representation of the manipulation angle θ_z in the upper-right corner and the MVF as a function of time. The center of rotation, (C), is 30 cm above the center of the palm. The bottom of Figure 10 shows the angle θ_z as a function of time; the two graphs share the x-axis.

Figure 9. Rendered images of spheres, from left to right the diameter of the sphere is respectively three, four, and six times the length of Polyps phalanx.

The diameter of the sphere affects the required vacuum force during the manipulation; the MVF increases with the radius of the sphere especially at the beginning of the manipulation. In the accelerating section, the inertia force is synchronized with the weight force, making the most critical angle the end of the acceleration ($\theta_z = 22$ deg). The MVF associated with this angle is 32.6, 24.6, and 21.2 N considering the spheres respectively from the biggest to the smallest. In addition, in the middle section, where angular velocity is constant, the influence of the sphere diameter is noticeable, increasing the maximum registered value of MVF from 16 to 24 N. In the last manipulation section, the diameter has almost no influence on the MVF due to the cancellation of tangential effects being inertia and weight opposite in direction. To hold the object with $\theta_z = 90$ deg, the MVF almost doubled when considering the biggest sphere compared with the smallest. These results show the improvement in performance if the object is wrapped by the gripper, which is the main feature of Polypus where suction cups adhere to the object with different orientation. For $R = 4L$ and $R = 3L$, 30 mm suction cups are sufficient, while for the biggest sphere ($R = 6L$), suction cups of 40 mm diameter should be considered.

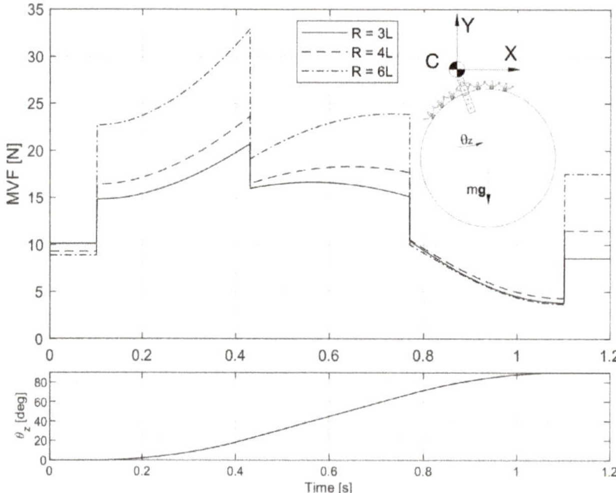

Figure 10. (**Upper plot**): Minimum vacuum force required during the manipulation of spheres with different dimension as a function of time. (**Bottom plot**): Angle θ_z that defines the rotation of the manipulation.

5. Optimization of the Grasping Location

The last set of simulations mimics a typical pick and place operation in a warehouse. Figure 11 explains the desired trajectory: starting from point A, there is a first translation of 1 m up to point B to remove the object from the shelf. Then, the object rotates (BC) about the vertical axis with a radius of 2 m. It is followed by a rotation of the wrist to place the object in point D rotated by 90 deg from the initial configuration. In this last sub-trajectory, a radius of 1 m is considered. This operation has been inspired by the short video that presents the Stretch on the Boston Dynamics website [24] and the Amazon Picking Challenge 2016 [11]. The same trajectory has been repeated considering five different grasping locations, as shown in Table 1, along with a side view of the object (bottom right) that shows the CoM location. The whole movement is achieved in 3 s, and the total time is divided in 20% for AB, 50% for BC, and 30% for CD. Figure 12 reports the value of the MVF as a function of the time. The most important aspect of this simulation is the difference between grasping locations. Here, we can see that the required vacuum force (i.e., the maximum value of MVF during the manipulation) has a strong dependence on the grip location. Locations 3 and 4 have almost the same MVF and are the best, followed by location 5, then location 2, whereas the worst is location 1. As expected, moving away from the CoM, the required MVF increases. It should be noticed that locations 3 and 4 can be achieved thanks to the specific feature of Polypus to have an articulated frame equipped with suction cups.

Table 1. This table reports the grasping locations used with the irregular rectangular prism. On the right side of the bottom row, there is a schematic representation that shows the center of mass G and the object dimensions ($L_1 = L_2 = 34$ cm, $h_1 = h_2 = 68$ cm, the width of the object is 68 cm).

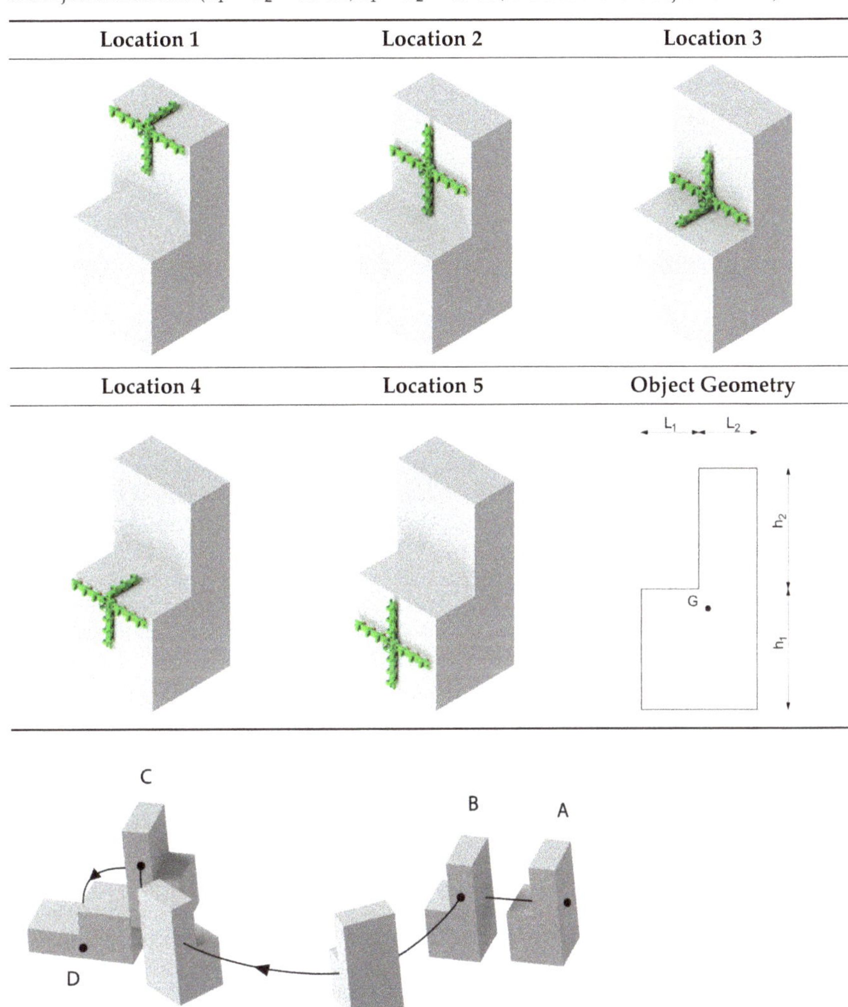

Location 1	Location 2	Location 3
Location 4	**Location 5**	**Object Geometry**

Figure 11. Schematic representation of the trajectory followed by the irregular prism. It simulates a robot that picks an object from a shelf (AB); then, after a rotation of the whole robot (BC), the object is placed rotated 90 degrees (CD).

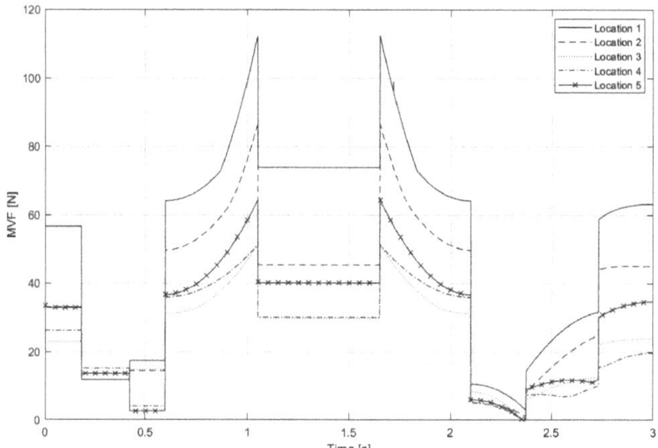

Figure 12. Minimum vacuum force required during the trajectory of Figure 11 as a function of time.

6. Discussion and Conclusions

This paper presents an analytical model to predict the vacuum force a gripper needs to grasp and rotate an object, including the inertia effect. Several simulations have been provided to show how the required vacuum force is affected by friction, manipulation time, and gripper location. Moreover, a brief comparison with a suction cup catalog has been included to indicate the suction cup dimension. From the results explained in the last section, a few considerations can be drawn. In the first two use cases, the grasped objects have the same mass and CoM distance from Polypus palm. In the first case, the gripper holds the object from only one face, while in the second case, Polypus adheres to two faces of the object bending one finger. Having suction cups in different planes is a unique feature of the proposed rigid gripper that allows a slight reduction of the vacuum force in fast manipulations with good friction coefficients (0.6) and a significant drop in the vacuum force while manipulating objects with low friction. Indeed, in the second case, 30 mm suction cups are enough even with low friction. On the other hand, in the first case, suction cups of at least 40 mm diameters are required for $\mu = 0.2$. In the third case, the object is a relatively big sphere, and it should be noticed that grasping the sphere could be a tricky task for grippers able only to power or that have a pinch grasping strategy as well as with an in-plane matrix of suction cups. Results from the last use case show that the proposed grasping model can be valuable to predict the best grasping configurations for objects with complex and irregular shapes. This can also be useful whenever specific requirements apply, namely the grasp can not be performed on one or more surfaces of the object that are delicate or fragile. Future research will investigate the impact on the grasping stability of the contact loss of one or more suction cups. One important development will also involve the experimental validation of the proposed approach via real experiments using a physical prototype of Polypus.

Author Contributions: All authors contributed equally to this work. All authors have read and agreed to the published version of the manuscript.

Funding: This research received no external funding.

Institutional Review Board Statement: Not applicable.

Informed Consent Statement: Not applicable.

Data Availability Statement: Data are available upon request to the corresponding author.

Conflicts of Interest: The authors declare no conflict of interest.

Abbreviations

The following abbreviations are used in this manuscript:

MVF   Minimum vacuum force
CoM   Center of mass
$F_{v_{i,j}}$ Vacuum force for the (i,j)-th phalanx
$F_{n_{i,j}}$ Normal force for the for the (i,j)-th phalanx
$F_{t_{i,j}}$ Tangential force for the (i,j)-th phalanx
$P_{i,j}$ Location of contact point of the (i,j)-th phalanx
$r_{G,P_{i,j}}$ Position vector from $P_{i,j}$ to G

References

1. Birglen, L.; Schlicht, T. A statistical review of industrial robotic grippers. *Robot. Comput.-Integr. Manuf.* **2018**, *49*, 88–97. [CrossRef]
2. Piazza, C.; Grioli, G.; Catalano, M.; Bicchi, A. A century of robotic hands. *Annu. Rev. Control Robot. Auton. Syst.* **2019**, *2*, 1–32. [CrossRef]
3. Watanabe, T.; Yamazaki, K.; Yokokohji, Y. Survey of robotic manipulation studies intending practical applications in real environments-object recognition, soft robot hand, and challenge program and benchmarking. *Adv. Robot.* **2017**, *31*, 1114–1132. [CrossRef]
4. Samadikhoshkho, Z.; Zareinia, K.; Janabi-Sharifi, F. A brief review on robotic grippers classifications. In Proceedings of the 2019 IEEE Canadian Conference of Electrical and Computer Engineering (CCECE), Edmonton, AB, Canada, 5–8 May 2019; pp. 1–4.
5. Borisov, I.I.; Borisov, O.I.; Gromov, V.S.; Vlasov, S.M.; Kolyubin, S.A. Versatile gripper as key part for smart factory. In Proceedings of the 2018 IEEE Industrial Cyber-Physical Systems (ICPS), St. Petersburg, Russia, 15–18 May 2018; pp. 476–481.
6. Mantriota, G.; Messina, A. Theoretical and experimental study of the performance of flat suction cups in the presence of tangential loads. *Mech. Mach. Theory* **2011**, *46*, 607–617. [CrossRef]
7. Mantriota, G. Theoretical model of the grasp with vacuum gripper. *Mech. Mach. Theory* **2007**, *42*, 2–17. [CrossRef]
8. Guo, J.; Leng, J.; Rossiter, J. Electroadhesion technologies for robotics: A comprehensive review. *IEEE Trans. Robot.* **2019**, *36*, 313–327. [CrossRef]
9. Glick, P.; Suresh, S.A.; Ruffatto, D.; Cutkosky, M.; Tolley, M.T.; Parness, A. A soft robotic gripper with gecko-inspired adhesive. *IEEE Robot. Autom. Lett.* **2018**, *3*, 903–910. [CrossRef]
10. Song, S.; Majidi, C.; Sitti, M. Geckogripper: A soft, inflatable robotic gripper using gecko-inspired elastomer micro-fiber adhesives. In Proceedings of the 2014 IEEE/RSJ International Conference on Intelligent Robots and Systems, Chicago, IL, USA, 14–18 September 2014; pp. 4624–4629.
11. Correll, N.; Bekris, K.E.; Berenson, D.; Brock, O.; Causo, A.; Hauser, K.; Okada, K.; Rodriguez, A.; Romano, J.M.; Wurman, P.R. Analysis and observations from the first amazon picking challenge. *IEEE Trans. Autom. Sci. Eng.* **2016**, *15*, 172–188. [CrossRef]
12. Eppner, C.; Höfer, S.; Jonschkowski, R.; Martín-Martín, R.; Sieverling, A.; Wall, V.; Brock, O. Lessons from the amazon picking challenge: Four aspects of building robotic systems. In Proceedings of the 2016 Robotics: Science and Systems Conference, Ann Arbor, MI, USA, 18–22 June 2016.
13. Bryan, P.; Kumar, S.; Sahin, F. Design of a soft robotic gripper for improved grasping with suction cups. In Proceedings of the 2019 IEEE International Conference on Systems, Man and Cybernetics (SMC), Bari, Italy, 6–9 October 2019; pp. 2405–2410.
14. Xie, Z.; Domel, A.G.; An, N.; Green, C.; Gong, Z.; Wang, T.; Knubben, E.M.; Weaver, J.C.; Bertoldi, K.; Wen, L. Octopus arm-inspired tapered soft actuators with suckers for improved grasping. *Soft Robot.* **2020**, *7*, 639–648. [CrossRef] [PubMed]
15. Mazzolai, B.; Mondini, A.; Tramacere, F.; Riccomi, G.; Sadeghi, A.; Giordano, G.; Del Dottore, E.; Scaccia, M.; Zampato, M.; Carminati, S. Octopus-Inspired Soft Arm with Suction Cups for Enhanced Grasping Tasks in Confined Environments. *Adv. Intell. Syst.* **2019**, *1*, 1900041. [CrossRef]
16. Takahashi, T.; Suzuki, M.; Aoyagi, S. Octopus bioinspired vacuum gripper with micro bumps. In Proceedings of the 2016 IEEE 11th Annual International Conference on Nano/Micro Engineered and Molecular Systems (NEMS), Sendai, Japan, 17–20 April 2016; pp. 508–511.
17. Wei, Y.; Zhang, W. OS hand: Octopus-inspired self-adaptive underactuated hand with fluid-driven tentacles and force-changeable artificial muscles. In Proceedings of the 2017 IEEE International Conference on Robotics and Biomimetics (ROBIO), Macau, Macao, 5–8 December 2017; pp. 7–12.
18. Pi, J.; Liu, J.; Zhou, K.; Qian, M. An Octopus-Inspired Bionic Flexible Gripper for Apple Grasping. *Agriculture* **2021**, *11*, 1014. [CrossRef]
19. Maggi, M.; Mantriota, G.; Reina, G. Introducing POLYPUS: A novel adaptive vacuum gripper. *Mech. Mach. Theory* **2022**, *167*, 104483. [CrossRef]
20. Aukes, D.M.; Heyneman, B.; Ulmen, J.; Stuart, H.; Cutkosky, M.R.; Kim, S.; Garcia, P.; Edsinger, A. Design and testing of a selectively compliant underactuated hand. *Int. J. Robot. Res.* **2014**, *33*, 721–735. [CrossRef]
21. Mantriota, G. Optimal grasp of vacuum grippers with multiple suction cups. *Mech. Mach. Theory* **2007**, *42*, 18–33. [CrossRef]
22. Paul, R. Manipulator Cartesian path control. *IEEE Trans. Syst. Man Cybern.* **1979**, *9*, 702–711. [CrossRef]
23. Farin, G. *Curves and Surfaces for CAGD: A Practical Guide*; Morgan Kaufmann Publishers: San Francisco, CA, USA, 2001.
24. Boston Dynamics. Available online: https://www.bostondynamics.com/products/stretch (accessed on 20 January 2022).

 *actuators*

Article

Simulation and Model-Based Verification of an Emergency Strategy for Cable Failure in Cable Robots

Roland Boumann * and Tobias Bruckmann *

Chair of Mechatronics, University of Duisburg-Essen, Forsthausweg 2, 47057 Duisburg, Germany
* Correspondence: roland.boumann@uni-due.de (R.B.); tobias.bruckmann@uni-due.de (T.B.)

Abstract: Cable failure is an extremely critical situation in the operation of cable-driven parallel robots (CDPR), as the robot might be instantly outside of its predefined workspace. Therefore, the calculation of a cable force distribution might fail and, thus, the controller might not be able to master the guidance of the system anymore. However, as long as there is a remaining set of cables, the dynamic behavior of the system can be influenced to prevent further damage, such as collisions with the ground. The paper presents a feasible algorithm, introduces the models for dynamical multi-body simulation and verifies the algorithm within control loop closure.

Keywords: cable-driven parallel robot; cable failure; cable break; modeling; simulation; nonlinear model prediction; emergency strategies; force distribution

1. Introduction

Cable-driven parallel robots (CDPR) use a set of cables that are coiled on digitally controlled winches to move a payload, such as a platform. CDPRs can be designed very lightweight, which makes them fast and efficient. Moreover, they can cover large workspaces in comparison to serial manipulators, as the cables can span over huge distances [1]. The range of applications for CDPRs investigated is increasing, including, e.g., intralogistics [2] or automation in construction [3]. Further applications are, e.g., vertical green maintenance [4] or lamella cleaning in sedimentation tanks [5]. Additionally, research has been conducted to improve the performance of CDPRs in underwater operations [6]. In [7], the accuracy, repeatability and long-term running of a CDPR is investigated. A recent project in which the authors are involved is a CDPR for automated masonry construction [8]. This implies a rapid progression of the future industrial utilization of CDPRs.

Consequently, aiming at practical applications, industrial safety requirements must be met in the future to avoid damage to the robot and its environment. Even if there are commonly known rules and guidelines for fundamental components, such as pulleys and cables, a cable might still fail [9]. Within literature, the detection of failures and their outcome in CDPRs has been broadly discussed. Several methods have been presented to handle such situations. In [10], initial thoughts on potential errors and error tolerance in CDPRs were provided, and the consequences of a failed cable on the nullspace of the robot's Jacobian were examined. The impact of a failed or slack cable resulting in improper cable forces is considered in [11]. It aims to recover the force and momentum capacity of the robot's platform. The conditions of practicability for this purpose are suggested. In [12], a strategy for recovery after a cable failure, where the end effector can be outside of its workspace, is proposed. It is based on the planning of an elliptical path within a dynamic trajectory. In [13], the workspace of a predefined system after cable failure is investigated. Moreover, an algorithm for the planning of a feasible trajectory along a straight line path back into the workspace is introduced. This work was extended in [14] by including drivetrain models. In [15], the approach was implemented on a prototype with three degrees-of-freedom (DOF) of the mobile end effector and four cables.

Citation: Boumann, R.; Bruckmann, T. Simulation and Model-Based Verification of an Emergency Strategy for Cable Failure in Cable Robots. *Actuators* **2022**, *11*, 56. https://doi.org/10.3390/act11020056

Academic Editor: Gianluca Palli

Received: 31 December 2021
Accepted: 1 February 2022
Published: 14 February 2022

Publisher's Note: MDPI stays neutral with regard to jurisdictional claims in published maps and institutional affiliations.

In [16], algorithms to stop a cable-driven camera system after a cable failure are proposed. Dynamic trajectories into the post-failure workspace are planned. Collision avoidance is considered and preferred target positions are identified. In [17], the effects of cable anchor point alignment on post-failure behavior is studied for an underwater CDPR. The winches are installed on marine barges and their optimal positions are determined to increase the systems fault tolerance. In [18], safety concepts for CDPRs are presented. This features, e.g., approaches to shorten the stopping distance during emergency braking and the monitoring of the workspace.

The strategies after cable failure described so far mainly focus on trajectory planning back into the remaining workspace, or cable force redistribution when the platform remains in the workspace after failure. On the contrary, the authors of this work have proposed two strategies to guard the platform back into the post-failure workspace without a predefined trajectory [9]. One strategy presented is based on potential fields employment, whereas the other aims to minimize the systems' kinetic energy [9].

In more recent works, the authors also proposed including actuated moving pulleys in cable failure strategies. The strategy of minimizing the systems' kinetic energy is enhanced and simulated using ideal assumptions in [19].

However, the proposed methods of the authors have neither been validated in loop-closure of a control algorithm, nor been tested in detailed multi-body simulations. Moreover, the presumed failure of a control algorithm commonly used by the authors was not demonstrated before. The aim of this contribution is to close this gap. Following a strictly model-based testing scheme, the examined behavior of the introduced emergency strategy is analyzed, discussed and compared to the commonly used control algorithm. This paves the way for future experimental validation on a prototype.

The structure of the paper is given in the following: An introduction on CDPRs is given first, focusing especially on cable failure and methods to cope with this issue. Section 2 introduces the kinematic and dynamic robot models, as well as the cable and drivetrain model applied within this work. In Section 3, an emergency strategy aiming to minimize the system's kinetic energy is introduced. This strategy involves nonlinear model predictive control. The dynamic multi-body simulation environment based on the modeling fundamentals used within this work is shown in Section 4, and also includes a standard position controller. For the chosen set of robot parameters, a brief examination of the post-failure static equilibrium workspace is carried out in Section 5. Subsequently, dynamic simulations within the chosen environment are performed, comparing the standard controller approach and the presented emergency controller within a cable failure scenario. This analysis is extended regarding the computation time and real-time suitability of the emergency algorithm, considering an emergency controller restricted in iterations. Finally, a summary and outlook on future work towards implementation within real hardware are given.

2. Modeling Fundamentals

In this section, the modeling equations and approaches used for this work will be introduced. These models form the basis for simulative studies in the upcoming sections.

2.1. Kinematics

The robot platform is the robot's end effector. It has m cables and n DOF. This results in the kinematic redundancy $r = m - n$. The coordinate system $\underline{P}$ is fixed to the platform, whereas the latter is referenced in the inertial system $\underline{B}$. The position of the platform ${}^B r_P$ and its orientation $\mathbf{\Phi}$ are denoted as row vectors. They are merged in the pose ${}^B x_P = [{}^B r_P\ \mathbf{\Phi}]^T$. The rotation matrix ${}^B R_P$ describes the platform orientation with respect to $\underline{B}$ with the use of yaw–pitch–roll angles.

The fixed platform anchor point of each cable is ${}^P p_i$; see Figure 1. They are joined in $P = [{}^P p_1 \ldots {}^P p_i]$.

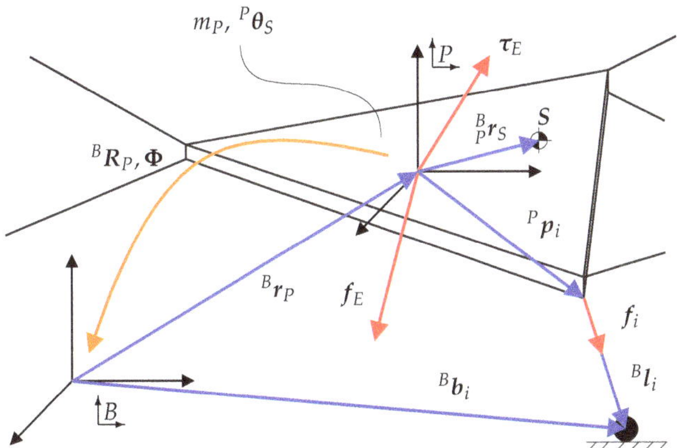

Figure 1. Cable-robot model parameters.

Furthermore, the base vectors $^B b_i$ describe the frame geometry. Here, the cables are led into the robot's workspace. These vectors are denoted in $B = \begin{bmatrix} ^B b_1 \dots ^B b_i \end{bmatrix}$. From inverse kinematics, the cable vectors $^B l_i$ can be derived as follows:

$$^B l_i = {}^B b_i - \underbrace{(^B r_P + {}^B R_P {}^P p_i)}_{^B p_{B_i}}, \quad 1 \leq i \leq m. \tag{1}$$

2.2. Forces and Dynamics

All vectors are decomposed and referenced within the inertial coordinate system. Note that the top-left index is left out for simplicity. The force of each cable in its direction v_i is f_i. The cable force vector $f \in \mathbb{R}^{m \times 1}$ is composed by concatenating all m cable forces. The force in each ith cable that acts on the platform is f_i. Each is computed by

$$f_i = f_i \cdot \frac{l_i}{\|l_i\|_2} = f_i \cdot v_i, \quad 1 \leq i \leq m. \tag{2}$$

All other forces f_E and torques τ_E that act on the end effector are summarized within w_E. The transpose of the robot's Jacobian is A^{T}, which is named as the structure matrix. Hence, the equilibrium of forces and torques at the end effector is

$$-w_E = \begin{bmatrix} -f_E \\ -\tau_E \end{bmatrix} = \begin{bmatrix} v_1 & \dots & v_m \\ p_1 \times v_1 & \dots & p_m \times v_m \end{bmatrix} \begin{bmatrix} f_1 \\ \vdots \\ f_m \end{bmatrix} = A^{\text{T}} f. \tag{3}$$

The inertia tensor of the platform is denoted as $^B\boldsymbol{\theta}_P$ and its mass is m_P. $^B\boldsymbol{\theta}_P$ needs to be determined by applying the Steiner–Huygens relation using the given tensor within the platform's gravitational center $^P\boldsymbol{\theta}_S$. Now, $^B_P r_S$ is the vector pointing from $\underline{P}$ to the gravitational center S. Accordingly, the skew–symmetric matrix $^B_P R_S$ is

$$^B_P R_S = \begin{pmatrix} 0 & -^B_P r_{S,z} & ^B_P r_{S,y} \\ ^B_P r_{S,z} & 0 & -^B_P r_{S,x} \\ -^B_P r_{S,y} & ^B_P r_{S,x} & 0 \end{pmatrix}. \tag{4}$$

Now the Newton-Euler equations [20] can be expressed in $\uparrow\underline{B}$ which leads to the dynamics model equations

$$\underbrace{\begin{bmatrix} m_P\,E_3 & -m_P\,{}^B_P R_S \mathcal{H} \\ m_P\,{}^B_P R_S & {}^B\theta_P \mathcal{H} \end{bmatrix}}_{M(x_P)} \underbrace{\begin{bmatrix} {}^B\ddot{r}_P \\ \ddot{\phi} \end{bmatrix}}_{\ddot{x}_P} + \dots$$

$$+ \underbrace{\begin{bmatrix} m_P\left[(\dot{\mathcal{H}}\dot{\phi}) \times {}^B_P r_S + (\mathcal{H}\dot{\phi}) \times ((\mathcal{H}\dot{\phi}) \times {}^B_P r_S)\right] \\ {}^B\theta_P \dot{\mathcal{H}}\dot{\phi} + (\mathcal{H}\dot{\phi}) \times ({}^B\theta_P \mathcal{H}\dot{\phi}) \end{bmatrix}}_{K(x_P,\dot{x}_P)} \dots$$

$$\dots + \underbrace{-{}^B w_E}_{Q(x_P,\dot{x}_P)} = {}^B A^\mathsf{T} f. \tag{5}$$

Here, $\mathcal{H}$ and $\dot{\mathcal{H}}$ can be determined from kinematic Kardan equations; see [21]. E_3 is a 3×3 identity matrix. $\dot{x}_P$ and $\ddot{x}_P$ are the first and second time derivatives of the platform pose. $K(x_P, \dot{x}_P)$ contains centrifugal and Coriolis forces and torques. The platforms mass matrix is $M(x_P)$. Within vector $Q(x_P, \dot{x}_P)$, all remaining forces and torques are included. These are, in particular, friction, disturbances and gravitational forces. For simplicity, cables are considered to be non-sagging and massless, following straight lines. Since they can never push but only pull, a minimal cable tension $f_{\min}$ needs to be kept at all times during regular operation. Along with this, the maximum tension limit $f_{\max}$ should not be surpassed due to the necessary limitations of the drive capabilities and the mechanical loads on the robot components. To follow a given trajectory, the required forces and torques at the platform can be determined in each time step. Consequently, required forces f can be obtained by solving Equation (5). For computation, several methods exist that are commonly known. They differ, e.g., in the required maximum or average computation time or in the resulting force level [1,22].

2.3. Cables

As force transmitting elements between the end effector and the cable-drum, the cables need to be considered within the modeling. In practice, cables show different dynamic effects depending e.g. on their material, tension or length. Such effects include e.g. creeping, sagging, vibration or elastic deformation. Depending on the modeling approach, some or all effects of the above can be displayed within simulation, varying in complexity and calculation time costs [1]. In this work, the cables are modeled as a parallel system of a spring and a damper without a mass. The transmitted force therefore depends only on the elongation of the cable, i.e. eigendynamics of the cables are neglected. This approach is rather simple and efficient in calculation time but is still able to display a linear-elastic deformation of the cable and allows for a description of platform vibrations [1]. Since the cables cannot push, a piecewise defined function is used to model the cable force magnitude f_i for each i-th cable.

$$f_i(x_P, \theta_i, \dot{x}_P, \dot{\theta}_i) = \begin{cases} k_{l,i}(l_{i,tot})\,\Delta l_i(x_P, \theta_i) + d_i\,\Delta \dot{l}_i(\dot{x}_P, \dot{\theta}_i) & \text{for } \Delta l_i > 0 \text{ and } k_i \Delta l_i > -d_i \Delta \dot{l}_i \\ 0 & \text{for } \Delta l_i \leq 0 \text{ or } k_i \Delta l_i \leq -d_i \Delta \dot{l}_i \end{cases} \tag{6}$$

Note that a negative damping force cannot exceed the applied spring force in order to avoid pushing forces within the cable model when a cable is compressed. The damper coefficient d_i is assumed to be constant, whereas the cable stiffness $k_{l,i}$ is dependent on the current total cable length $l_{i,tot}$. This is a function of the cable length at starting position $l_{i,0}$ and the unwound cable length of the drum with radius r_d and angle θ_i. $l_{i,0}$ is the sum of the free cable length within the workspace $||l_i(x_{P,0})||_2$ at the starting position $x_{P,0}$ and the cable length between the drum and the deflection point of the cable $l_{d,i}$, which is assumed to be constant for simplicity. Moreover, the wound cable length on the drum is neglected, as well as the cable length covering the deflection elements, since pulleys are not considered in this work.

$$l_{i,tot}(\theta_i, l_{i,0}) = \theta_i\, r_d + \underbrace{||l_i(x_{P,0})||_2 + l_{d,i}}_{l_{i,0}} \tag{7}$$

From catalog data, there is a linear relationship of 5.8% between the minimum breaking force f_b and the corresponding cable length, which is proposed to be used to determine the current cable stiffness.

$$k_{l,i}(l_{i,tot}(\theta_i, l_{i,0})) = \frac{f_b}{0.058\, l_{i,tot}(\theta_i, l_{i,0})} \tag{8}$$

The cable length difference Δl_i describes whether the cable is tensed or slack. Thus, the total cable length dependent on the wound or unwound cable $l_{i,tot}$ is compared to the kinematic cable length at the current pose x_P. Note that this approach is only possible since the sagging of the cable is neglected in the modeling for simplicity and the cables are assumed to be straight lines.

$$\Delta l_i(x_P, \theta_i) = (||l_i(x_P)||_2 + l_{d,i}) - l_{i,tot}(\theta_i, l_{i,0}) \tag{9}$$

For the cable velocity difference $\Delta \dot{l}_i$, the approach is similar.

$$\Delta \dot{l}_i(\dot{x}_P, \dot{\theta}_i) = \underbrace{(A\dot{x}_P)_i}_{\dot{l}_i(x_P, \dot{x}_P)} - \dot{\theta}_i r_d \tag{10}$$

2.4. Drivetrain

Let J_d be the total inertia of the drivetrain system about the rotation axis, including the cable drum and motor inertia for one joint. $\ddot{\theta}_i$ is the angular acceleration of one cable drum. Depending on the angular velocity $\dot{\theta}_i$ of one cable drum and its sign, a coulomb friction with coefficient μ_c and a static friction μ_s is assumed. The backlashing force acting on the cable drum by the tensed cable converted into the respective torque is $f_i\, r_d$. From a desired torque command, the motor controller is assumed to process this command with the dead time T_t and is able to produce it with a delay. This is modeled by a PT1 system with time constant T_1. The produced torque is denoted as $\tau_{c,i}$. Finally, the equation of motion of the drivetrain system can be described by

$$J_d\, \ddot{\theta}_i = \tau_{c,i} - f_i\, r_d - (\mu_s\, \text{sign}(\dot{\theta}_i) + \mu_c\, \dot{\theta}_i) \tag{11}$$

3. Emergency Strategy Based on Energy Minimization

Cable failure during the operation of CDPRs can lead to severe outcomes. As the end effector might move in an uncontrolled manner, collisions can occur, possibly leading to substantial damage to the machine, employees or the surroundings. Moreover, the carried payload might be lost. Consequently, the necessity for action strategies after a cable failure is given, especially during the transfer of this technology into industry. When a cable fails, the end effector can suddenly get beyond the borders of its pre-failure workspace. If this is the case, standard methods for cable force calculation are expected to fail [9]. Two strategies of action after a failure have been proposed by the authors in previous work [9]. One of those is presented in particular as follows.

For simplicity, it is assumed now that just one cable fails. Hence, a CDPR with a number of $m - 1$ cables remains. In the rest of the paper, the force distribution of the remaining cables is referred to as $f^* \in \mathbb{R}^{(m-1) \times 1}$. Accordingly, the corresponding structure matrix, reduced by the column of the broken cable, is $A^{T*} \in \mathbb{R}^{n \times m-1}$. As long as at least a single cable remains after a cable break, the movement of the end effector can still be influenced. In case of CDPR with higher redundancies r, there might even be a good chance to stop the movement in a controlled way, even after the loss of multiple cables.

However, the post-failure workspace can be significantly smaller than the original one. This promotes an unfavorable scenario where the platform suddenly gets beyond the borders of its pre-failure workspace, as mentioned above.

A possible approach to safeguard the end effector in a controlled way is to incorporate model predictive control [23] in order to minimize the systems' kinetic energy after a cable break [9]. Within this strategy, the state of the system is predicted utilizing the dynamical and nonlinear model of the robot to compute cable force distributions, minimizing the end effector's kinetic energy over time. Consequently, this minimizes the system's velocity, which leads to a system stop. Hence, if the system is finally able to stop, the robot will automatically have been led into the remaining post-failure workspace. This strategy does not require the definition of a goal pose, which is highly advantageous. A rigid body is assumed in order to model the moving masses of platform and payload. Its kinetic energy is defined as

$$E_{Kin} = \frac{1}{2} m_p \, {}^{B}\dot{r}_P^{\mathrm{T}} \, {}^{B}\dot{r}_P + \frac{1}{2} \Omega^{\mathrm{T}B} \theta_P \Omega. \tag{12}$$

The angular velocity $\Omega = \mathcal{H}\dot{\phi}$ of the body is directly dependent on $\dot{\phi}$ [21]. Therefore, the magnitude of $\dot{x}_P = [\dot{r}_P \; \dot{\phi}]^{\mathrm{T}}$ needs to be be minimized in order to decrease the system's kinetic energy. The system is described by the nonlinear discrete state equations

$$\begin{aligned} x(k+1) &= f\big(x(k), u(k)\big) \\ y(k+1) &= C\, x(k+1). \end{aligned} \tag{13}$$

It is based upon the nonlinear system function $f\big(x(k), u(k)\big)$ derived from Equation (5). $x(k)$ is the state vector containing position and velocity. $y(k)$ describes the output of the system. Both are defined as

$$\begin{aligned} x(k) &= [\dot{r}_p(k), \dot{\phi}(k), r_p(k), \phi(k)]^{\mathrm{T}} \\ y(k) &= [\dot{r}_p(k), \dot{\phi}(k)]^{\mathrm{T}}. \end{aligned} \tag{14}$$

Applying the 3×3 identity matrix K, matrix C describes the output:

$$C = \begin{bmatrix} K & 0 & 0 & 0 \\ 0 & K & 0 & 0 \end{bmatrix}. \tag{15}$$

The system's input vector contains the remaining cable forces

$$u(k) = f^*(k). \tag{16}$$

The state vector $x(k+1)$ for the next time step is obtained from $f\big(x(k), u(k)\big)$ applying numerical integration with a fixed prediction step size Δt_c. For this purpose, the Euler–Cromer scheme is used:

$$\begin{aligned} \begin{bmatrix} \dot{r}_p(k+1) \\ \dot{\phi}(k+1) \end{bmatrix} &= \begin{bmatrix} \dot{r}_p(k) \\ \dot{\phi}(k) \end{bmatrix} + \left[M(x_P)^{-1} \big(A^{\mathrm{T}*} f^* - K(x_P, \dot{x}_P) - Q(x_P, \dot{x}_P) \big) \right](k) \Delta t_c \\ \begin{bmatrix} r_p(k+1) \\ \phi(k+1) \end{bmatrix} &= \begin{bmatrix} r_p(k) \\ \phi(k) \end{bmatrix} + \begin{bmatrix} \dot{r}_p(k+1) \\ \dot{\phi}(k+1) \end{bmatrix} \Delta t_c. \end{aligned} \tag{17}$$

Now, a cost function J is set up as follows:

$$\begin{aligned} J = \sum_{i=1}^{n_p} \big(y(k+i) - y_r(k+i)\big)^{\mathrm{T}} Q \big(y(k+i) - y_r(k+i)\big) + \dots \\ \sum_{i=i}^{n_c} \big(f^*(k+i-1) - f^*(k+i-2)\big)^{\mathrm{T}} r_1 \big(f^*(k+i-1) - f^*(k+i-2)\big) \end{aligned} \tag{18}$$

Herein, the number of steps n_p is the prediction horizon, whereas n_c is the control horizon. To obtain set-point cable force distributions f^*, Equation (18) is minimized in each time step, considering the given limitations for cable forces. In order to stop the system, the goal velocity is set to $y_r = 0$. As soon as the platform gets to a full stop and the cable force distribution remains steady, the costs J will be minimal. The weighting parameters Q and r_1 can be used to scale the optimization problem.

4. Simulation Environment

In order to perform dynamic simulation experiments, a multi-body simulation environment needs to be set up. This environment and a commonly used position controller are described in this section.

4.1. System Structure

Based on the modeling fundamentals described in section 2, an environment is set up to carry out simulative experiments. The resulting system structure used within this work is shown in Figure 2.

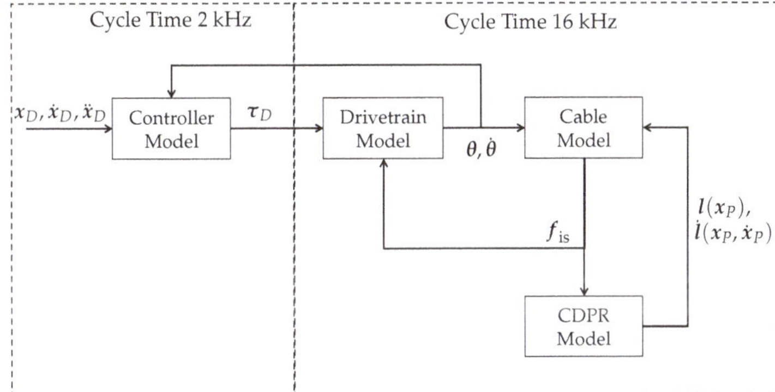

Figure 2. System structure of the simulation environment with two differing fixed sample times.

The simulation is split up into two areas with differing fixed sample times. The controller model is sampled with 2 kHz, whereas the physical system, including the drivetrains, cables and a robot model, is simulated with 16 kHz. The lower sample time corresponds to the cycle time used within the control of prototypes such as the SEGESTA [24], whereas the higher sample time is intended to create an adequate representation of the physics, aligning to typical clock frequencies of motor current controllers. From a desired trajectory given by $x_D, \dot{x}_D, \ddot{x}_D$ and $\theta, \dot{\theta}$ as a feedback from the drives, the position controller model delivers set torques τ_D to the drivetrain model. Details on the incorporated controller are given in the upcoming section.

Depending on the current cable forces f_{is} and the torque command τ_D, the internal drivetrain control accelerates the winch according to Equation (11). This results in current winch velocities $\dot{\theta}$ and angles θ by integration. To determine the current cable force according to Equation (6), the cable model receives θ and $\dot{\theta}$, as well as the current kinematic cable length and velocities $l, \dot{l}$ from the CDPR model. Depending on the current cable forces, the acceleration of the end effector is calculated using Equation (5). The position and velocity of the end effector are also obtained by integration. Finally, from the end effector state, the current kinematic cable length and velocities of the robot model are derived, employing the inverse kinematic; see Section 2. Note that the system states in the drivetrain model, as well as in the robot model, are time-discretely integrated using the trapezoidal integration method.

4.2. Standard Controller Model

Since CDPRs have a strongly nonlinear character, a position control of them is non-trivial. As described in [25,26], an augmented PD controller is a highly suitable choice for this task. Figure 3 displays the structure of this controller. As a central component, a PD position controller in joint space is acting, which controls the desired cable length l_D and cable velocity $\dot{l}_D$ obtained from the desired trajectory using inverse kinematics. Using the wrench w_{ff}, this linear controller is extended by a model-based feed forward path based on Equation (5) to handle the strong nonlinear character of the robot. The controller law is given as follows:

$$w = \underbrace{A^{\mathrm{T}}(K_P\, e + K_D\, \dot{e})}_{w_{\mathrm{PD}}} + w_{ff} \tag{19}$$

K_P and K_D are diagonal gain matrices, whereas e and $\dot{e}$ are the errors in cable length and velocities, respectively. The measured length l and velocities $\dot{l}$ are determined by the feedback of the measured drum angles θ, velocities $\dot{\theta}$ and the drum radii r_d (see also Section 2.3). A brief discussion on the stability and parametrization of the controller is given in [27]. Inserting the controller output wrench in Equation (2), a cable force distribution can be found using well-known methods; see Section 2.2. The desired force distribution f_D is converted to torques τ_D and can be fed to the drives.

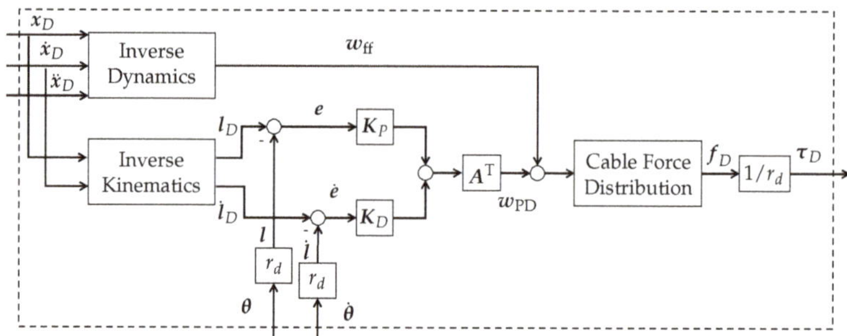

Figure 3. Diagram of the augmented PD controller [26].

5. Dynamic Simulation—Experiments and Results

Within this work, a planar dynamic CDPR model using two translational DOF and four cables ($n = 2$, $m = 4$) is considered. In comparison to former work of the authors featuring models with two DOF (see, e.g., [9]), this model is extended by drive-train modeling and basic cable modeling, as well as a standard control system model (see Sections 2 and 4). Note, that for the model at hand, the rotational components are set to zero in Equations (5) and (17). For simplicity, all elements of parameters and variables that go beyond the chosen two DOF are not displayed in the following.

The model is subject to gravity in the negative y-direction. The cables are massless but can become slack (see Section 2.3). Modeling of pulleys is neglected for simplicity, as well as collisions between cables or disturbances. The chosen set of parameters for the simulation model is as follows. The cable and drivetrain parameters are inspired by the SEGESTA prototype.

$$P = [{}^P p_1 \ldots {}^P p_4] = \begin{bmatrix} 0\,0\,0\,0 \\ 0\,0\,0\,0 \end{bmatrix}\, \mathrm{m},\ B = [{}^B b_1 \ldots {}^B b_4] = \begin{bmatrix} -0.5001\ 0.4999\ -0.4998\ 0.5002 \\ 1.0000\ \ 1.0000\ \ 0.0000\ \ 0.0000 \end{bmatrix}\, \mathrm{m},$$

$$f_{\min} = 10\,\mathrm{N},\ f_{\max} = 150\,\mathrm{N},\ {}^B_P r_S = [0\,0]^{\mathrm{T}}\,\mathrm{m},\ f_E = [0\,0]^{\mathrm{T}}\,\mathrm{N},\ {}^B \theta_S = \begin{bmatrix} 0\,0 \\ 0\,0 \end{bmatrix}\,\mathrm{kg\,m^2},$$

$$J_d = 2.25 \times 10^{-4}\,\mathrm{kg\,m^2},\ r_d = 0.015\,\mathrm{m},\ f_b = 4\,\mathrm{kN},\ d = 78\,\mathrm{Ns/m},\ m_P = 1\,\mathrm{kg},\ \mu_s = 7 \times 10^{-4}\,\mathrm{Nm},$$

$$\mu_c = 0.05\,\mathrm{Nms/rad},\ T_1 = 4.3 \times 10^{-4},\ T_t = 1.875 \times 10^{-4}\,\mathrm{s},\ [l_{d,1} \ldots l_{d,4}] = [0.25, 0.25, 0.25, 0.25]\,\mathrm{m}. \tag{20}$$

Note, that only results for the failure of an cable coming from above are introduced in this work, due to lack of space.

5.1. Analysis of the Post-Failure Workspace

As demonstrated in the former work of the authors [9,19], the workspace of a CDPR can change significantly after a cable failure. This increases the risk of the platform being outside the workspace for static force equilibrium ($\mathcal{SEW}$) post-failure, leading to an inevitable movement of the platform.

For the model parameters at hand, the $\mathcal{SEW}$ for static equilibrium pre- and post-failure under influence of gravity is displayed in Figures 4 and 5. The workspace was calculated using a discrete point grid and checking for a feasible cable force distribution at each point while applying the closed-form method [1]. Clearly, it is visible that the workspace is approximately halved post-failure for the given model, promoting the risk of the platform being outside the workspace after cable failure.

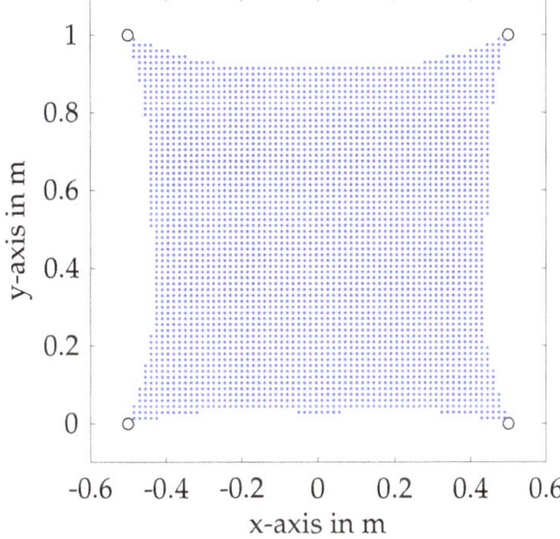

Figure 4. $\mathcal{SEW}$ of the CDPR pre cable failure. Only gravitational forces are considered. The positions, where the cables are fed into the workspace are displayed by circles in the corners.

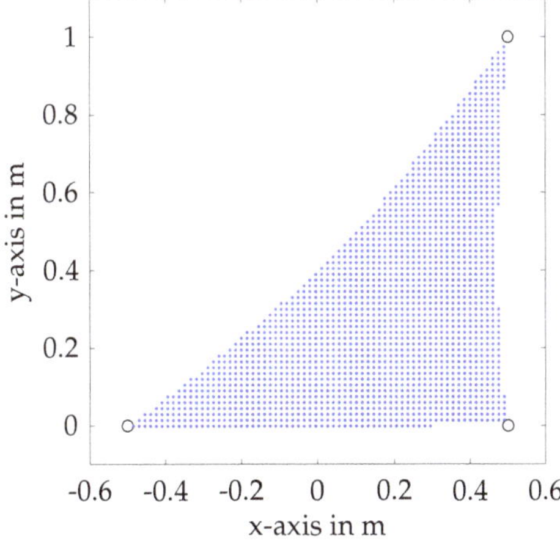

Figure 5. $\mathcal{SEW}$ of the CDPR post failure of cable 1. Only gravitational forces are considered. The positions, where the cables are fed into the workspace are displayed by circles in the corners.

5.2. Multi-Body Cable Failure Simulation with Standard Controller

To perform dynamic cable failure simulations, the intervention points of the cable failure within the model need to be defined. At a given time, the cable failure is induced by setting the output force of the cable model to zero for the broken cable. Therefore, only the remaining cables can impose a dynamic effect on the robot platform within the robot model; see Section 4.1. Moreover, due to the missing reactive force from this cable, the associated drivetrain can easily accelerate the winch only bounded by friction. However, the winch movement does not impose any effect on the cable model anymore. Still, due to the feedback of motor angles and velocities, the controller senses the winch movement.

For a first experiment, the platform is set into a position at $r_P = [-0.35, 0.65]^\mathsf{T}$ m, while the APD controller aims to hold that position as goal pose. Desired velocity and acceleration are set to zero. Note, that this position is outside of the $\mathcal{SEW}$ after cable failure. The parameter are set to $K_P = 40,000 \,\mathrm{diag}(1,1,1,1)$ and $K_D = 50 \,\mathrm{diag}(1,1,1,1)$. To generate a desired cable force distribution from the controller output (see section 4.2), the closed-form method is used [1]. Subsequently, the model behavior after cable break is investigated, as if the controller would have no inherent emergency strategy and no detection of the cable failure. The cable failure is induced at 0.25 s. Figure 6 shows the trajectory of the platform and the drives after cable failure employing the augmented PD controller without emergency strategy. As the APD controller has no knowledge of the cable failure, it tries to hold the old desired position, leading to an undesired movement of the platform. The velocity in the x-direction reaches $1.9\frac{\mathrm{m}}{\mathrm{s}}$, and reaches $-1.8\frac{\mathrm{m}}{\mathrm{s}}$ in the y-direction. The drives coil the cables in an undesired manner, trying to maintain the desired cable force given from the augmented PD controller. The drive velocities reach up to $\sim 30\frac{\mathrm{rad}}{\mathrm{s}}$ when the platform is falling down. When the cables get back in tension, as shown in the right side of Figure 7, the conducted impulses in cable forces can also be seen within the drive velocities. They are caused whenever the slack cables catch the falling platform. After being pulled into the reduced $\mathcal{SEW}$ with the remaining cables and some uncontrolled movement at the workspace border, the joint errors exceed a certain level, resulting in non-feasible cable force distributions (see Figure 7, left). As displayed on the right side of Figure 7, the remaining cables stay in tension after failure of the cable force calculation at approximately 2.25 s due to inertia and friction in the drivetrains and the mass of the falling platform, which finally crashes to the ground. This underlines the necessity for an emergency strategy, guiding the end effector back into the static equilibrium workspace in a controlled manner, where it can be stopped in a statically stable position without collision with the ground.

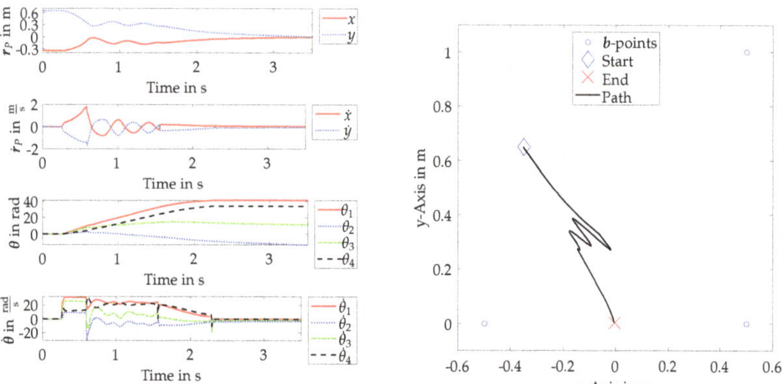

Figure 6. Trajectory of end effector and drives (**left**) and path (**right**) after cable failure of cable 1 using the augmented PD controller.

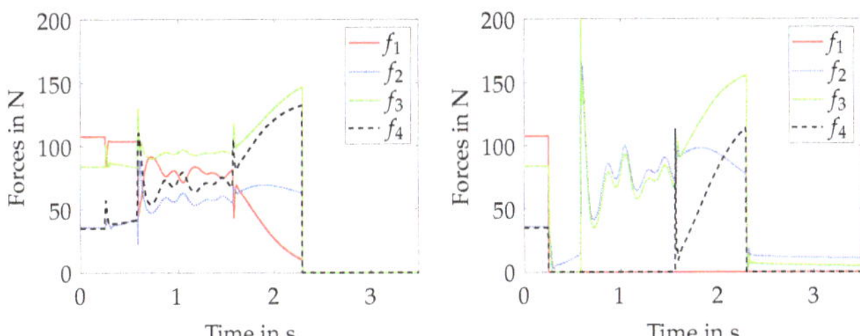

Figure 7. Commanded forces (**left**) and measured cable forces (**right**) after cable failure using the augmented PD controller.

5.3. Implementation of the Emergency Strategy

To implement the proposed emergency strategy (see Section 3), a few changes are applied to the system structure, as displayed in Figure 8. As soon as the cable fails, the information about the failure is directly fed to the controller for simplicity. At that point, the APD controller gets bypassed and the desired force from the last calculation cycle serves as a warm start point for the optimization-based emergency controller. Of course, for the application on real hardware, the cable failure needs to be detected. Potential points of detection that need to be investigated in future work are, e.g., cable force measurement or an unexpected deviation of the winch angle measurement. As the emergency strategy of kinetic energy minimization needs to have knowledge on the CDPR system state, an ideal pose measurement is assumed in the simulation. On real hardware, this task needs to be accomplished in real time by a suitable measurement unit, e.g., based on laser tracking. A second option would be the use of a forward kinematic algorithm. As a result of the risk of slack cables and uncertain movement of the end effector, implementing a cable failure resistant forward kinematic is nontrivial and part of future work.

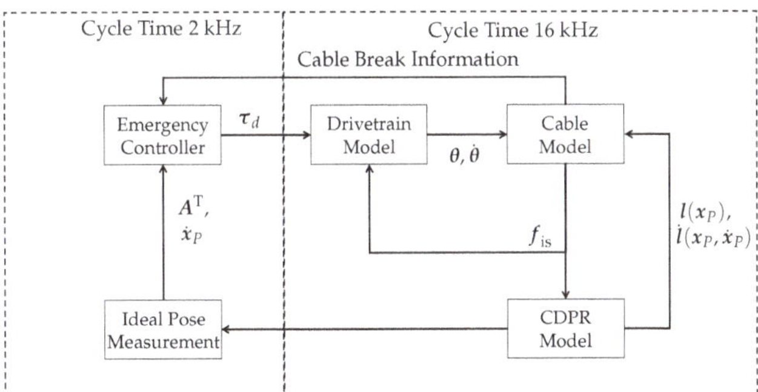

Figure 8. System structure of the simulation environment with two differing fixed sample times. Implementation of the NMPC-based emergency strategy in closed loop.

After an induced cable break, the emergency controller will start to calculate desired cable forces while minimizing the cost function Equation (18). The forces are converted to torques and commanded to the drivetrain system. Note that the NMPC algorithm is capable of predicting the system behavior over several time steps and, moreover, can obtain a more complex model, including, e.g., drivetrain behavior and friction. Due to the expected shortage in the available computation time when using the algorithm on prototype hardware later on, a rather simple model based on Equation (17) is included in this work, and the prediction is carried out only over one time step, i.e., $n_p = n_c = 1$. The optimization problem is solved using MATLAB's `fmincon()` and an interior point algorithm. The output of the last cycle is fed back in order to enable a warm start of the optimization.

5.4. Multi-Body Cable Break Simulation with Emergency Strategy

The simulation is started with the same initial conditions as the experiment in Section 5.2. Subsequently, as soon as the cable breaks, the APD controller is bypassed and the NMPC-based emergency controller starts. The NMPC parameters are set as follows: $r_1 = 4 \times 10^{-4}$, $Q = \text{diag}(1, 5)$, $\Delta t_c = 0.5$ ms, $y_r = 0$. $\texttt{fmincon()}$ is used with default parameters first. Note that Q is designed for less desired vertical movement, since it produces five times more costs in Equation (18) than horizontal movements.

Figure 9 displays the trajectory of the end effector and the drives after cable failure while the emergency controller is commanding desired cable forces. It is worth noting that the platform travels a slight diagonal path into the remaining $\mathcal{SEW}$, covering double the distance in the horizontal direction than in the vertical direction. In contrast to Figure 6, the velocity in the x-direction is slightly higher, whereas it is nearly halved in the y-direction. This evidences that the chosen parameter Q indeed leads to less vertical movement. The progressions of θ and $\dot{\theta}$ show how the drives react properly to the commanded torque coming from the emergency method. The drives reach velocities up to $\sim 50 \frac{\text{rad}}{\text{s}}$. Drive 1, which corresponds to the broken cable, stops movement, as there is no torque commanded to it. The platform can be dragged into the remaining workspace. Then, it is orbiting near the position where it entered the workspace. Finally, and in contrast to the experiment in Section 5.2, it gets stopped without being subjected to collision with the robot boundaries.

The commanded and measured forces in this experiment are shown in Figure 10. The platform can be stopped in a statically stable position after approx. 2 s. The algorithm fully takes advantage of the given cable force boundaries, but does not exceed it within the commanded values. However, as displayed in the right hand figure, the current cable forces progress differently to the set forces throughout the trajectory. Since the cable force is not controlled directly, this is not unexpected. In the first segment after the cable failure, only cable 2 is held in tension while the platform starts to move. Since vertical movement is not desired within the optimization, cables 3 and 4 are commanded to minimum tension, which causes them to go slack in the model. Roughly 0.25 s after the failure, the platform reaches a pose where cable 3 gets back in tension, creating an impulse that slightly exceeds the desired cable force boundaries by 20 N. From there on, the platform starts to swing near the workspace border until cable 4 is brought back into tension by the algorithm. In that phase, the platform movement causes this cable to go in and out of tension a few times, creating some impulses in cable force 4. Notably, the second impulse also exceeds the cable force boundaries by 50 N. After all three cables are back in tension, the movement is stopped immediately. Even though the observed impulses exceed the desired cable force limits, which can be problematic on real hardware, the cable forces that occur are still far below the cable breaking force f_b. In summary, the experiment gives evidence that the chosen emergency method successfully works in a multi-body simulation environment and can stop the platform without collision in a fast manner.

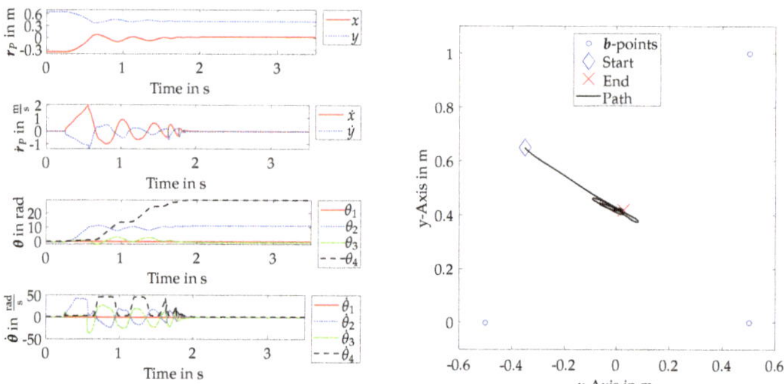

Figure 9. Trajectory of end effector and drives (**left**) and path (**right**) after cable failure of cable 1 using the emergency controller with default optimizer settings.

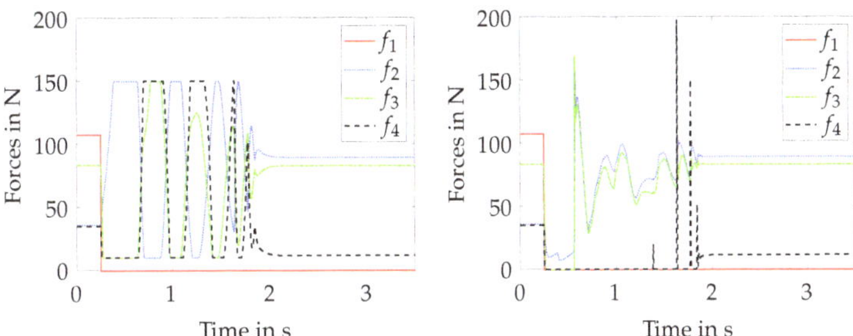

Figure 10. Commanded forces (**left**) and measured cable forces (**right**) after cable failure using the emergency controller with default optimizer settings.

5.5. Investigation on Computational Efficiency

Since the solution of numerical optimization problems on real-time hardware might lead to problems because of their iterative structure, a consideration of the computation time of the algorithm is crucial. The experiment from the last section is now evaluated in computation time and compared to an experiment with more restriction on the maximum allowed iterations of the optimizer. The calculations were carried out within Matlab R2019a using an Intel CPU i7-8700 with 3.2 GHz on a Windows 10 System. The maximum allowed iterations are now reduced from 1000 (default) to 10. Limiting the number of iterations within the optimizer may lead to a premature abortion of the optimization before reaching a cost function minimum. This impacts the cyclic computation time in a positive manner in terms of predictability and constancy. Nonetheless, the quality of the optimization result is expected to be negatively affected by this limitation.

In comparison to Section 5.4, the trajectory of the platform and the drives after cable failure is quite similar, as displayed in Figure 11.

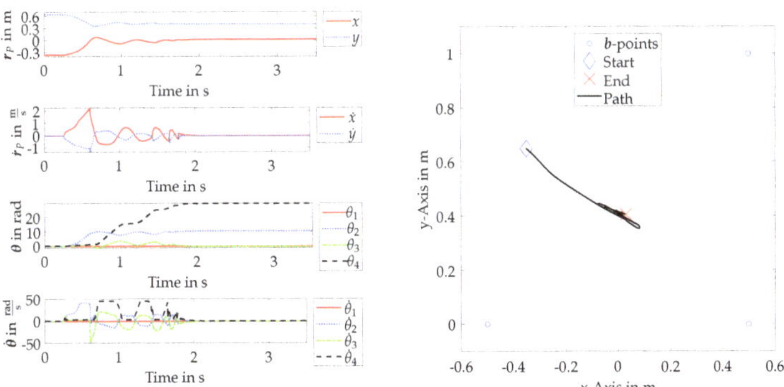

Figure 11. Trajectory of end effector and drives (**left**) and path (**right**) after cable failure of cable 1 using the emergency controller with constrained iterations of 10.

In addition, the progression of the cable forces, as displayed in Figure 12, is comparable; however, some differences can be highlighted. The progression of the controlled forces is slower when a different extreme cable force distribution near the force boundaries is found. This can be explained by the assumption that the algorithm tends to stay in local minima or runs into the iteration restriction while propagating to a new optimal solution. This is also displayed by θ and $\dot{\theta}$. It is worth noting that the drives tend to react slower due to the commanded torque values. Drive 2, for example, is not accelerated as fast as in Figure 9, while the first impulse in cable force 3 results in a higher negative velocity on drive 3. The platform is led to a full stop after ~ 2.15 s, which is a little slower than with default parameters. Within the measured forces, it can be observed that the first impulse—when cable 3 gets back into tension—is roughly 30 N higher in comparison. In addition, the first impulse on cable 4 is 80 N higher, whereas the second impulse is 10 N lower. All in all, the height of the impulses when the slack cables go back into tension tends to rise.

The computation time per cycle step of the optimization algorithm throughout both experiments is displayed in Figure 13. The mean times have been determined starting from 0.25 s, where the cable failure is induced and the optimizer is switched on. For default optimization parameters, the cycle times range from 2.5 ms to 25 ms while the platform is in motion, with a mean time of 9.24 ms. After the stop of the platform, the cycle times drop to a quite continuous level of 2.48 ms. The total mean computation time per cycle is 6.26 ms. As expected, restricting the maximum allowed iterations leads to more constant cycle times, as displayed on the right side of Figure 13. Here, the cycle times range from 7.25 ms to 20.4 ms while the platform is in motion. The mean is 8.14 ms, which is approximately 1 ms lower than with default parameters. After stopping, the times drop to roughly 2.8 ms, leading to a total mean of 5.84 ms. Note that both cases have a peak in cycle time within the first iteration of the optimizer at 0.25 s due to initialization and memory preallocation. Restricting the algorithm further to 5 iterations maximum, the range of calculation times drops to 5 ms minimum and 8.5 ms maximum. However, in that case, the platform cannot be fully stopped within the simulation time, and the peak force that is induced when one cable get back into tension first is 260 N. The minimum number of iterations to stop the platform successfully in the present simulation example is 6. Still, in that case, a peak tension of 250 N occurs. Due to a lack of space, those experiments are not displayed here. This indicates that further restrictions of the maximum iterations are not useful. In summary, a compromise between the quality of the optimization result and computation time needs to be found when restricting the maximum iterations of the algorithm.

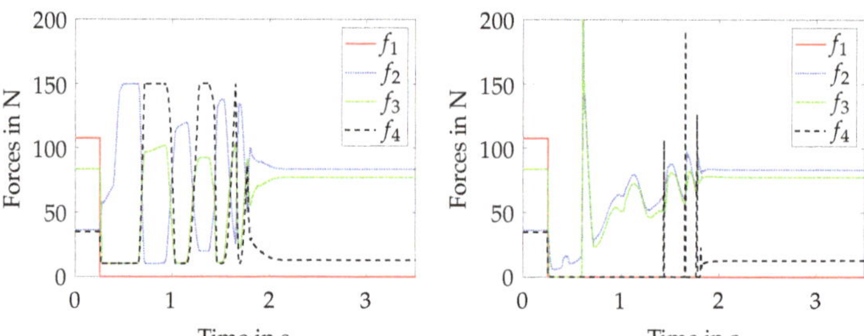

Figure 12. Commanded forces (**left**) and measured cable forces (**right**) after cable failure using the emergency controller with constrained iterations.

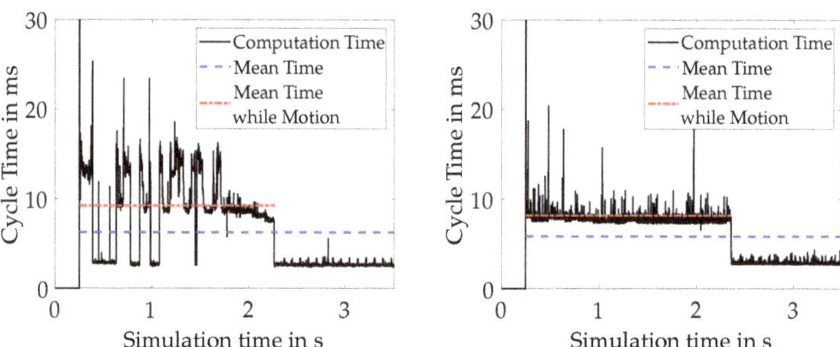

Figure 13. Computation time of the NMPC algorithm along the emergency trajectory. **Left:** default parameters of `fmincon()`. **Right:** iterations constrained to 10.

The experiments show that the maximum iterations can be reduced, which leads to a lower mean computation time, while maintaining comparable results. Even though the results look promising, the resulting computation times are still far away from the 0.5 ms time frame available in the hardware controller running at 2 kHz. To apply the algorithms on real hardware, this issue needs to be solved in future work. Moreover, as cables tend to become slack within the simulation, an additional position control or force control in joint space might be needed for experiments on hardware. As mentioned in Section 5.3, the NMPC has a rather simple model of the robot. In future work, it might be reasonable to investigate the behavior of an extended algorithm through also incorporating winch dynamics and cable behavior. This will be especially interesting when carrying out tests on real hardware in terms of the quality of the results versus computation costs. Moreover, it needs to be investigated if a larger prediction horizon of the NMPC algorithm leads to better results in terms of experienced cable forces, cable slackness and platform movement. Finally, it is worth noting that the path of the platform and force progressions can be adjusted with the weighting parameters of the cost function—e.g., for less desired vertical movement or slower progression in the cable forces—which is not further investigated here.

6. Conclusions and Outlook

In this work, an emergency strategy after the cable failure of a CDPR based on kinetic energy minimization was presented and implemented within a multi-body simulation environment with control loop closure. The proposed simulation environment was briefly described, and also featuring modeling of the drivetrains, as well as a simple cable model. The chosen two DOF robot was analyzed regarding a pre- and post-cable failure static equilibrium workspace. A well-known position control approach was described and implemented within the simulation environment as a standard controller. To display the event of a cable failure, the dynamic simulation was modified accordingly. Within the simulation, the standard control approach was proven to fail. In contrast, utilizing the proposed emergency strategy, the platform was successfully recovered and stopped within the post-failure workspace. This paves the way for the application of the proposed method on robot hardware. However, some issues have been identified and discussed that need to be resolved in order to carry out hardware experiments. This includes, e.g., the reliable detection of cable failure and a real-time position determination in the emergency situation. For the latter, in order to close the control loop, a measurement unit or a cable-failure-resistant forward kinematic code is needed.

Given the simulation scenario, the emergency controller was not able to hold tension in all cables throughout the whole recovery phase. This indicates a possible demand for an additional controller in joint space in order to maintain the desired force. An alternative option would be to extend the NMPC algorithm to incorporate the platform position and cable length, which might be used for position control in order to maintain the desired cable length, preventing the cables from going slack. Finally, the computational costs of the NMPC optimization algorithm were shown. Despite fast computing times in the range of milliseconds, the algorithm does not reach the desired computation time for a 2 kHz control system. This is planned to be resolved in future work using either real-time hardware with high computation power or by altering the algorithm. It is worth noting that a large amount of damage potential coming, for example, from undesired rotations is not addressed using a two DOF model, which requires further consideration.

All in all, this work contributes to an improvement in safety and reliability for CDPRs, helping to pave the way for CDPRs in industrial applications.

Author Contributions: Conceptualization and methodology, R.B. and T.B.; software, R.B.; validation, R.B.; formal analysis, R.B. and T.B.; investigation, R.B.; resources, T.B.; data curation, R.B.; writing—original draft preparation, R.B.; writing—review and editing, T.B.; visualization, R.B.; supervision, T.B.; project administration, T.B.; funding acquisition, T.B. All authors have read and agreed to the published version of the manuscript.

Funding: This work was supported by the Federal Ministry of Economic Affairs and Energy (BMWi) via the Arbeitsgemeinschaft industrieller Forschungsvereinigungen "Otto von Guericke" e.V. (AiF) within the program Industrial Collective Research (IGF), funding no. 20061 BG—"Entwicklung von Seilrobotern für die Erstellung von Kalksandstein-Mauerwerk auf der Baustelle", based on a resolution of the German Bundestag.

Acknowledgments: The authors would like to thank the funding institutions for their financial support.

Conflicts of Interest: The authors declare no conflict of interest.

Abbreviations

The following abbreviations are used in this manuscript:

CDPR	cable-driven parallel robot
$\mathcal{SEW}$	static equilibrium workspace
APD	augmented PD
NMPC	nonlinear-model-based control
DOF	degrees of freedom

References

1. Pott, A. *Cable-Driven Parallel Robots: Theory and Application*; Springer Tracts in Advanced Robotics; Springer: Cham, Switzerland, 2018. [CrossRef]
2. Reichert, C.; Bruckmann, T. Optimization of the Geometry of a Cable-Driven Storage and Retrieval System. In *Robotics and Mechatronics, Proceedings of the Fifth IFToMM International Symposium on Robotics & Mechatronics (ISRM 2017), Sydney, Australia, 29 November–1 December, 2017*; (Chunhui) Yang, R., Takeda, Y., Zhang, C., Fang, G., Eds.; Springer: Cham, Switzerland, 2019; pp. 225–237. [CrossRef]
3. Iturralde, K.; Feucht, M.; Hu, R.; Pan, W.; Schlandt, M.; Linner, T.; Bock, T.; Izard, J.B.; Eskudero, I.; Rodriguez, M.; et al. A Cable Driven Parallel Robot with a Modular End Effector for the Installation of Curtain Wall Modules. In *Proceedings of the 37th International Symposium on Automation and Robotics in Construction and Mining (ISARC), Kitakyushu, Japan, 27–28 October 2020*; Osumi, H., Furuya, H., Tateyama, K., Eds.; The International Association for Automation and Robotics in Construction (IAARC): Munich, Germany, 2020; pp. 1472–1479. [CrossRef]
4. Schröder, S. Under Constrained Cable-Driven Parallel Robot for Vertical Green Maintenance. In *Cable-Driven Parallel Robots*; Gouttefarde, M., Bruckmann, T., Pott, A., Eds.; Springer: Cham, Switzerland, 2021; pp. 389–400.
5. Leung, C.M.; Lam, W.Y.; Kwok, C.K.; Lau, D. Real-World Development of a Cleaning CDPR for Primary Lamella Sedimentation Tanks. In *Cable-Driven Parallel Robots*; Gouttefarde, M., Bruckmann, T., Pott, A., Eds.; Springer: Cham, Switzerland, 2021; pp. 401–412.
6. Rodriguez-Barroso, A.; Saltaren, R. Cable-Driven Robot to Simulate the Buoyancy Force for Improving the Performance of Underwater Robots. In *Cable-Driven Parallel Robots*; Gouttefarde, M., Bruckmann, T., Pott, A., Eds.; Springer: Cham, Switzerland, 2021; pp. 413–425.
7. Métillon, M.; Pedemonte, N.; Caro, S. Evaluation of a Cable-Driven Parallel Robot: Accuracy, Repeatability and Long-Term Running. In *Cable-Driven Parallel Robots*; Gouttefarde, M., Bruckmann, T., Pott, A., Eds.; Springer: Cham, Switzerland, 2021; pp. 375–388.
8. Bruckmann, T.; Boumann, R. A Simulation and Optimization Framework for Cable Robots in Automated Construction. *Adv. Eng. Informatics* 2021, *Submitted*.
9. Boumann, R.; Bruckmann, T. Development of Emergency Strategies for Cable-Driven Parallel Robots after a Cable Break. In *Cable-Driven Parallel Robots: Proceedings of the 4th International Conference on Cable-Driven Parallel Robots*; Pott, A., Bruckmann, T., Eds.; Springer: Cham, Switzerland, 2019; pp. 269–280. [CrossRef]
10. Roberts, R.G.; Graham, T.; Lippitt, T. On the inverse kinematics, statics, and fault tolerance of cable-suspended robots. *J. Robot. Syst.* **1998**, *15*, 581–597. [CrossRef]
11. Notash, L. Wrench Recovery for Wire-Actuated Parallel Manipulators; In *RoManSy 19*; Padois, V., Bidaud, P., Khatib, O., Eds.; Springer: Vienna, Austria, 2013; pp. 201–208. [CrossRef]
12. Berti, A.; Gouttefarde, M.; Carricato, M. Dynamic Recovery of Cable-Suspended Parallel Robots After a Cable Failure. In *Advances in Robot Kinematics 2016*; Lenarčič, J., Merlet, J.P., Eds.; Springer: Cham, Switzerland, 2018; pp. 331–339. [CrossRef]
13. Boschetti, G.; Passarini, C.; Trevisani, A. A Strategy for Moving Cable Driven Robots Safely in Case of Cable Failure. In *Advances in Italian Mechanism Science*; Boschetti, G., Gasparetto, A., Eds.; Springer: Cham, Switzerland, 2017; pp. 203–211.
14. Boschetti, G.; Minto, R.; Trevisani, A. Improving a Cable Robot Recovery Strategy by Actuator Dynamics. *Appl. Sci.* **2020**, *10*, 7362. [CrossRef]
15. Boschetti, G.; Minto, R.; Trevisani, A. Experimental Investigation of a Cable Robot Recovery Strategy. *Robotics* **2021**, *10*, 35. [CrossRef]
16. Passarini, C.; Zanotto, D.; Boschetti, G. Dynamic Trajectory Planning for Failure Recovery in Cable-Suspended Camera Systems. *J. Mech. Robot.* **2019**, *11*. [CrossRef]

17. Ghaffar, A.; Hassan, M. Failure Analysis of Cable Based Parallel Manipulators. In *Recent Trends in Materials, Mechanical Engineering, Automation and Information Engineering*; Applied Mechanics and Materials; Trans Tech Publications Ltd.: Frienbach, Switzerland, 2015; Volume 736, pp. 203–210. [CrossRef]
18. Winter, D.L.; Ament, C. Development of Safety Concepts for Cable-Driven Parallel Robots. In *Cable-Driven Parallel Robots*; Gouttefarde, M., Bruckmann, T., Pott, A., Eds.; Springer: Cham, Switzerland, 2021; pp. 360–371.
19. Boumann, R.; Bruckmann, T. An Emergency Strategy for Cable Failure in Reconfigurable Cable Robots. In *Cable-Driven Parallel Robots: Proceedings of the 5th International Conference on Cable-Driven Parallel Robots*; Pott, A., Bruckmann, T., Eds. Springer: Cham, Switzerland, 2021; pp. 217–229. [CrossRef]
20. Hahn, H. *Rigid Body Dynamics of Mechanisms: 1 Theoretical Basis*, 1st ed.; Springer: Berlin/Heidelberg, Germany, 2002. [CrossRef]
21. Schramm, D.; Hiller, M.; Bardini, R. *Modellbildung und Simulation der Dynamik von Kraftfahrzeugen*, 3rd ed.; Springer Vieweg: Berlin/Heidelberg, Germany, 2018. [CrossRef]
22. Müller, K.; Reichert, C.; Bruckmann, T. Analysis of a real-time capable cable force computation method. In *Cable-driven Parallel Robots: Proceedings of the Second International Conference on Cable-Driven Parallel Robots*; Bruckmann, T., Pott, A., Eds.; Mechanisms and machine science; Springer: Cham, Switzerland, 2015; Volume 32, pp. 227–238. [CrossRef]
23. Grüne, L.; Pannek, J. Nonlinear Model Predictive Control. In *Nonlinear Model Predictive Control: Theory and Algorithms*; Springer: Cham, Switzerland, 2017; pp. 45–69. [CrossRef]
24. Reichert, C.; Bruckmann, T. Unified contact force control approach for cable-driven parallel robots using an impedance/admittance control strategy. In Proceedings of the 14th IFToMM World Congress, Taipei, Taiwan, 25–30 October 2015; pp. 645–654. [CrossRef]
25. Reichert, C.; Unterberg, U.; Bruckmann, T. Energie-optimale Trajektorien für seilbasierte Manipulatoren unter Verwendung von passiven Elementen. In Proceedings of the First IFToMM D-A-CH Conference 2015, Duisburg, Germany, 11 March 2015; TU Dortmund. DuEPublico, Universität Duisburg-Essen: Duisburg, Germany, 2015. [CrossRef]
26. Reichert, C.; Glogowski, P.; Bruckmann, T. Dynamische Rekonfiguration eines seilbasierten Manipulators zur Verbesserung der mechanischen Steifigkeit. In *Fachtagung Mechatronik*; Bertram, T., Corves, B., Janschek, K., Eds.; Inst. für Getriebetechnik und Maschinendynamik: Dortmund, Germany, 2015; pp. 91–96. [CrossRef]
27. Hufnagel, T. Theoretische und praktische Entwicklung von Regelungskonzepten für redundant angetriebene parallelkinematische Maschinen. Ph.D. Thesis, University of Duisburg-Essen, Duisburg, Germany, 2014.

Article

Robotic Sponge and Watercolor Painting Based on Image-Processing and Contour-Filling Algorithms

Lorenzo Scalera [1,*], Giona Canever [2], Stefano Seriani [2], Alessandro Gasparetto [1] and Paolo Gallina [2]

[1] Polytechnic Department of Engineering and Architecture, University of Udine, 33100 Udine, Italy; alessandro.gasparetto@uniud.it

[2] Department of Engineering and Architecture, University of Trieste, 34127 Trieste, Italy; gionacanever96@gmail.com (G.C.); sseriani@units.it (S.S.); pgallina@units.it (P.G.)

* Correspondence: lorenzo.scalera@uniud.it

Abstract: In this paper, the implementation of a robotic painting system using a sponge and the watercolor painting technique is presented. A collection of tools for calibration and sponge support operations was designed and built. A contour-filling algorithm was developed, which defines the sponge positions and orientations in order to color the contour of a generic image. Finally, the proposed robotic system was employed to realize a painting combining etching and watercolor techniques. To the best of our knowledge, this is the first example of robotic painting that uses the watercolor technique and a sponge as the painting media.

Keywords: robotic art; watercolor; sponge painting; etching; image processing

Citation: Scalera, L.; Canever, G.; Seriani, S.; Gasparetto, A.; Gallina, P. Robotic Sponge and Watercolor Painting Based on Image-Processing and Contour-Filling Algorithms. *Actuators* **2022**, *11*, 62. https:// doi.org/10.3390/act11020062

Academic Editors: Marco Carricato and Edoardo Idà

Received: 29 December 2021
Accepted: 16 February 2022
Published: 19 February 2022

Publisher's Note: MDPI stays neutral with regard to jurisdictional claims in published maps and institutional affiliations.

1. Introduction

At the present time, there is a growing interest in creating artworks using machines and robotic systems, and the introduction of machine creativity and intelligence in art fascinates both engineers and artists [1,2]. Robotic art is a niche sector of robotics, which includes different types of performance such as theater, music, and painting. Robotic art painting is a sector in which artists coexist with experts in robotics, computer vision, and image processing. Robotic art projects can be seen as a new art sector in which art and technology mix together and the exact representation of the world is not the main focus.

In recent years, technology has entered our lives more and more, and the use of robots in art projects can be seen as the natural progression of this continuous increase of the presence of technology in our daily lives. Robotic art should no longer be seen simply as an exercise to see how robots can be moved freely or a test of robots' abilities, but also as a novel art sector with its syntax and artistic characteristics. Currently, several drawing machines and robotic systems capable of producing artworks can be found in the present literature. The majority of these systems adopt image-processing and non-photorealistic-rendering techniques to introduce elements of creativity into the artistic process [3,4]. Furthermore, most robotic machines capable of creating artworks are focused on drawing and brush painting as artistic media. *eDavid* is a robotic system based on non-photorealistic rendering techniques, which produces impressive artworks with dynamically generated strokes [5]. *Paul the robot* is an installation capable of producing sketches of people guided by visual feedback with impressive results [6]. Further examples of artistically skilled machines are given by the compliant robot able to draw dynamic graffiti strokes shown in [7] and by the robot capable of creating full-color images with artistic paints in a human-like manner described in [8]. In [9], non-photorealistic rendering algorithms were applied for the realization of artworks with the watercolor technique using a robotic brush painting system. In [10], the authors presented a Cartesian robot for painting artworks by interactive segmentation of the areas to be painted and the interactive design of the orientation of the brush strokes.

Furthermore, in [11], the authors adopted a machine learning approach for realizing brushstrokes as human artists, whereas in [12], a drawing robot, which can automatically transfer a facial picture to a vivid portrait and then draw it on paper within two minutes on average, was presented. More recently, in [13], a humanoid painting robot was shown, which draws facial features extracted robustly using deep learning techniques. In [14], a fast robotic pencil drawing system based on image evolution by means of a genetic algorithm was illustrated. Finally, in [15], a system based on a collaborative robot that draws stylized human face sketches interactively in front of human users by using generative-adversarial-network-style transfer was presented.

There are also examples of painting and drawing on 3D surfaces, such as in [16], where an impedance-controlled robot was used for drawing on arbitrary surfaces, and in [17], where a visual pre-scanning of the drawing surface using an RGB-D camera was proposed to improve the performance of the system. Moreover, in [18], a 3D drawing application using closed-loop planning and online picked pens was presented.

In addition to brush painting and drawing, other artistic media have been investigated in the context of robotic art. Interesting examples are the spray painting systems based on a serial manipulator [19] or on an aerial mobile robot [20], the robotic system based on a team of mobile robots equipped for painting [21], and the artistic robot capable of creating artworks using the palette-knife technique proposed in [22]. Other applications of robotic systems to art include light painting, performed with a robotic arm [23] or with an aerial robot [24], as well as the stylized water-based clay sculpting realized with a robot with six degrees of freedom in [25].

Elements of creativity in the creation process can be also acquired by means of the interaction between the painting machine and human artist. Examples are given by the human–machine co-creation of artistic paintings presented in [26] and by the Skywork-daVinci, a painting support system based on a human-in-the-loop mechanism [27]. Furthermore, in [28], a flexible force-vision-based interface was employed to draw with a robotic arm in telemanipulation, whereas the authors in [29] presented a system for drawing portraits with a remote manipulator via the 5G network. Further examples of human–machine interfaces for robotic drawing are given by the use of the eye-tracking technology, which is applied to allow a user paint using the eye gaze only, as in [30,31].

In this paper, we present a robotic sponge and watercolor painting system based on image-processing and contour-filling algorithms. To the best of our knowledge, this is the first example of robotic painting that uses the watercolor technique and a sponge as the painting media. In this work, we focused on filling continuous, irregular areas of a processed digital image with uniform color through the use of a sponge. More in detail, an input image was pre-processed before the execution of a custom-developed contour-filling algorithm, with the main focus being to find the best position of the sponge imprint inside the contour of a figure in order to color it.

In summary, the main contributions of this paper are the following: (a) the implementation of a robotic sponge and watercolor painting system; (b) the development of an algorithm for the image processing and the contour filling of an image to be painted using a brush with a pre-defined shape; (c) the experimental validation of the proposed method with the realization of two artworks also using a robotic etching technique.

This work fits into the context of figurative art, which, in contrast to abstract art, refers to artworks (paintings and sculptures) that are derived from real object sources. Within this expressive set, there are situations in which, in order to represent the subject, the activity of reproducing the contours is separated from the activity of filling in the backgrounds with colors or grey-scale tints. For example, it is well known that, in the case of comic books, the artist who is in charge of making the images of the panels (penciller) is different from the artist who is later in charge of coloring the backgrounds (inker or colorist). Another example is given by the engravings by Henry Fletcher (1710–1750). The artist realized the contours of the subject by means of etching; in a second step, he colored the flower petals [32].

In this work, we present two artistic subjects as test cases: the "Palazzo della Civiltà Italiana" and "Wings". The former is a benchmark example to test the performance of the contour-filling algorithm. The latter was realized with a combination of etching and the watercolor technique. The etching was an intaglio printmaking process in which lines or areas are incised using acid into a metal plate in order to hold the ink [33]. In this case, the proposed algorithm was adopted to color an image for which the contours were first obtained with the etching technique.

The paper is organized as follows: Section 2 describes the image-processing and contour-filling algorithm. In Section 3, the watercolor technique, the experimental setup, and the software implementation are illustrated, followed by a description of the calibration process. The experimental results are shown in Section 4. Finally, the conclusions of the paper are given in Section 5.

2. Theoretical Framework

As explained in the Introduction, the purpose of the research was to identify an appropriate strategy to evenly fill an area of the canvas using a sponge as the painting tool. The problem involved a complex solution since the layout of the sponge, unlike that of a brush, can have a non-circular shape; in fact, often, the sponge has a rectangular design. We call the algorithm that fulfills this task the *contour-filling algorithm*. The input of the algorithm was an RGB image. The contour-filling algorithm was composed of four steps:

- *Image preparation*: The image was analyzed, and a color reduction was performed. After that, the contours of the uniformly colored areas were carried out;
- *Area division*: This passage can be skipped according to the artist that uses the software tool; if selected, it performs a Voronoi partition of the area to be painted. This possibility was introduced for two reasons: (1) in order to break up the excessive regularity of a large background, which is usually not aesthetically pleasing; (2) to stop the painting process in case the partial results are not as expected;
- *Image erosion*: The obtained contours were eroded in such a way to prevent the sponge from painting beyond the area bounded by the edges;
- *Contour-filling algorithm*: This is the heart of the algorithm, where sponge positions and orientations (poses) are defined.

2.1. Image Preparation

In this first step, the image to be reproduced was selected and saved in a proper format: a matrix with dimensions *height* × *length* × 3. In this work, RGB images were considered, as the one of Figure 1a. The number of colors of the image was reduced according to the user-defined parameter n_{colors}, since the number of colors that the robot can manage was limited. Furthermore, the mixing of multiple colors to obtain particular shades has not yet been implemented. The resulting image was an RGB image, with pixel values px in the range $0 \leq px \leq n_{colors}$, as in Figure 1b.

During the whole process, only one color at a time was analyzed. The considered color image was converted to a black and white image (addressed as L in the following), and the obtained areas were labeled and converted into MATLAB `polyshape` objects for computational reasons. The area of possible internal holes was considered negligible with respect to the larger one, and for this reason, it was removed using a simple filter.

2.2. Area Division

The image can now be divided into smaller areas. This division was implemented to break the whole robotic task into several sub-processes in order to carry out eventual on-the-fly adjustments during the painting process. There are several ways to subdivide an area. We discarded the simplest ones, which consisted of dividing the area into regular subareas (such as squares), because graphic regularity, when perceived by the observer's eye, is seen as aesthetically poor.

(a)   (b)

Figure 1. Original image (**a**); image after reducing its number of colors (**b**).

The image was therefore divided by applying a Voronoi diagram to the considered shape to paint. In the present work, the Euclidean distance and a 2D space were considered. In usual terms, the Voronoi diagram resulted in a series of regions R_k in which every point inside was closer to P_k, the center of the considered region k, than to the generic P_j. The points P_j that define the Voronoi cells were random ones that fell inside the area to paint.

The area division was performed by superimposing the Voronoi cells lines on the original image. In this way, the original image area was divided into smaller subareas. Formally, V is the image matrix that represents the lines that define the Voronoi cells ($V_{ij} = 0$ at the pixel belonging to the lines; $V_{ij} = 1$ elsewhere). L is the image matrix to be divided ($L_{ij} = 1$ at pixels of the area to paint; $L_{ij} = 0$ elsewhere). The final image F is defined by the logical *and* operator: $F = and(V, L)$. An example of Voronoi area division is shown in Figure 2, where, for the sake of clarity, each separated area is represented with a different color.

Figure 2. Voronoi area division of the bigger contour of the image of Figure 1b.

2.2.1. Image Erosion

As described below, the algorithm for filling an area aimed at generating a sequence of positions and orientations of the sponge (poses) in such a way that the envelope of all sponge imprints covered the entire area to be painted without exiting excessively from the area edges. To do this, some points were selected on the image where the center of the sponge would be sequentially positioned during the painting process. To facilitate the point selection, the area to be painted was divided into subsets of pixels through an erosion process applied to the initial area, as shown in Figure 3. An example of a typical binary erosion can be found in [34].

The original contours (referred to as C_j) of the colored areas inside the image F were considered. The divided areas were eroded, and three regions were created:

- Am *innermost area I* (red in Figure 3a);
- An *outermost area O* (blue in Figure 3a);
- A *median area M* between the innermost and the outermost ones (green in Figure 3a).

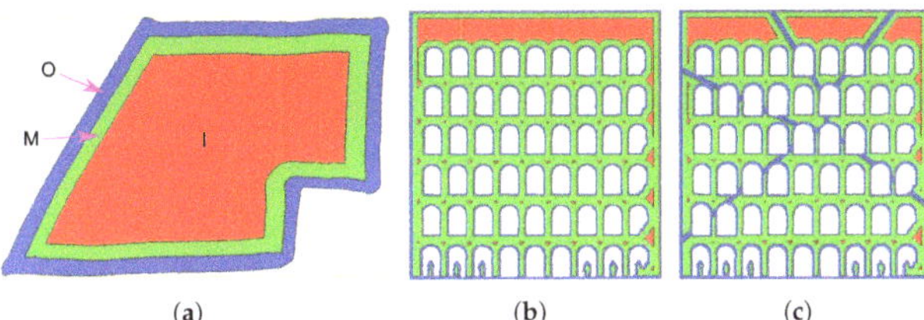

Figure 3. Example of image erosion. (**a**) Example of the eroded contour of a region and the definition of the obtained areas. (**b**) Eroded image obtained without dividing the area. (**c**) Eroded image obtained by dividing the area.

A lower point density was set in the innermost area than in the median area. Indeed, in the innermost area, all points were valid positions for the sponge, since its imprint always fell inside the contour C_j, whatever the sponge orientation was. On the other side, in the median area, a higher density of points was necessary to better follow the small details of the contour by properly defining the sponge orientation. Finally, in the outermost area, no points were placed because the sponge imprint in this area would surely fall out of the contour to paint. The acceptable distance of going beyond the contour was a user-defined percentage of the sponge area (e.g., 5%). This parameter also influenced the amount of erosion computed by the algorithm.

2.2.2. Contour-Filling Algorithm

The contour-filling algorithm was the core of the proposed strategy for robotic sponge and watercolor painting. In the following, it is established how the sponge positions were selected and how they were considered valid or not.

There are two possible approaches the user can choose to define the center sponge points p_i: random or based on a grid. For each generated point p_i, the algorithm evaluates if it falls inside the contour to paint. Let us define α as the angular orientation of the sponge with respect to the horizontal axis (see Figure 4a). Inside the innermost area I, for every value of α, the sponge imprint completely belongs to the area to be painted ($\forall p_i \in I$ and $\forall \alpha \Rightarrow B \subseteq I$, where B is the sponge imprint); therefore, a random value α was assigned to the points belonging to I.

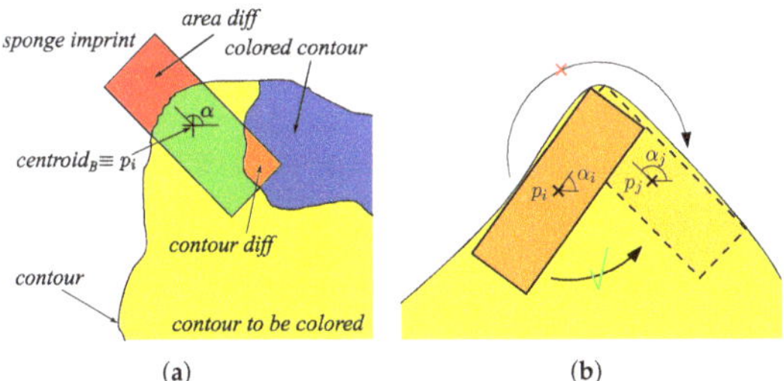

Figure 4. Variables introduced in the algorithm (**a**); sponge rotation in the dragging technique (**b**).

No points were placed in the outermost area O since $\forall p_i \in O$ and $\forall \alpha \Rightarrow B \not\subseteq I$. However, there could be a rare situation in which the sponge imprint corresponding to a particular point would fall within the area to be painted, for a particular value of α. This could happen because image erosion is a coarse way to define the positions of the points. However, this strategy has the merit of speeding up the algorithm.

In the median region M, the sponge is rotated by an angle α to define the best orientation that makes the sponge imprint fall as much as possible inside the contour to paint. This can be seen in Figure 4a, where the area to be minimized is the red one (*area diff*). To better follow the contour, a series of points were also placed on the line C that divides the outermost and the median regions. These points were chosen among the ones that define the contour vertices: if two consecutive vertices p_i and p_j are too far apart, a series of intermediate points p_k are automatically added.

To find the best angular rotation of the sponge (or the brush) in every defined point, the algorithm flowchart shown in Figure 5 is provided. The variables and parameters used in the flowchart are illustrated in Figure 4a. Their definitions are the following:

- $centroid_B$: coordinates of the sponge centroid;
- p_i: point in which the sponge is placed. There are 3 possibilities:
    - $p_i \in I$;
    - $p_i \in M$;
    - $p_i \in C$;
- α: angle of rotation of the sponge around its centroid;
- *contour to color*: contour yet to be colored;
- *ang incr*: angular increment used in the for cycle;
- *contour*: external contour to color;
- *area diff*: area of the sponge imprint that falls outside the contour to color; this is the variable to minimize;
- *% sponge out*: maximum user-defined percentage of the sponge size accepted to be outside of the contour to color;
- *contour diff*: area where the already painted contour and the rotated sponge overlap.

As can be seen in Figure 5, if $p_i \in M \vee p_i \in C$, the sponge is at first rotated clockwise and then counterclockwise. It was also evaluated if it fell sufficiently inside the image contour. If there was a valid position, the one that paints most of the yet-to-be-colored area was saved. In Figure 6, an example of a sequence of sponge positions resulting from the processing of Figure 2 with the proposed algorithm is provided.

Once the sequence of sponge positions is defined, two different painting technique can be selected:

- *Dabbing technique*: The sponge is moved between positions p_i on the canvas by raising the sponge in the passage between points. This technique does not require particular care since the sponge positions are already defined;
- *Dragging technique*: The sponge is moved between positions p_i on the canvas without being raised. This technique is more complex to simulate. Considering two defined points, p_i and p_j, if $\|(p_j - p_i)\| > d_{max}$, with d_{max} a user-defined maximum distance, a series of n intermediate points p_k that connect p_i and p_j are created. In these intermediate points p_k, the sponge imprint has to be verified; if it falls outside the area to paint, the sponge is raised. Furthermore, during the movement through the intermediate points p_k, the sponge is smoothly rotated between the rotation configuration of p_i and p_j (the sequence of consecutive poses is interpolated). If the starting angular position is α_i and the ending one is α_j, the intermediate points' angular positions are $\alpha_k = (alpha_j - \alpha_i)/n$. It was also evaluated if the rotation was convenient to be clockwise or counterclockwise, with the goal of maximizing the coloring contribution to the area yet to be painted (Figure 4b). This process substantially corresponds to the addition of several poses to the original ones.

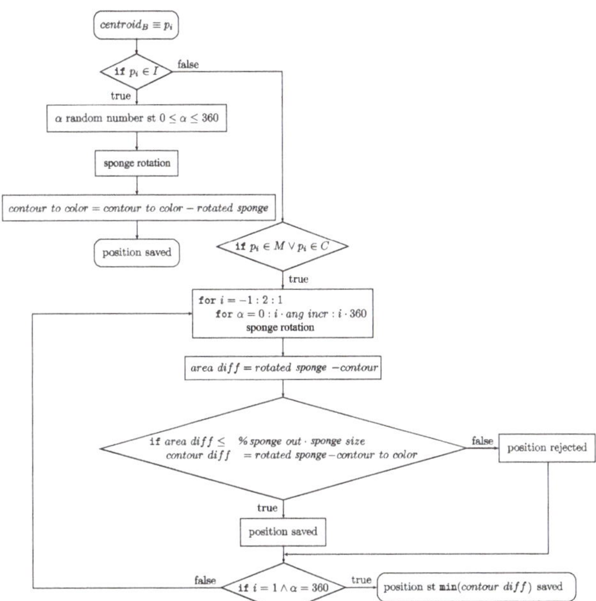

Figure 5. Flowchart of the algorithm for the definition of sponge angular rotation.

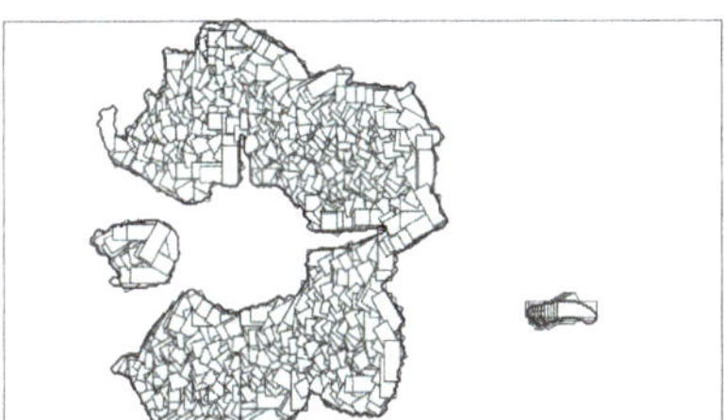

Figure 6. Sequence of sponge positions for the image of Figure 2.

3. Materials and Methods

In this section, we first briefly recall the watercolor painting technique. Then, the robotic system used in the experimental tests and the calibration operations needed for sponge painting are described.

3.1. Watercolor Painting

In the watercolor painting technique, ground pigments are suspended in a solution of water, binder, and surfactant [35]. This technique exhibits textures and patterns that show the movements of the water on the canvas during the painting process. The main uses have been on detailed pen-and-ink or pencil illustrations. Great artists of the past such as J. M. W. Turner (1775–1851), John Constable (1776–1837), and David Cox (1783–1859) began to investigate this technique and the unpredictable interactions of colors with the canvas, which produce effects such as dry-brush, edge darkening, back runs, and glazing. In more recent years, authors, such as Curtis et al. [36], Lum et al. [37], and Van Laerhoven et al. [38], have investigated the properties of watercolor to create digital images by simulating the interactions between pigment and water.

Great importance in this technique is given to the specific paper and pigment used: the canvas should be made of linen or cotton rags pounded into small fibers, and the pigment should be formed by small separate particles ground in a milling process. In the tests performed in the present work, the Fabriano Rosaspina paper was adopted. This is an engraving paper made of cotton fibers and characterized by a high density per square meter. Color pigments in tube format specifically made for watercolor were used as well.

This complex and unpredictable painting technique combined with the use of a sponge as the painting media is challenging. In fact, a trial and error process to fine-tune the system was necessary to find the perfect ratio between color and water and the best compression of the sponge during the releasing of color on the canvas.

3.2. Robotic Painting System

The painting system was built on a KUKA LBR iiwa 14 R820 robotic platform, shown in Figure 7a. This 7-axis robot arm offers a payload of 14 kg and a reach of 820 mm. The repeatability of the KUKA LBR iiwa 14 R820 robot is ± 0.1 mm [39]. The manipulator was equipped with a Media flange Touch electrical, a universal interface that enables the user to connect electrical components to the robot flange (Figure 7b). The flange position is described by its tool center point, which is user defined and can be personalized depending on the tool attached to the robot flange.

In order to equip the robot for sponge painting, three components were designed (Figure 8): a connector, a calibration tip, and a sponge support. These components were fabricated using an Ultimaker 2+ Extended 3D printer. The connector was fixed to the robot flange with four screws. The calibration tip and the sponge support can be easily interfaced with the connector thanks to a pair of magnets with a 4 mm diameter, which were inserted into the printed material. The sponge used for painting was a make-up sponge (20×4 cm) with a piece of fabric glued on its top (Figure 8e). The size of the sponge was chosen empirically. However, the algorithm works with any size and section of sponge, but the larger the sponge, the less details can be painted and the less time it takes for the robot to paint. The piece of fabric on the tool was used to improve the dosage of water: the sponge works as an absorber, and the piece of fabric reduces the quantity of water that is deposited on the canvas. The main problems are related to the great variability given by the sponge behavior, by the color that can be less or more diluted, but also by the canvas paper that, after some time, becomes soaked by the water contained in the pigment solution.

The KUKA LBR iiwa was programmed in the Java environment, and the Sunrise Workbench application by KUKA was used to develop the applications needed to control the robot and set the software parameters. The image-processing and contour-filling algorithms (described in Section 2) were developed in MATLAB. To facilitate the software interface, a user-friendly MATLAB app was developed, which allows the user to set the parameters for the non-photorealistic rendering techniques, as well as to calibrate the painting setup and send commands to the manipulator (Figure 9).

A client–server socket communication based on the TCP protocol was implemented to establish the communication between the MATLAB app and the Java program loaded on the robot, which interprets the MATLAB commands sent by the user. In this manner, the coordinates of the points resulting from the image processing can be fed to the robot controller and executed by the manipulator. The motion of the robot was planned using a trapezoidal speed profile for each couple of subsequent points. The painting operation was performed by limiting the joint speed of the manipulator to 30% of its maximum value, since abrupt movements of the sponge could damage the paper, especially during dragging.

3.3. Calibration

The robotic sponge painting required a calibration of the painting surface, as well as a calibration of the height of the sponge during color discharge. The canvas surface was calibrated using a procedure similar to the one implemented in [9,22]. More in detail, the robot equipped with the calibration tip was manually moved in contact with the canvas surface in five calibration points, as shown in Figure 10. The z position of the robot tool on each of these points was acquired using the MATLAB app. Actually, only three points could be sufficient, but a higher number was considered to improve accuracy. The interpolating plane was then generated starting from the five calibration coordinates and used to define the corresponding z coordinate for each (x, y) planar position generated by the image-processing algorithm.

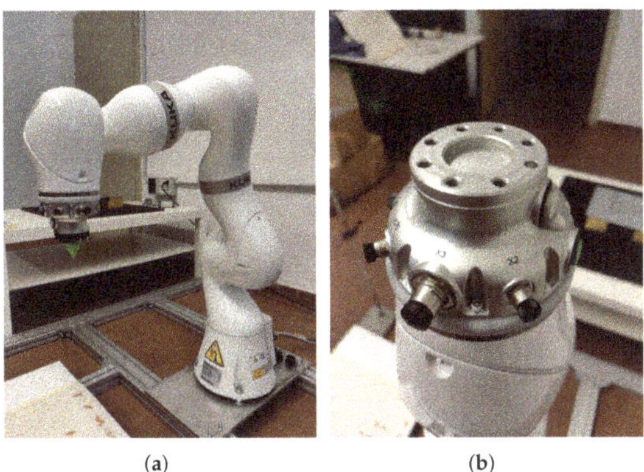

Figure 7. KUKA LBR iiwa 14 R820 manipulator (**a**); Media flange Touch electrical (**b**).

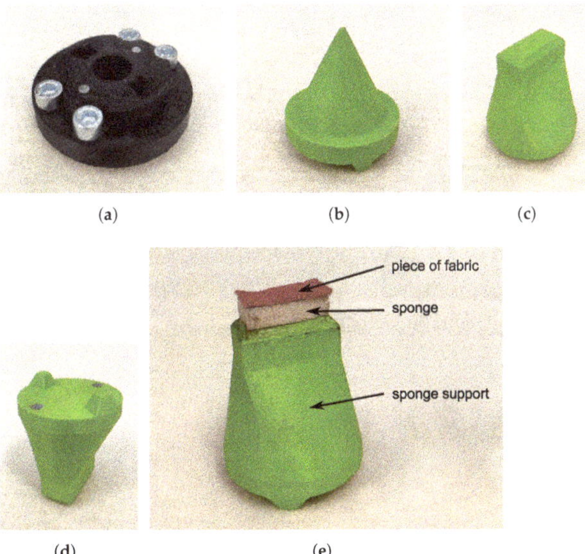

Figure 8. The 3D-printed parts for robotic sponge painting: connector (**a**); calibration tip (**b**); sponge support (**c–e**).

A calibration of the sponge height during color discharge was also needed to adjust the sponge compression during painting and, therefore, the quantity of color released. This operation was performed with a trial-and-error approach, since in this work, force control was not implemented on the robot. Figure 11 shows a scheme of the sponge discharge and some examples of sponge imprints for different heights. As can be seen from the figure, a compromise between a complete sponge imprint and a good amount of released water was needed. Indeed, if too much water were released on the canvas, the wet paper would produce ripples and undulations that would affect the quality of the results.

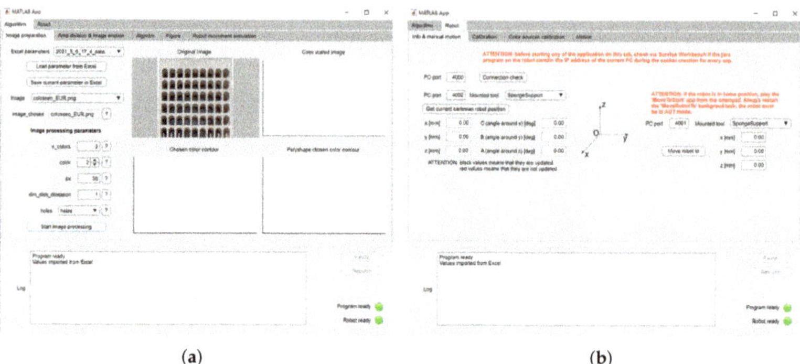

(a)   (b)

Figure 9. MATLAB app: image-processing section (**a**); robot-control section (**b**).

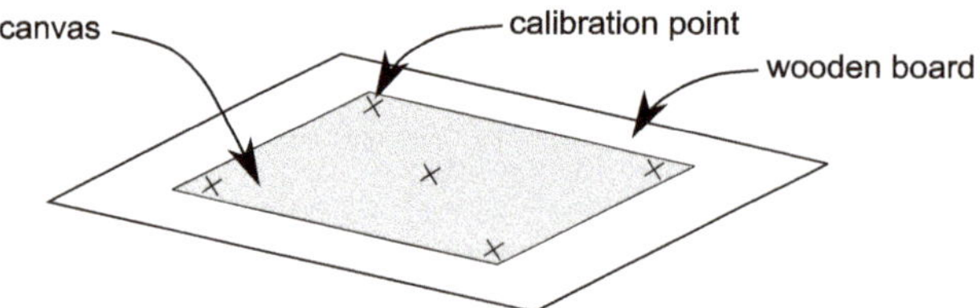

Figure 10. Calibration points on the painting canvas.

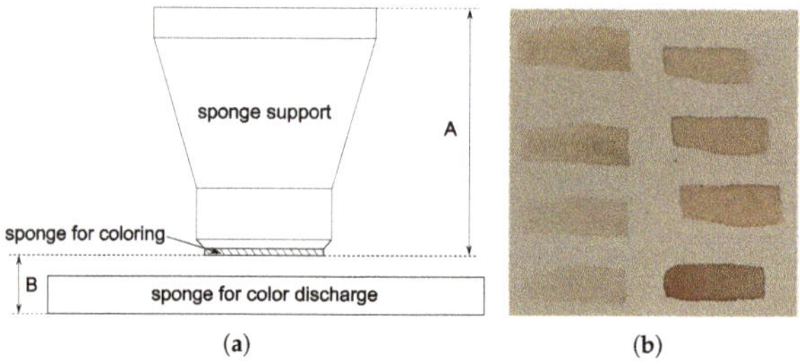

(a)   (b)

Figure 11. Sponge discharge scheme (**a**); examples of sponge imprints for different heights (**b**).

4. Experimental Results and Discussion

In order to provide a validation of the algorithm presented in Section 2, we set up an experimental campaign using two artistic subjects as test cases. The first was the "Palazzo della Civiltà Italiana", a monument by architects Guerrini, La Padula, and Romano and located in Rome (1938). It was chosen for its regular geometric shape and high contrast. The second artwork was "Wings", which was exposed at the *Arte Fiera* public exhibition in Padova, Italy, in 2021. In the following, we report the results of the pre-processing and painting of the two subjects together with a brief discussion of the results.

4.1. The "Palazzo Della Civiltà Italiana"

The original image used for the processing of this subject is shown in Figure 12. To paint it, we used the dabbing technique and the tool shown in Figure 8e. The results obtained with this subject are shown in Figure 13, where the figures on the left show the simulated painting and those on the right show the real painting. The overlapping between the windows and the wall area was due to the tolerance set on the sponge area, which can exit the contour to color (see Figure 13a), as previously explained.

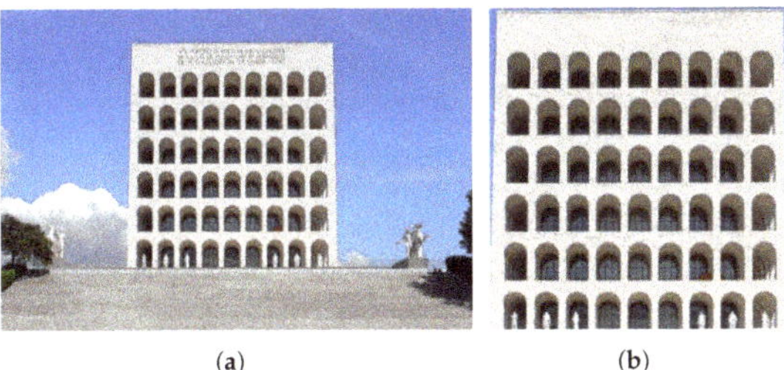

Figure 12. Palazzo della Civiltà Italiana in Eur District in Rome, Italy. Source: [40]. (**a**) Original image. (**b**) Image used in the proposed algorithm: the original one was cropped, and the top inscription was removed.

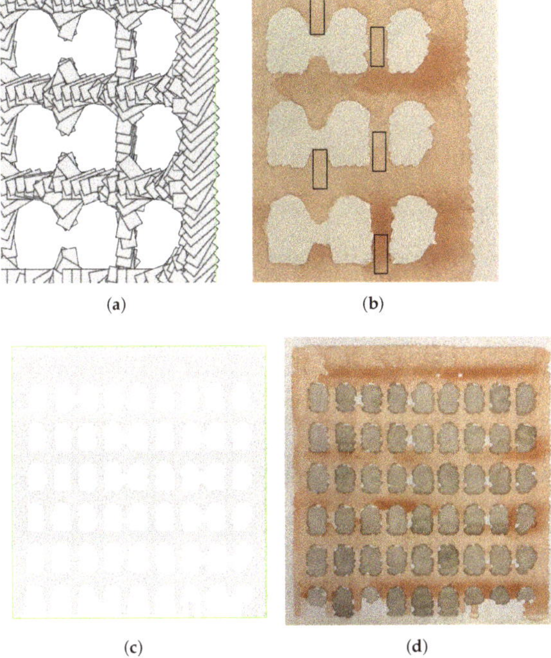

Figure 13. Processing of the "Palazzo della Civiltà Italiana": in (**a**) the planned poses of the sponge on the canvas; in (**b**), a detailed view of some poses on the real painting; in (**c**), the simulated painting; in (**d**), the experimental result (dimensions: 35 × 35 cm.)

There were still some creases caused by the water content in the pigment solution, which tended to accumulate inside them, but this problem can hardly be mitigated because water naturally soaks the canvas. The texture of the wall was still visible in some areas, but with a reduced entity. Some areas, for example the one between some windows, were not painted, since the points generated by the algorithm did not satisfy the constraints on the percentage of brush area exceeding the contour. This is visible in Figure 13a,b.

The variation of the intensity in the experimental results (as in Figure 13b,d) was due to the excess of water on the canvas and to the wet paper wrinkles. To eliminate the unwanted color variations, the amount of water adsorbed by the sponge and the dipping height should be reduced, and the robot should be moved more slowly during dipping and color releasing.

This was also granted by the calibration process. In fact, during the pre-incision calibration, only three calibration points were used: the darker crosses visible in the three corners of the final painting. In the algorithm used for this work piece, we used five calibration points; for this reason, two calibration points (the lighter crosses in the lower left corner and in the center) were added to the three initial ones. By having the same calibration points as a reference, the canvas was correctly positioned with respect to the robot base (the canvas was fixed at the same height of the robot base and at a distance of 60 cm from the robot base reference frame).

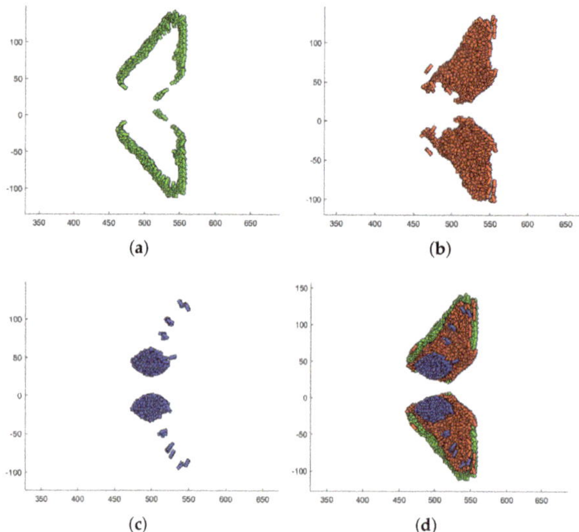

Figure 17. Simulated results for "Wings". (**a**) First threshold image simulation. (**b**) Second threshold image simulation. (**c**) Third threshold image simulation. (**d**) Final image simulation.

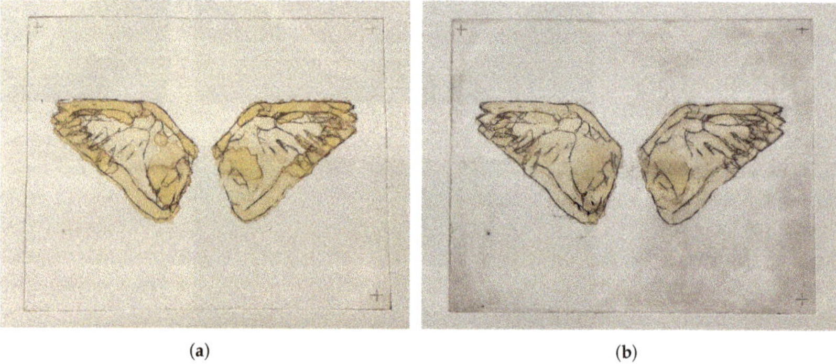

Figure 18. Final results for "Wings". (**a**) First artwork. (**b**) Second artwork, with more diluted colors than in the first one.

The different color intensities were obtained by adding multiple passes over the same area; the first threshold image contours were colored with two color passes; the second one was obtained with one color pass and the third with two color passes. The overlap of the different areas was visible because of the stronger color intensities in these overlapping areas; the edge darkening effects were clearly visible as well.

A second test with the same settings was carried out and is shown in Figure 18b; this time, the pigment solution was slightly more diluted. The first and third threshold layers were obtained with three color passes and the second with two; the result was more uniform, and the edge darkening effects were mitigated.

5. Conclusions

In this paper, the implementation of a robotic system using a sponge tool and the watercolor painting technique was presented. A collection of tools for calibration operations and sponge support were designed and realized. A contour-filling algorithm was developed, which defines the sponge positions and orientations in order to color the contours of a generic image. The system was also employed to realize artworks combining etching and watercolor techniques.

The results of this work highlighted the challenges that arise in the robotic sponge and watercolor painting. First of all, defining a sequence of positions and rotations of the sponge to paint the contours of an image requires the definition of an appropriate strategy that has to consider the shape of the available sponge, as well as the desired detail level. Furthermore, a proper calibration is needed for both the painting surface and the sponge height during color discharge. The quantity of water adsorbed by the sponge and released on paper is indeed fundamental to avoid ripples and undulations of the wet paper, which produced a variation of intensity in the experimental results. Care should be also taken when adopting the dragging technique, since abrupt movements of the sponge could damage the paper. On the other hand, the dabbing technique does not require particular care, since the sponge is raised by the robot in the passage between each couple of points. However, this approach is more time consuming and does not take advantage of the previous position of the sponge on the canvas.

Future works will include the improvement of the contour-filling algorithm and the optimization of the sponge placements to prevent repeated application of paint at the same location and to ensure a better coverage of the area to be painted without putting too much water on the paper. The sponge shape will be also optimized with respect to the size of the details to be painted, and alternative strategies for the coverage of an area will be performed in future developments of this work.

In future developments of this work, we will also consider objective criteria and quantitative metrics to evaluate the results of the study according to subjective data. Objective criteria could be, for example, the painting rate of the picture, the amount of painted area outside the borders of the picture, the number of points applied by the sponge, and the amount of overlapped sponge imprints. However, it has to be noted that these criteria are useful from a scientific point of view, but not from an artistic point of view, since the appreciation of an artwork is subjective and personal.

Furthermore, visual feedback will be implemented to allow the robot to identify uncolored areas or defects of the painting operation. Finally, force control will be introduced to ensure a better uniformity of the color deposition, and a suction table will be adopted to keep the paper flat, avoiding ripples and undulations produced by the wet paper.

Author Contributions: Conceptualization, P.G.; methodology, software, validation, G.C. and P.G.; formal analysis, investigation, resources, data curation, and writing—original draft preparation, L.S., G.C., S.S. and P.G.; writing—review and editing, L.S., G.C., S.S., A.G. and P.G.; visualization and supervision, A.G. and P.G.; project administration, P.G. All authors have read and agreed to the published version of the manuscript.

Funding: This research was partially funded by the Laboratory for Advanced Mechatronics—LAMA FVG, an international research center for product and process innovation of the three Universities of Friuli Venezia Giulia Region (Italy).

Institutional Review Board Statement: Not applicable.

Informed Consent Statement: Not applicable.

Conflicts of Interest: The authors declare no conflict of interest.

References

1. Gomez Cubero, C.; Pekarik, M.; Rizzo, V.; Jochum, E. The Robot is Present: Creative Approaches for Artistic Expression with Robots. *Front. Robot.* **2021**, *8*, 1–19. [CrossRef] [PubMed]
2. Leymarie, F.F. Art. Machines. Intelligence. *Enjeux Numer.* **2021**, *13*, 82–102.
3. Scalera, L.; Seriani, S.; Gasparetto, A.; Gallina, P. Non-photorealistic rendering techniques for artistic robotic painting. *Robotics* **2019**, *8*, 10. [CrossRef]
4. Karimov, A.; Kopets, E.; Kolev, G.; Leonov, S.; Scalera, L.; Butusov, D. Image Preprocessing for Artistic Robotic Painting. *Inventions* **2021**, *6*, 19. [CrossRef]
5. Deussen, O.; Lindemeier, T.; Pirk, S.; Tautzenberger, M. Feedback-guided stroke placement for a painting machine. In *Proceedings of the Eighth Annual Symposium on Computational Aesthetics in Graphics, Visualization, and Imaging*; Eurographics Association: Geneve, Switzerland 2012; pp. 25–33.
6. Tresset, P.; Leymarie, F.F. Portrait drawing by Paul the robot. *Comput. Graph.* **2013**, *37*, 348–363. [CrossRef]
7. Berio, D.; Calinon, S.; Leymarie, F.F. Learning dynamic graffiti strokes with a compliant robot. In Proceedings of the Intelligent Robots and Systems (IROS), 2016 IEEE/RSJ International Conference, Daejeon, Korea, 9–14 October 2016; pp. 3981–3986.
8. Karimov, A.I.; Kopets, E.E.; Rybin, V.G.; Leonov, S.V.; Voroshilova, A.I.; Butusov, D.N. Advanced tone rendition technique for a painting robot. *Robot. Auton. Syst.* **2019**, *115*, 17–27. [CrossRef]
9. Scalera, L.; Seriani, S.; Gasparetto, A.; Gallina, P. Watercolor robotic painting: A novel automatic system for artistic rendering. *J. Intell. Robot. Syst.* **2019**, *95*, 871–886. [CrossRef]
10. Igno-Rosario, O.; Hernandez-Aguilar, C.; Cruz-Orea, A.; Dominguez-Pacheco, A. Interactive system for painting artworks by regions using a robot. *Robot. Auton. Syst.* **2019**, *121*, 103263. [CrossRef]
11. Bidgoli, A.; De Guevara, M.L.; Hsiung, C.; Oh, J.; Kang, E. Artistic Style in Robotic Painting; a Machine Learning Approach to Learning Brushstroke from Human Artists. In Proceedings of the 2020 29th IEEE International Conference on Robot and Human Interactive Communication (RO-MAN), Naples, Italy, 31 August–4 September 2020; pp. 412–418.
12. Gao, F.; Zhu, J.; Yu, Z.; Li, P.; Wang, T. Making robots draw a vivid portrait in two minutes. In Proceedings of the 2020 IEEE/RSJ International Conference on Intelligent Robots and Systems (IROS), Las Vegas, NV, USA, 25–29 October 2020; pp. 9585–9591.
13. Chen, G.; Sheng, W.; Li, Y.; Ou, Y.; Gu, Y. Humanoid Robot Portrait Drawing Based on Deep Learning Techniques and Efficient Path Planning. *Arab. J. Sci. Eng.* **2021**, 1–12. [CrossRef]
14. Adamik, M.; Goga, J.; Pavlovicova, J.; Babinec, A.; Sekaj, I. Fast robotic pencil drawing based on image evolution by means of genetic algorithm. *Robot. Auton. Syst.* **2022**, *148*, 103912. [CrossRef]
15. Wang, T.; Toh, W.Q.; Zhang, H.; Sui, X.; Li, S.; Liu, Y.; Jing, W. RoboCoDraw: Robotic Avatar Drawing with GAN-based Style Transfer and Time-efficient Path Optimization. *Proc. AAAI Conf. Artif. Intell.* **2020**, *34*, 10402–10409. [CrossRef]
16. Song, D.; Lee, T.; Kim, Y.J.; Sohn, S.; Kim, Y.J. Artistic Pen Drawing on an Arbitrary Surface using an Impedance-controlled Robot. In Proceedings of the 2018 IEEE International Conference on Robotics and Automation (ICRA), Brisbane, Australia, 21–25 May 2018.
17. Song, D.; Kim, Y.J. Distortion-free robotic surface-drawing using conformal mapping. In Proceedings of the 2019 International Conference on Robotics and Automation (ICRA), Montreal, QC, Canada, 20–24 May 2019; pp. 627–633.
18. Liu, R.; Wan, W.; Koyama, K.; Harada, K. Robust Robotic 3-D Drawing Using Closed-Loop Planning and Online Picked Pens. *IEEE Trans. Robot.* **2021**. doi:10.1109/TRO.2021.3113996. [CrossRef]
19. Scalera, L.; Mazzon, E.; Gallina, P.; Gasparetto, A. Airbrush Robotic Painting System: Experimental Validation of a Colour Spray Model. In *International Conference on Robotics in Alpe-Adria Danube Region*; Springer: Berlin/Heidelberg, Germany 2017; pp. 549–556.
20. Vempati, A.S.; Siegwart, R.; Nieto, J. A Data-driven Planning Framework for Robotic Texture Painting on 3D Surfaces. In Proceedings of the 2020 IEEE International Conference on Robotics and Automation (ICRA), Paris, France, 31 May–31 August 2020; pp. 9528–9534.
21. Santos, M.; Notomista, G.; Mayya, S.; Egerstedt, M. Interactive Multi-Robot Painting Through Colored Motion Trails. *Front. Robot.* **2020**, *7*, 143. [CrossRef] [PubMed]
22. Beltramello, A.; Scalera, L.; Seriani, S.; Gallina, P. Artistic Robotic Painting Using the Palette Knife Technique. *Robotics* **2020**, *9*, 15. [CrossRef]
23. Huang, Y.; Tsang, S.C.; Wong, H.T.T.; Lam, M.L. Computational light painting and kinetic photography. In Proceedings of the Joint Symposium on Computational Aesthetics and Sketch-Based Interfaces and Modeling and Non-Photorealistic Animation and Rendering, Victoria, BC, Canada, 17–19 August 2018; pp. 1–9.
24. Ren, K.; Kry, P.G. Single stroke aerial robot light painting. In Proceedings of the 8th ACM/Eurographics Expressive Symposium on Computational Aesthetics and Sketch Based Interfaces and Modeling and Non-Photorealistic Animation and Rendering, Genoa, Italy, 5–6 May 2019; pp. 61–67.
25. Ma, Z.; Duenser, S.; Schumacher, C.; Rust, R.; Baecher, M.; Gramazio, F.; Kohler, M.; Coros, S. Stylized Robotic Clay Sculpting. *Comput. Graph.* **2021**, *98*, 150–1664. [CrossRef]
26. Zhuo, F. Human-machine Co-creation on Artistic Paintings. In Proceedings of the 2021 IEEE 1st International Conference on Digital Twins and Parallel Intelligence (DTPI), Beijing, China, 15 July–15 August 2021; pp. 316–319.

27. Guo, C.; Bai, T.; Lu, Y.; Lin, Y.; Xiong, G.; Wang, X.; Wang, F.Y. Skywork-daVinci: A novel CPSS-based painting support system. In Proceedings of the 2020 IEEE 16th International Conference on Automation Science and Engineering (CASE), Hong Kong, China, 20–21 August 2020; pp. 673–678.
28. Quintero, C.P.; Dehghan, M.; Ramirez, O.; Ang, M.H.; Jagersand, M. Flexible virtual fixture interface for path specification in tele-manipulation. In Proceedings of the 2017 IEEE International Conference on Robotics and Automation (ICRA), Singapore, 29 May–3 June 2017; pp. 5363–5368.
29. Chen, L.; Swikir, A.; Haddadin, S. Drawing Elon Musk: A Robot Avatar for Remote Manipulation. In Proceedings of the 2021 IEEE/RSJ International Conference on Intelligent Robots and Systems (IROS), Prague, Czech Republic, 27 September–1 October 2021; pp. 4244–4251.
30. Dziemian, S.; Abbott, W.W.; Faisal, A.A. Gaze-based teleprosthetic enables intuitive continuous control of complex robot arm use: Writing & drawing. In Proceedings of the 2016 6th IEEE International Conference on Biomedical Robotics and Biomechatronics (BioRob), Singapore, 26–29 June 2016; pp. 1277–1282.
31. Scalera, L.; Maset, E.; Seriani, S.; Gasparetto, A.; Gallina, P. Performance evaluation of a robotic architecture for drawing with eyes. *Int. J. Mech. Control.* **2021**, *22*, 53–60.
32. Art Institute Chicago. Henry Fletcher. Available online: https://www.artic.edu/artists/34495/henry-fletcher (accessed on 24 December 2021).
33. The Metropolitan Museum of Art. Etching. Available online: https://www.metmuseum.org/about-the-met/collection-areas/drawings-and-prints/materials-and-techniques/printmaking/etching (accessed on 24 December 2021).
34. van den Boomgaard, R.; van Balen, R. Methods for fast morphological image transforms using bitmapped binary images. *Cvgip Graph. Model. Image Process.* **1992**, *54*, 252–258. [CrossRef]
35. Mayer, R. *The Artist's Handbook of Materials and Techniques*; Viking: New York, NY, USA, 1991.
36. Curtis, C.J.; Anderson, S.E.; Seims, J.E.; Fleischer, K.W.; Salesin, D.H. Computer-generated watercolor. In Proceedings of the 24th Annual Conference on Computer Graphics and Interactive Techniques, Los Angeles, CA, USA, 3–8 August 1997; pp. 421–430.
37. Lum, E.B.; Ma, K.L. Non-photorealistic rendering using watercolor inspired textures and illumination. In Proceedings Ninth Pacific Conference on Computer Graphics and Applications. Pacific Graphics, Tokyo, Japan, 16–18 October 2001; pp. 322–330.
38. Van Laerhoven, T.; Liesenborgs, J.; Van Reeth, F. Real-time watercolor painting on a distributed paper model. In Proceedings of the Computer Graphics International, Crete, Greece, 16–19 June 2004; pp. 640–643.
39. KUKA. LBR Iiwa. Available online: https://www.kuka.com/en-us/products/robotics-systems/industrial-robots/lbr-iiwa (accessed 24 December 2021).
40. Uozzart. Colosseo Quadrato, il Capolavoro dell'Eur e il Riferimento al Duce. Available online: https://uozzart.com/2020/06/14/palazzo-della-civilta-italiana-colosseo-quadrato-eur/ (accessed 24 December 2021).
41. Pixabay. Blue Wings. Available online: https://pixabay.com/photos/blue-wings-at-liberty-blue-bird-2471094/ (accessed 24 December 2021).

 *actuators*

Article

Control Design for CABLEankle, a Cable Driven Manipulator for Ankle Motion Assistance

Idumudi Venkata Sai Prathyush [1], Marco Ceccarelli [2,*] and Matteo Russo [3]

[1] Department of Mechanical Engineering, Birla Institute of Technology and Science
   Pilani, Hyderabad 500078, India; prathyushivs@gmail.com
[2] LARM2: Laboratory of Robot Mechatronics, University of Rome "Tor Vergata", 00133 Rome, Italy
[3] The Rolls-Royce UTC in Manufacturing and On-Wing Technology, Faculty of Engineering,
   University of Nottingham, Nottingham NG8 1BB, UK; matteo.russo@nottingham.ac.uk
* Correspondence: marco.ceccarelli@uniroma2.it

Abstract: A control design is presented for a cable driven parallel manipulator for performing a controlled motion assistance of a human ankle. Requirements are discussed for a portable, comfortable, and light-weight solution of a wearable device with an overall design with low-cost features and user-oriented operation. The control system utilizes various operational and monitoring sensors to drive the system and also obtain continuous feedback during motion to ensure an effective recovery. This control system for CABLEankle device is designed for both active and passive rehabilitation to facilitate the improvement in both joint mobility and surrounding muscle strength.

Keywords: cable-driven robots; control design; motion assistance

Citation: Venkata Sai Prathyush, I.; Ceccarelli, M.; Russo, M. Control Design for CABLEankle, a Cable Driven Manipulator for Ankle Motion Assistance. *Actuators* **2022**, *11*, 63. https://doi.org/10.3390/act11020063

Academic Editors: Marco Carricato and Edoardo Idà

Received: 27 January 2022
Accepted: 18 February 2022
Published: 21 February 2022

Publisher's Note: MDPI stays neutral with regard to jurisdictional claims in published maps and institutional affiliations.

1. Introduction

The ankle is a complex joint that forms a kinematic linkage between the lower limb and the foot, allowing day-to-day tasks. It is under high compressive and shear forces during gait, but due to its structure, it functions with a high degree of stability [1–3]. Unfortunately, this joint is very prone to acute and long-term injuries in physically active individuals. Thus, there is a need for rehabilitation to ensure the injured joint returns to its complete functionality [4–8].

Effective rehabilitation is a long process and also requires a qualified physiotherapist to help bring back the joint mobility. Hence, in this regard, robotics has been involved in rehabilitation therapy to constantly monitor the patient and help in movement execution [9–11]. Most of the past robotic designs are based on a static platform design, where a non-portable device requires the patient to keep his foot on a grounded platform (usually while sitting down) to perform rehabilitation [12–21]. A similar design involves a mechanism driven by pneumatic muscles to help exercise a single leg [22,23]. Most of such designs are not only bulky, but also fixed to the ground. As such, patients are required to travel to the hospital or facility where the rehabilitation device is installed despite a potential mobility impairment [24].

To overcome this issue, cable-driven parallel robots (CDPR) were introduced into the field of rehabilitation, owing to their lighter weight, safe nature, and better payload-to-weight ratio [25]. Devices such as the ones developed by Aggogeri et al. [26] and Dai et al. [27] adopted the CDPR models for leg and elbow rehabilitation. For the case of ankle rehabilitation, a CDPR with a grounded base platform was proposed by Liu et al. [28]. A portable CDPR solution for ankle motion assistance is the CABLEankle introduced in [29].

Previous work regarding CABLEankle [29] describes the working of the model along with the necessary kinematics, static, and force closure analysis to evaluate the performance based on parameter such as maximum cable tension, load on the ankle joint, and range of motion. However, whereas previous work discusses the CABLEankle's mechanical design, a control system design is needed to integrate the sensors and motors with proper motion capabilities, which are required to perform a desired motion assistance by exerting forces and tensions according to the requirements of different stages of rehabilitation. This is usually achieved by developing three different operational modes, namely active training, passive training, and assistive training. Preliminary work on such control schemes has been presented in [30,31], but it is only applied to designs with a static platform, rather than wearable ones.

Therefore, in this work, such a motion controller is developed to ensure the proper functioning of a wearable ankle rehabilitation device. In this regard, this article is organized as follows. After introducing the problem requirement, the working of the model, the important equations, and results of CABLEankle from [27] are summarized. Section 4, which explains in detail, including with simulations, the control system that is developed as a solution to make the CABLEankle controllable and adapt to user requirements. In Section 5, the results of the simulations report cable tensions and motor torques during different rehabilitation modes, as well as motor input and platform orientation. Finally, the discussion sums up the work performed and gives insights into future research directions.

2. Requirements and Problems

Rehabilitation is a very important and necessary component to fully heal an injured joint/limb to ensure recovery of its functional abilities and range of motion. There are two main types of rehabilitation—active and passive rehabilitation. In passive rehabilitation, an external force is applied to move the injured joints/muscles to reduce the localized stiffness and to also regain the range of motion. On the other hand, active rehabilitation requires the patient to use their own muscles to work the injured joint to recover its functional ability, endurance, and strength. Hence, there is a need for the design and development of more sophisticated and flexible rehabilitation robots.

Problems in designing and operating a cable-driven system for motion assistance can be identified along the lines of (i) mechanical design with proper features that can also provide comfort to the wearer and (ii) control design for a user-oriented motion-controlled operation. These design problems bring upon design requirements such as a lightweight design with comfort operation, a user-oriented operation for user feedback, a flexible controller that can switch between from active to passive rehabilitation modes and vice-versa, limited power consumption for the duration of usage, and, finally, the necessary safety conditions of design and operation. A complete control system considers all the above-mentioned requirements in designing and operating the cable-driven robot with proper limits and requirements that are adjustable as a function of the user.

Depending on the type of injury, the status of the patient and the recommendation of a physiotherapist, both active and passive or either one of the rehabilitation modes might be required. Hence, it is necessary to make both these modes available to the patient to ensure their speedy and efficient recovery. In this regard, this work has developed a motion controller for the CABLEankle that can accommodate both active and passive rehabilitation modes. The control system has sensors for both monitoring and operation to ensure that the rehabilitation process proceeds smoothly even in the absence of physiotherapists. Additionally, the controller can both assist the motion by using the servomotors to generate an external moment to drive the foot platform and generate an external moment against the intended direction of motion to increase the strength and endurance of the joint. The detailed solution to the problem statement along with the description of the motion controller and the sensors used is provided in the following section.

3. CABLEankle, an Ankle Assisting Device

3.1. Mechanism Design of a Cable-Driven Assistive Device

The CABLEankle is a S-4SPS cable-driven parallel mechanism that functions as a lightweight wearable robot for ankle rehabilitation [32,33]. The ankle joint has three main motion modes—dorsiflexion/plantarflexion, abduction/adduction, and inversion/eversion.

This device can accommodate the ankle joint motion along all the three motion modes but within a limited range, as described in Table 1 [27]. Four cables are used to achieve this triaxial motion, with the help of motors, according to the requirement of the patient. The motors help orient the foot platform with respect to the stationary shank platform, as shown in Figures 1 and 2, and help provide motion assistance or guidance to the user.

Table 1. Ranges of motion of the human ankle joint [1–3].

Motion	Dorsiflexion	Plantarflexion	Abduction/Addjuction	Inversion/Eversion
Range Limits	20 deg	50 deg	$+/-10$ deg	$+/-12$ deg

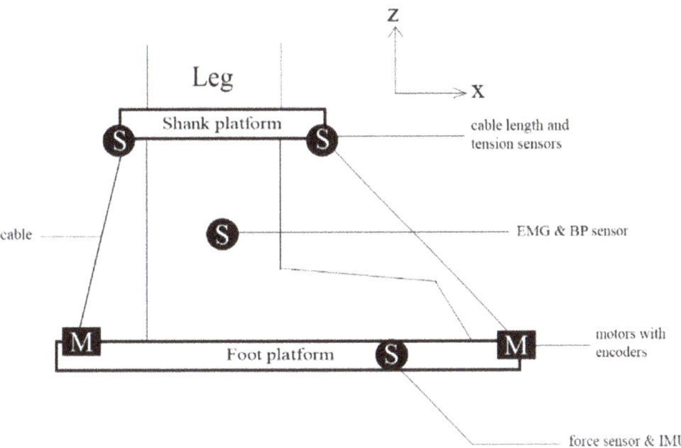

Figure 1. Conceptual design of CABLEankle.

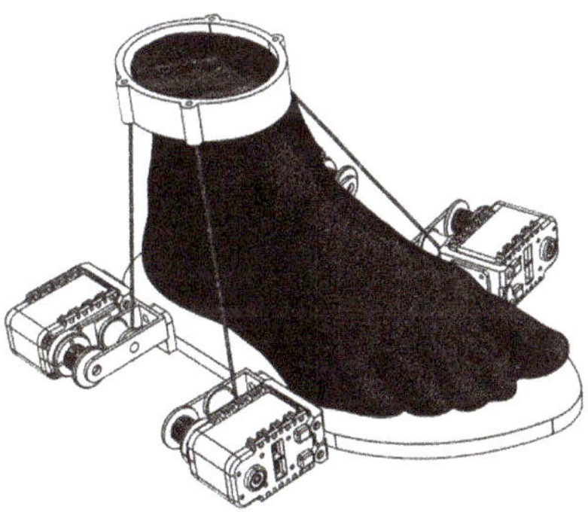

Figure 2. CAD design of CABLEankle as in [27].

3.2. Kinematic Analysis

The CABLEankle device consists of two platforms, one at the base of the foot and the other at the shank of the patient. They are fastened to the patient by means of straps and can be considered as fixed to the patient's leg. Their relative motion is here analysed by assuming the shank platform as fixed, while the foot platform can orient itself relative to the shank by varying the length of the cables connecting the two platforms. This relative motion of the foot platform happens about the ankle joint, whose kinematic behavior can be assumed to be a passive spherical joint with a restricted range of angular motion.

The shank platform is represented by a reference frame A contains $Oxyz$ frame, while the reference frame B of the foot platform is given by $Ox'y'z'$, as seen in Figure 3. The relative motion of the foot platform with respect to the shank platform is a roll–pitch—yaw motion that can be expressed as $^{AB}R = R_z(\alpha)\,R_y(\phi)\,R_x(\theta)$, where ϕ measures dorsiflexion/plantarflexion, θ measures inversion/eversion, and α measures abduction/adduction. During the motion, three assumptions are followed: (i) all cables are always under tension, (ii) the points where the cable is attached to the platform act as spherical joints, and (iii) the cables are considered to be prismatic joints with negligible axial deformation.

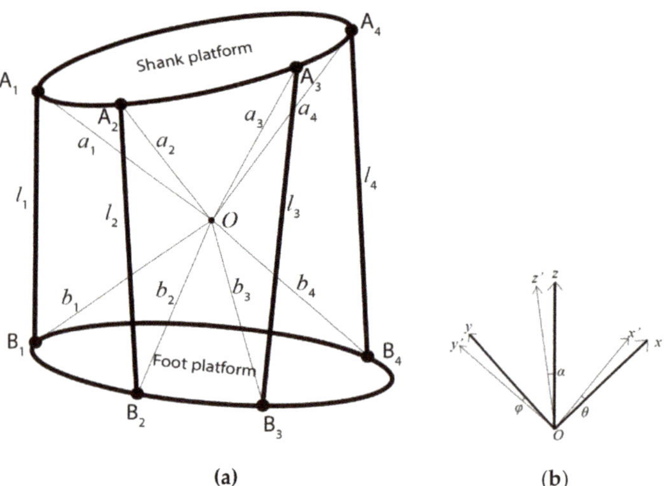

Figure 3. (**a**) Kinematic scheme with motion variables and (**b**) reference frames.

The positions where the cables are attached can be generalized as $A_{a_i} = \begin{pmatrix} a_{ix} & a_{iy} & a_{iz} \end{pmatrix}^T$ at the shank platform and $B_{b_i} = \begin{pmatrix} b_{ix} & b_{iy} & b_{iz} \end{pmatrix}^T$ at the foot platform. Using the model in Figure 3, the vectorial loop-closure equation for each cable can be written as

$$^A l_i = {}^{AB}R\,{}^B b_i - {}^A a_i \tag{1}$$

where, l_i is the distance between the points A_i and B_i for the ith cable. The scalar product of either side of the (1) is multiplied by itself, and the length of each cable is expressed as

$$l_i = \sqrt{{}^A a_i^{T\,A} a_i + {}^B b_i^{T\,B} b_i - 2{}^A a_i^{T\,AB}R^B b_i} \tag{2}$$

3.3. Static Analysis

Rehabilitation requires a smooth, slow, and controlled motion to ensure that minimal stress or pain is experienced by the patient. Since this mechanism is designed to operate with limited speed, inertial effects and dynamics can be neglected. Hence, a static analysis can be adopted for performance evaluation.

When an external wrench defined by force F_{ext} and moment M_{ext} is applied to the foot platform, static equilibrium is achieved through reactions such as tension T, reaction force F_R, and reaction moment M_R, as shown in Figure 4.

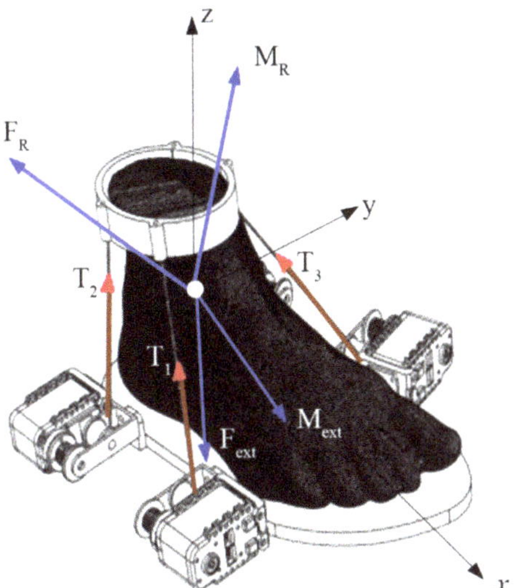

Figure 4. Static model of CABLEankle.

The equilibrium condition for translation in frame A can be expressed as

$$\sum_{i=1}^{n} T_i + F_R + F_{ext} = 0 \tag{3}$$

while the equilibrium condition for rotation is given by

$$\sum_{i=1}^{n} b_i \times T_i + M_R + M_{ext} = 0 \tag{4}$$

where T_i is the tension in the ith cable. The tension in each cable is expressed as the product of it's magnitude and cable unit vector as $T_i = -T_i\, p_i$, where the actuation vector T is defined as $(T_1\ T_2\ T_3\ T_4)^T$. Using this, Equations (3) and (4) can, respectively, be rewritten as

$$P^T \cdot T - F_R = F_{ext} \tag{5}$$

and,

$$Q^T \cdot T - M_R = M_{ext} \tag{6}$$

where, $P^T = [p_1\ p_2\ p_3\ p_4]$ and $Q^T = [b_1 \times p_1 \ldots b_4 \times p_4]$.

As discussed previously, since the ankle joint's kinematic behavior is assumed to be that of a spherical joint, rotational motion is not constrained within the limited range of the ankle joint and hence the reaction moment M_R is considered a null vector. Hence, Equations (5) and (6) can be combined to represent the full equilibrium given by

$$\begin{bmatrix} P^T & -I_3 \\ Q^T & 0_3 \end{bmatrix} \begin{pmatrix} T \\ F_R \end{pmatrix} = \begin{pmatrix} F_{ext} \\ M_{ext} \end{pmatrix} \tag{7}$$

The operation problem can be characterized by solving the actuating torque as function of a prescribed ankle motion in controlled assisted exercise.

4. Solution for Control Design Unit

A complete control system for CABLEankle would involve both position and force control for achieving a desired position of the platform and a desired force and torque from the motors based on user input. The flowchart of such a control system design is provided in Figure 5. It is designed to incorporate two modes of rehabilitation, active and passive, for the efficient recovery of the patient.

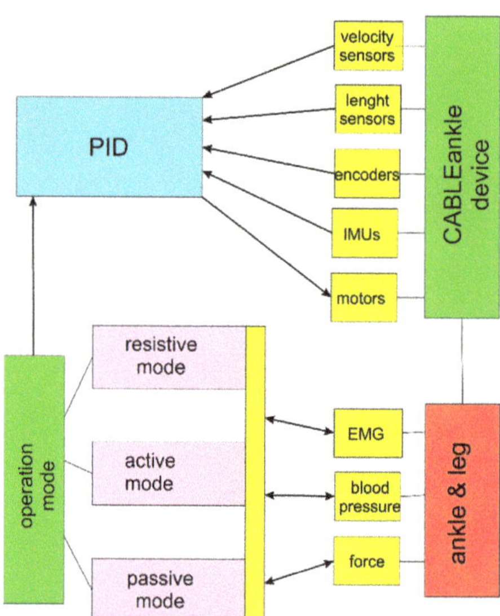

Figure 5. A scheme for control design in CABLEankle device.

Resistive mode and co-operative mode fall under the category of active rehabilitation. In resistive mode, a resistive torque is applied by the motors forcing the user to apply additional effort to move the platform. This mode causes a moment that opposes the force exerted by the muscles. Hence, it is especially used for increasing the strength of the joint. For the case of co-operative mode, any force exerted by the patient is understood as an intention to perform the motion in that direction, and torque is applied to assist that motion. On the other hand, passive mode rehabilitation is where only the motors perform all the work required to move the leg rested on the platform and no muscular activity is involved from the user's side. These types of exercises are aimed at improving the flexibility of the joint. There are two main subcategories of passive exercises, relaxed and forced passive exercises. As noted earlier, joints have a set range of motion, and forced passive exercises force the joint to move a little beyond the limited range. Here, relaxed passive exercises that simply move the joint with the help of a motor are adopted since this work focuses on functioning within the limited range.

The remaining blocks in the flowchart in Figure 5 represent sensors that can be divided into two categories, namely, operation control and monitoring. Sensors such as motor encoder, IMU, and cable-length and cable-tension sensors fall under operation control sensors. They help to obtain information such as angular position/velocity, pose of the platform, length of the cable measured in real-time, and tension values in the cables to ensure they always satisfy the force-closure equation, respectively. In the other category are the force sensors that are placed on the platform to measure the force exerted by the patient and to determine the direction in which the user wants to rotate the platform. The EMG sensor is used to monitor the muscle activity in the region and provide information on the healing status of the ankle. Prior research in rehabilitation robotics has shown various successful ways in which EMG sensors could be used to monitor the muscular activity in the muscles surrounding the joint to measure the progress of the strength of the joint. A blood pressure sensor is used to monitor the blood pressure levels in the ankle region to monitor both for indicating patient reaction and patient safety since active exercises can cause a pressure increase that could be either a good signal or harmful after an ankle injury (e.g., avoid harmful increase in blood pressure in individuals with edema from the ankle injury as pointed out in [34]). In addition, this sensor is used only in the case of active rehabilitation because, in general, the mean average pressure does not change during passive rehabilitation.

In this work, a control framework is designed as in Figure 6 for passive and active-resistive modes with cable length sensors and cable velocity sensors as per the control feedback during dorsiflexion/plantarflexion. From the kinematic and static analysis in Section 2, it can be noted that cable length influences the tension in the cables and the load on the ankle and is also directly related to the flexion angle. Hence, it is chosen as the primary control input to achieve position control of the foot platform during ankle exercise. During the motion, errors could arise from physical inaccuracies in the mechanical components that are due mainly to parameter identification and tolerance in construction and assembly, and to deal with such scenarios, this framework is equipped with PID controllers to ensure that the actual cable length and velocity values are close enough to the desired values. Referring to Figure 6, the Inverse Kinematics block calculates the cable lengths taking the flexion angle as the input by using formulation in Section 2 with Equations (2) and (3). The Cable Velocity block outputs the cable velocities based on the rate of ankle pose, which can be computed yet by using a formulation. The data necessary to generate results for control design refer to the configuration in Table 2, with the design parameters that have been obtained through an optimization procedure that considers both design and path [33], and the control parameters tuned through Matlab Simulink. When a velocity control is implemented, the PID controller thus operates as per Figure 6 according to

$$\dot{l}_i(t) = K_p l_{i,err}(t) + K_i \int l_{i,err}(t)dt + K_d \frac{dl_{i,err}(t)}{dt} \tag{8}$$

where $\dot{l}_i$ represents the velocity of the ith cable (i.e., its length variation over time), $l_{i,err}$ is the difference between the current and desired cable configuration, K_p is the proportional gain of the controller, K_i is the integral gain of the controller, and K_d is the derivative gain of the controller. While this represents a way of controlling the proposed device, when other sensors are included (e.g., EMG, blood pressure, and force), their inputs can be implemented in the control scheme to better react to the patient's behavior and improve the rehabilitation therapy.

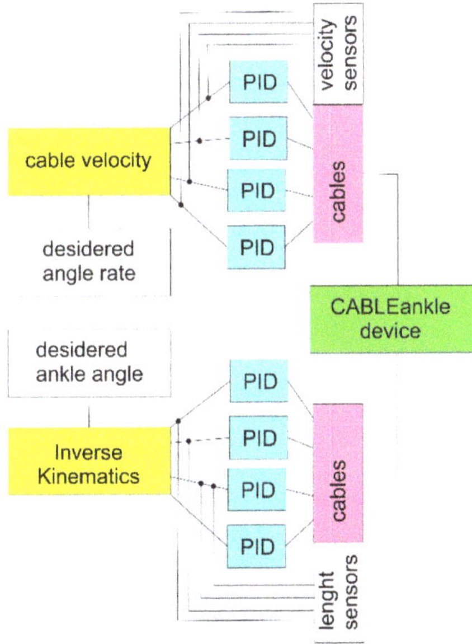

Figure 6. Flowchart of the designed control system for CABLEankle device.

Table 2. Parameters of performance evaluation for the control design.

Shank Platform (mm)	Foot Platform (mm)	Control Parameters
$^{A}a_1 = (20\ 30\ 50)^{\mathrm{T}}$	$^{B}b_1 = (80\ 75\ {-40})^{\mathrm{T}}$	$l_i \in [20\ \mathrm{mm},\ 200\ \mathrm{mm}]$
$^{A}a_2 = (-20\ 30\ 50)^{\mathrm{T}}$	$^{B}b_2 = (0\ 60\ {-40})^{\mathrm{T}}$	$\phi \in [-50°,\ 20°]$
$^{A}a_3 = (20\ {-30}\ 50)^{\mathrm{T}}$	$^{B}b_3 = (10\ 60\ {-40})^{\mathrm{T}}$	$K_p = 0.1$
$^{A}a_4 = (-20\ {-30}\ 50)^{\mathrm{T}}$	$^{B}b_4 = (90\ 60\ {-40})^{\mathrm{T}}$	$K_i = 1$
		$K_p = 0.001$

5. Performance Analysis

Based on the kinematic and static analysis performed in Section 2, the controller was designed referring to the scheme in Figure 6. Using Equation (2), the length of each cable can be computed for the simulated motion modes. The static analysis gives information about the cable tensions and reaction forces. These values have been computed for different inclinations of the mobile platform within the motion range of the ankle joint and for different rehabilitation modes such as resistive and passive. For the passive rehabilitation scenario, the cable tensions and motor torque are computed by assuming no external wrench being applied on the foot except for its own mass of 1 Kg. For the active resistive rehabilitation case, an external moment of 0.5 Nm is applied about the y-axis. The motor torques for the ith motor was calculated by multiplying the tension of the ith cable by the pulley diameter, which was assumed to be 0.02 m. The results of cable tensions and motor torques during dorsiflexion/plantarflexion are provided in Figures 7 and 8.

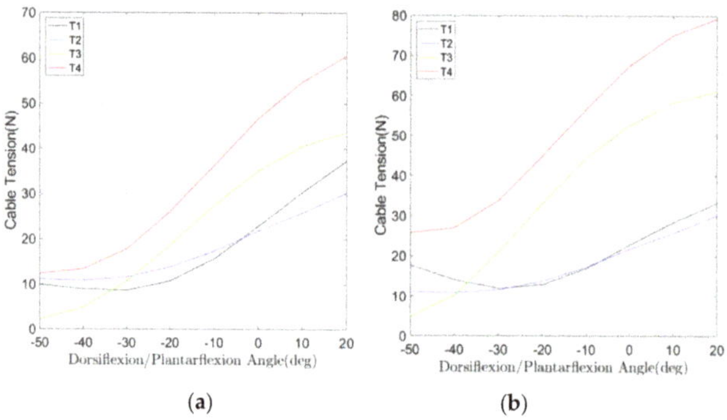

(a)   (b)

Figure 7. Computed cable tensions during a simulated controlled operation: (**a**) in passive mode and (**b**) in active resistive mode.

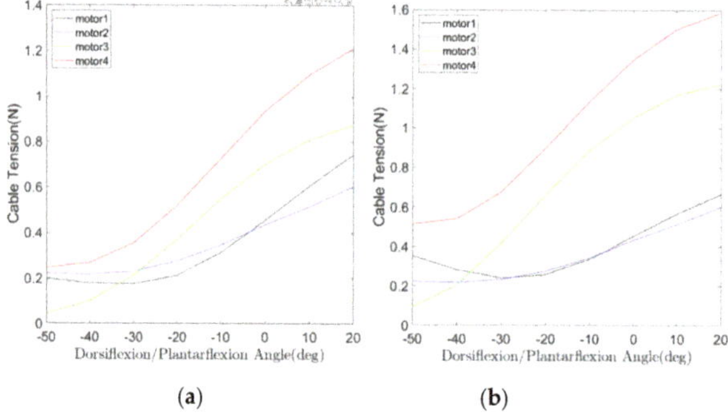

(a)   (b)

Figure 8. Computed motor torques during a simulated controlled operation: (**a**) in passive mode and (**b**) in active resistive mode.

The developed PID controller ensures that the position of the platform always reaches the desired inclination of the user by taking control variables such as cable length and velocity as inputs. The kinematic and static analysis shows a direct relationship between the control inputs and desired angle and rate of desired angle, respectively. Thus, it can be implied that reducing the error in the above control inputs would help achieve the desired flexion angle as well. Simulation has been carried out on a controlled operation for an example of dorsiflexion/plantarflexion assisted motion from $0°$ to $40°$. Results are reported in Figures 9–11 to characterize the designed control system for the device.

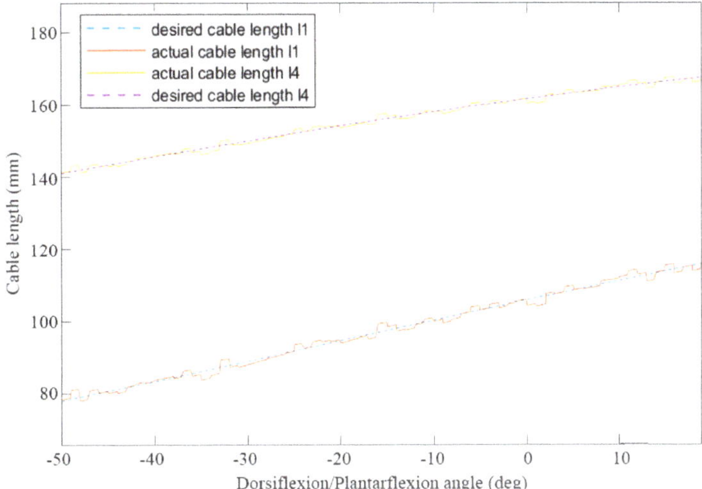

Figure 9. Desired and computed cable length values for a simulated operation of a controlled ankle assisted motion exercise.

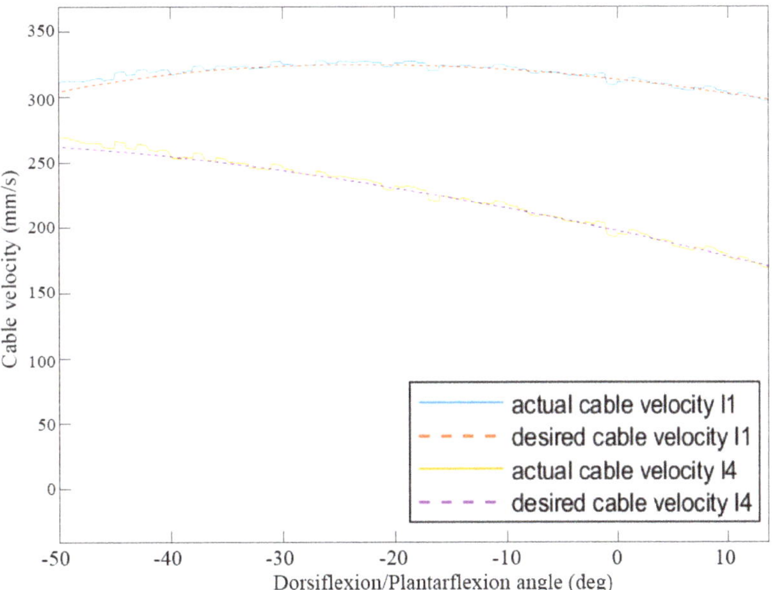

Figure 10. Desired and computed cable velocity values for a simulated operation of a controlled ankle assisted motion exercise.

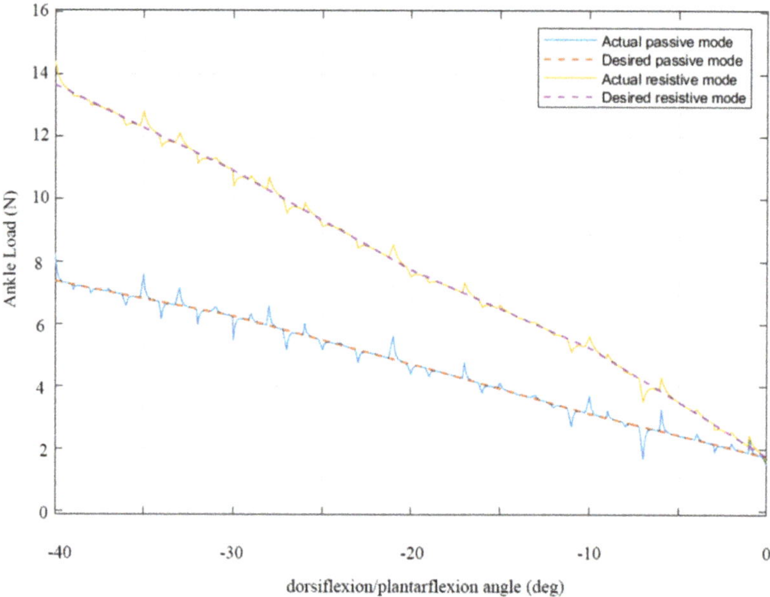

Figure 11. Desired and computed ankle load for simulated passive and resistive modes values of a controlled ankle assisted motion exercise.

In particular, plots representing the desired and actual cable length and velocity values are shown in Figures 9 and 10. It can be noted that, with the proposed controller, the system is able to achieve a stable behaviour, which closely follows the desired one, with a maximum error of 2 mm throughout the full range of motion over dorsiflexion/plantarflexion exercises.

Apart from achieving the desired angle of inclination of the foot platform, it is equally important to pay attention to the ankle load. The developed controller also provides information about the force on the ankle, as illustrated in the results in Figure 11 for the same exercise reported in Figures 9 and 10. The graph reports the results from two different control modes: the upper line represents the forces in resistive mode, when the user has to overcome an additional load from the rehabilitation device, while the lower line characterizes functioning in passive mode, with the device fully supporting and controlling the ankle of the patient throughout a rehabilitation exercise.

For the case of dorsiflexion/plantarflexion, it can be observed that, as the ankle moves from $0°$ to $-40°$, there is a visible increase in the load on the joint for both functioning modes. This is expected, as the foot moves from its "resting" position to a more "uncomfortable" one and thus requires effort to sustain the resulting bent position. For the passive control mode, this increase goes from 2 N to less than 8 N, which is a safe value for an injured ankle and can be used in early rehabilitation stages to avoid straining the joint further. Conversely, the increase is steeper for the resistive mode, imposing a load up to 14 N, suggesting that the ankle is being compelled to apply a force to counteract the resistive torque from the cables and motors. This higher load is more suited to later stages of rehabilitation or daily exercising.

When compared to the expected load for the ideal motion, which can be computed with Equation (7), the load on the joint relative to the control simulation in Figures 9 and 10 presents a less smooth behaviour with local force peaks. Nevertheless, these fluctuations are always within 1 N, and the overall load on the joint never reaches a critical threshold throughout the whole rehabilitation exercise, thus validating the proposed control.

6. Discussion

The results in Figures 7 and 8 show for the motion of dorsiflexion/plantarflexion that the maximum cable tension values are less than 90 N for both passive and active resistive rehabilitation cases. Thus, with an appropriate pulley design, the device can operate without reaching the servo motor's upper limit for the ankle's whole range of motion as planned in a full exercise [27].

Apart from the motor limits, the results in Figures 9 and 10 show the error between the computed and desired cable length and velocity values that implies achieving position control of the angle of inclination of the foot platform, which is necessary for the safe operation with a patient. These values, upon being extended to find the load on the ankle joint as reported in Figure 11, confirm the validity of both active and passive rehabilitation modes. From these results, it can be noted that this solution satisfies the requirement and solves the problems mentioned in Section 3. As such, the proposed robot can be successfully used to perform rehabilitation for both early treatment of an injury (in passive mode, thanks to the limited load imposed on the ankle in this more critical phase) and later daily exercises (in resistive mode, when the added load helps the patient exercise ankle muscles and recover full mobility and strength).

Even though the requirements have been met, advances can still be made to provide a user with more flexibility and to equip the controller with more control parameters for achieving improved monitoring and operation functionalities. The designed control framework presented in this work considers only cable length and cable velocity. However, as mentioned in the general design scheme in Section 4, future work could include utilizing IMUs, force sensors, blood pressure sensors, and EMG sensors in order to build a more complete controller. Typically, EMG sensors are used just for monitoring purposes, but it would be interesting to utilize them along with the feedback loop to take in motor control commands based on myoelectric signals and deliver them to the PID controller so that the rehabilitation device could better react to the patient's behaviour during therapy and provide an overall better rehabilitation experience.

Overall, the control system introduced in this manuscript integrates previous hardware with novel software capability and functionality, providing an adaptive control that can evaluate and apply actuation forces and torques as required from different stages of ankle rehabilitation and assistance while ensuring safe operation within the physical limits of a patient. In particular, the proposed controller is able to assist users in both passive and active exercise, by either controlling the load on the ankle, increasing it to a desired value for strengthening exercises, or supporting and guiding ankle motion in earlier rehabilitation phases. While the performance of the proposed system has been evaluated through simulations, future works will focus on prototype development and experimental validations on both healthy subjects and people with limited ankle mobility.

Author Contributions: Conceptualization, I.V.S.P., M.C. and M.R.; methodology, I.V.S.P. and M.C.; software, I.V.S.P.; validation, I.V.S.P.; formal analysis, I.V.S.P. and M.C.; investigation, I.V.S.P., M.C. and M.R.; resources, M.C.; data curation, I.V.S.P.; writing—original draft preparation, I.V.S.P. and M.C.; writing—review and editing, I.V.S.P., M.R. and M.C.; visualization, I.V.S.P.; supervision, M.C.; project administration, M.C.; funding acquisition, M.C. All authors have read and agreed to the published version of the manuscript.

Funding: This research received no external funding.

Conflicts of Interest: The authors declare no conflict of interest.

References

1. Wang, C.; Wang, L.; Wang, T.; Li, H.; Du, W.; Meng, F.; Zhang, W. Research on an Ankle Joint Auxiliary Rehabilitation Robot with a Rigid-Flexible Hybrid Drive Based on a 2-S′PS′ Mechanism. *Appl. Bionics Biomech.* **2019**, *2019*, 7071064. [CrossRef]
2. Brockett, C.L.; Chapman, G.J. Biomechanics of the ankle. *Orthop. Trauma* **2016**, *30*, 232–238. [CrossRef]
3. Knudson, D. *Fundamentals of Biomechanics*; Springer Science & Business Media: Berlin/Heidelberg, Germany, 2007.

4. Mattacola, C.G.; Dwyer, M.K. Rehabilitation of the Ankle After Acute Sprain or Chronic Instability. *J. Athl. Train.* **2002**, *37*, 413–429.

5. Chen, E.T.; McInnis, K.C.; Borg-Stein, J. Ankle sprains: Evaluation, rehabilitation, and prevention. *Curr. Sports Med. Rep.* **2019**, *18*, 217–223. [CrossRef]

6. Chinn, L.; Hertel, J. Rehabilitation of Ankle and Foot Injuries in Athletes. *Clin. Sports Med.* **2010**, *29*, 157–167. [CrossRef]

7. Nagaya, S.; Hayashi, H.; Fujimoto, E.; Maruoka, N.; Kobayashi, H. Passive ankle movement increases cerebral blood oxygenation in the elderly: An experimental study. *BMC Nurs.* **2015**, *14*, 14. [CrossRef]

8. Díaz, I.; Gil, J.J.; Sánchez, E. Lower-Limb Robotic Rehabilitation: Literature Review and Challenges. *J. Robot.* **2011**, *2011*, 759764. [CrossRef]

9. Zhang, M.; Davies, T.C.; Xie, S. Effectiveness of robot-assisted therapy on ankle rehabilitation–a systematic review. *J. Neuroeng. Rehabil.* **2013**, *10*, 30. [CrossRef]

10. Alvarez-Perez, M.G.; Garcia-Murillo, M.A.; Cervantes-Sánchez, J.J. Robot-assisted ankle rehabilitation: A review. *Disabil. Rehabil. Assist. Technol.* **2020**, *15*, 394–408. [CrossRef]

11. Miao, Q.; Zhang, M.; Wang, C.; Li, H. Towards Optimal Platform-Based Robot Design for Ankle Rehabilitation: The State of the Art and Future Prospects. *J. Health Eng.* **2018**, *2018*, 1534247. [CrossRef]

12. Yoon, J.; Ryu, J.; Lim, K.-B. Reconfigurable ankle rehabilitation robot for various exercises. *J. Robot. Syst.* **2006**, *22*, S15–S33. [CrossRef]

13. Jamwal, P.K.; Hussain, S.; Ghayesh, M.H.; Rogozina, S.V. Impedance Control of an Intrinsically Compliant Parallel Ankle Rehabilitation Robot. *IEEE Trans. Ind. Electron.* **2016**, *63*, 3638–3647. [CrossRef]

14. Saglia, J.A.; Tsagarakis, N.G.; Dai, J.S.; Caldwell, D.G. Control Strategies for Patient-Assisted Training Using the Ankle Rehabilitation Robot (ARBOT). *IEEE/ASME Trans. Mechatron.* **2013**, *18*, 1799–1808. [CrossRef]

15. Roy, A.; Krebs, H.I.; Williams, D.J.; Bever, C.T.; Forrester, L.W.; Macko, R.M.; Hogan, N. Robot-Aided Neurorehabilitation: A Novel Robot for Ankle Rehabilitation. *IEEE Trans. Robot.* **2009**, *25*, 569–582. [CrossRef]

16. Rodríguez-León, J.F.; Chaparro-Rico, B.D.M.; Russo, M.; Cafolla, D. An Autotuning Cable-Driven Device for Home Rehabilitation. *J. Health Eng.* **2021**, *2021*, 6680762. [CrossRef]

17. Sung, E.; Slocum, A.H.; Ma, R.; Bean, J.F.; Culpepper, M.L. Design of an Ankle Rehabilitation Device Using Compliant Mechanisms. *J. Med. Devices* **2011**, *5*, 011001. [CrossRef]

18. Fan, Y.; Yin, Y. Mechanism design and motion control of a parallel ankle joint for rehabilitation robotic exoskeleton. In Proceedings of the 2009 IEEE International Conference on Robotics and Biomimetics (ROBIO), Guilin, China, 19–23 December 2009; Institute of Electrical and Electronics Engineers (IEEE): Piscataway, NJ, USA; pp. 2527–2532.

19. Lin, C.C.; Ju, M.S.; Chen, S.M.; Pan, B.W. A specialized robot for ankle rehabilitation and evaluation. *J. Med. Biol. Eng.* **2008**, *28*, 79–86.

20. Chang, T.-C.; Zhang, X.-D. Kinematics and reliable analysis of decoupled parallel mechanism for ankle rehabilitation. *Microelectron. Reliab.* **2019**, *99*, 203–212. [CrossRef]

21. Nurahmi, L.; Caro, S.; Solichin, M. A novel ankle rehabilitation device based on a reconfigurable 3-RPS parallel manipulator. *Mech. Mach. Theory* **2019**, *134*, 135–150. [CrossRef]

22. Park, Y.-L.; Chen, B.-R.; Pérez-Arancibia, N.O.; Young, D.; Stirling, L.; Wood, R.J.; Goldfield, E.C.; Nagpal, R. Design and control of a bio-inspired soft wearable robotic device for ankle–foot rehabilitation. *Bioinspir. Biomim.* **2014**, *9*, 016007. [CrossRef]

23. Liu, Q.; Zuo, J.; Zhu, C.; Meng, W.; Ai, Q.; Xie, S.Q. Design and Hierarchical Force-Position Control of Redundant Pneumatic Muscles-Cable-Driven Ankle Rehabilitation Robot. *IEEE Robot. Autom. Lett.* **2021**, *7*, 502–509. [CrossRef]

24. Dong, M.; Zhou, Y.; Li, J.; Rong, X.; Fan, W.; Zhou, X.; Kong, Y. State of the art in parallel ankle rehabilitation robot: A systematic review. *J. Neuroeng. Rehabil.* **2021**, *18*, 52. [CrossRef] [PubMed]

25. Cafolla, D.; Russo, M.; Carbone, G. Design and Validation of an Inherently-Safe Cable-Driven Assisting Device. *Int. J. Mech. Control.* **2018**, *19*, 23–32.

26. Aggogeri, F.; Pellegrini, N.; Adamini, R. Functional Design in Rehabilitation: Modular Mechanisms for Ankle Complex. *Appl. Bionics Biomech.* **2016**, *2016*, 9707801. [CrossRef]

27. Dai, J.S.; Zhao, T.; Nester, C. Sprained Ankle Physiotherapy Based Mechanism Synthesis and Stiffness Analysis of a Robotic Rehabilitation Device. *Auton. Robot.* **2004**, *16*, 207–218. [CrossRef]

28. Liu, Q.; Wang, C.; Long, J.J.; Sun, T.; Duan, L.; Zhang, X.; Zhang, B.; Shen, Y.; Shang, W.; Lin, Z.; et al. Development of a New Robotic Ankle Rehabilitation Platform for Hemiplegic Patients after Stroke. *J. Health Eng.* **2018**, *2018*, 3867243. [CrossRef]

29. Russo, M.; Ceccarelli, M. Analysis of a Wearable Robotic System for Ankle Rehabilitation. *Machines* **2020**, *8*, 48. [CrossRef]

30. Wang, Y.; Wang, K.; Zhang, Z.; Mo, Z. Control strategy and experimental research of a cable-driven lower limb rehabilitation robot. *Proc. Inst. Mech. Eng. Part C J. Mech. Eng. Sci.* **2021**, *235*, 2468–2481. [CrossRef]

31. Oyman, E.L.; Korkut, M.Y.; Ylmaz, C.; Bayraktaroglu, Z.Y.; Arslan, M.S. Design and control of a cable-driven rehabilitation robot for upper and lower limbs. *Robotica* **2021**, *40*, 1–37. [CrossRef]

32. Russo, M.; Ceccarelli, M. A Wearable Device for Ankle Motion Assistance. In *Advances in Italian Mechanism Science*; Springer: Berlin, Germany, 2020; pp. 173–181. [CrossRef]

33. Russo, M.; Raimondi, L.; Dong, X.; Axinte, D.; Kell, J. Task-oriented optimal dimensional synthesis of robotic manipulators with limited mobility. *Robot. Comput. Manuf.* **2021**, *69*, 102096. [CrossRef]
34. Cloughley, W.B.; Mawdsley, R.H. Effect of Running on Volume of the Foot and Ankle. *J. Orthop. Sports Phys. Ther.* **1995**, *22*, 151–154. [CrossRef] [PubMed]

Article

Smooth-Switching Gain Based Adaptive Neural Network Control of n-Joint Manipulator with Multiple Constraints

Qing Yang [1,2], Haisheng Yu [1,2,*], Xiangxiang Meng [1,2], Wenqian Yu [3] and Huan Yang [4]

[1] College of Automation, Qingdao University, Qingdao 266071, China; 2020020573@qdu.edu.cn (Q.Y.); 2018020450@qdu.edu.cn (X.M.)
[2] Shandong Province Key Laboratory of Industrial Control Technology, Qingdao University, Qingdao 266071, China
[3] State Grid Dongping Power Supply Company, State Grid, Taian 271000, China; yuwenqiandp@163.com
[4] School of Mechanical and Automotive Engineering, Qingdao University of Technology, Qingdao 271000, China; 13675327120@139.com
*   Correspondence: yhsh_qd@qdu.edu.cn

Abstract: Modeling errors, external loads and output constraints will affect the tracking control of the n-joint manipulator driven by the permanent magnet synchronous motor. To solve the above problems, the smooth-switching for backstepping gain control strategy based on the Barrier Lyapunov Function and adaptive neural network (BLF-ANBG) is proposed. First, the adaptive neural network method is established to approximate modeling errors, unknown loads and unenforced inputs. Then, the gain functions based on the error and error rate of change are designed, respectively. The two gain functions can respectively provide faster response speed and better tracking stability. The smooth-switching for backstepping gain strategy based on the Barrier Lyapunov Function is proposed to combine the advantages of both gain functions. According to the above strategy, the BLF-ANBG strategy is proposed, which not only solves the influence of multiple constraints, unknown loads and modeling errors, but also enables the manipulator system to have better dynamic and steady-state performances at the same time. Finally, the proposed controller is applied to a 2-DOF manipulator and compared with other commonly used methods. The simulation results show that the BLF-ANBG strategy has good tracking performance under multiple constraints and model errors.

Keywords: manipulator; multiple constraints; adaptive neural network; smooth-switching for gain; Barrier Lyapunov Function

Citation: Yang, Q.; Yu, H.; Meng, X.; Yu, W.; Yang, H. Smooth-Switching Gain Based Adaptive Neural Network Control of n-Joint Manipulator with Multiple Constraints. *Actuators* **2022**, *11*, 127. https://doi.org/10.3390/act11050127

Academic Editors: Marco Carricato and Edoardo Idà

Received: 1 February 2022
Accepted: 25 April 2022
Published: 29 April 2022

Publisher's Note: MDPI stays neutral with regard to jurisdictional claims in published maps and institutional affiliations.

1. Introduction

The manipulator has been widely used in various scenarios such as medical treatment, automobile production and metal processing due to its strong safety, high precision and high efficiency [1–3]. The permanent magnet synchronous motor (PMSM) has the characteristics of small size, low loss and large starting torque [4,5], which is often used as the drive motor for the servo control of the manipulator [6,7]. The manipulator system driven by PMSM is a multi-variable, nonlinear and strongly coupled system. Therefore, as the production requirements increase, the rapidity, accuracy and stability of manipulator tracking have always been a research hotspot.

For manipulator tracking control, many scholars have proposed different control strategies. Traditional control strategies such as proportional integral derivative (PID) control, feedback linearization control, sliding mode control (SMC), adaptive control and backstepping control are commonly used. Intelligent control methods such as fuzzy control and neural network control are also widely used. Shojaei, Pradhan, and Kim respectively used self-tuning PID control, second-order PID control and PD control to effectively improve the steady-state tracking performance of the manipulator [8–10]. Feng, Yeh, and Huang respectively designed non-singular fast terminal SMC strategy [11], output feedback SMC strategy [12] and adaptive SMC strategy [13]. Each method optimizes the traditional SMC strategy, however, the chattering phenomenon still exists. Gabriele and Meng adopted the feedback linearization strategy [14,15], but this strategy requires an accurate mathematical model. In literature [16–18], adaptive control was designed to effectively estimate the uncertainty of the system. Kanellakopoulos proposed a recursion-based backstepping control [19], which was then widely used. Cheng and Farrell applied backstepping strategy to the control of the manipulator [20,21]. Chang, Yang and Song designed fuzzy backstepping, fuzzy adaptive and fuzzy command filter controllers to improve the stability of position tracking [22–24]. The neural network control has strong approximation ability, so it is used by many scholars to approximate the modeling errors and nonlinear terms [25–28].

The above methods have improved the dynamic and steady-state performances of manipulator tracking, respectively, however, it is difficult to guarantee better dynamic and steady-state performance at the same time. In addition, most of the existing strategies only consider the manipulator system and ignore the drive motor system, along with failing to consider the effects of multiple constraints, unknown loads and modeling errors at the same time. These problems often affect the safe and smooth operation of the manipulator in engineering practice. Many scholars have devised different solutions to these problems. Singh proposed the modeling concept of fractional calculus [29,30], and systematically described the fractional order model of the manipulator in the book [31]. The fractional order dynamic model can describe the system model more accurately. Meng and Liu adopted the coordination strategy of two controllers, combining the advantages of the two controllers to improve the dynamic and steady-state characteristics at the same time, but the use of two different controllers will increase the complexity of the control system [32,33]. Other studies [34–36] used the Barrier Lyapunov Function (BLF) to satisfy the output constraint problem. Sung and Cheng proposed a neural network strategy to approximate the model uncertainty [37,38]. Yang et al. designed a variable-gain backstepping strategy to improve the rapidity and stability of the controller [39–42].

In this paper, the smooth-switching for backstepping gain control strategy based on BLF and adaptive neural network (BLF-ANBG) is designed. Combined with the manipulator and the drive motor, the overall model of the manipulator control system is obtained. The adaptive radial basis function (RBF) neural networks are designed to approximate the modeling errors, unknown loads and unenforced inputs of the system. The gain function based on the error and the change rate of error is designed, and the Gaussian function is used as the switching function to design the method of smooth-switching for backstepping gain, which combines the advantages of the two gain functions. When the error is large, the gain function based on the error plays a major role, and the error is proportional to the gain, which shortens the rise time of the system. On the contrary, when the error is small, the gain function based on the change rate of error plays a major role, and the change rate of error is inversely proportional to the gain, which improves the stability of the steady-state of the system. The smooth-switching for backstepping gain controller is designed based on BLF (BLF-GSS) to realize the normal operation of the system under asymmetric or symmetric time-varying output limited. The BLF-ANBG strategy is proposed by combining the adaptive neural network strategy and the BLF-GSS strategy.

The main contents of this article are organized as follows. In Section 2, the overall model of the manipulator control system is provided by combining the manipulator system and the driven motor system. In Section 3, the BLF-ANBG controller based on an adaptive neural network and BLF-GSS is designed. In Section 4, the stability of the control strategy is proved by using the Lyapunov function [43]. In Section 5, the controller is applied to the 2-DOF manipulator, and the feasibility of the controller is verified by a simulation example. Some conclusions are summarized in Section 6.

2. The Overall System Model of n-Joint Manipulator Driven by PMSM

2.1. The Model of n-Joint Manipulator System

The system model of the n-joint manipulator considering the modeling error and unknown load is

$$(M(q)+\Delta M(q))\ddot{q}+(C(q,\dot{q})+\Delta C(q,\dot{q}))\dot{q}+(G(q)+\Delta G(q))=\tau_r-\tau_L-\tau_f-\Delta E \quad (1)$$

$$\tau_L = J^\mathrm{T}F, \quad \tau_f = R_f\dot{q} + F_c\mathrm{sgn}(\dot{q}) \quad (2)$$

where $q = [q_1, \cdots, q_n]^\mathrm{T}$ represents the position of each joint. $M(q)$ and $C(q,\dot{q}) \in R^{n\times n}$ are the positive-definite inertia matrix and Coriolis force matrix of the nominal model, respectively. $G(q) \in R^n$ is the system gravity vector of the nominal model. τ_r, $\tau_L \in R^n$ and τ_f are respectively expressed as the output torque, load torque and friction torque of the manipulator system. $\Delta M(q)$, $\Delta C(q,\dot{q})$ and $\Delta G(q)$ are the modeling error. J and F are the Jacobian matrix and load force of the manipulator, respectively. R_f and F_c are the diagonal viscous friction and Coulomb friction matrix. ΔE is the interference signal caused by position measurement error and velocity measurement noise.

2.2. The Model of Drive Motor System

The mathematical model of PMSM with modeling errors in the $d-q$ rotating coordinate system is described by

$$L_q\frac{di_q}{dt} = -n_p\Phi\omega - n_pBL_di_d - R_si_q + u_q \quad (3)$$

$$L_d\frac{di_d}{dt} = -R_si_d + n_pBL_qi_q + u_d \quad (4)$$

$$(J_m + \Delta J_m)\frac{d\omega}{dt} = \tau - \tau_{mL} - R_m\omega \quad (5)$$

$$\frac{d\theta}{dt} = \omega \quad (6)$$

$$\tau = n_p[(L_d - L_q)i_di_q + \Phi i_q] \quad (7)$$

where θ, $\omega \in R^n$ indicate the rotation angle and speed of the PMSM. L_d, L_q is the diagonal square matrix of $d-q$ axis inductance. $B = diag\{\omega_1, \cdots, \omega_n\}$, ω_i represents the ith component of the speed. i_d, i_q and u_d, u_q denote the $d-q$ axis stator current and voltage vector. n_p, Φ and $R_m \in R^{n\times n}$ are pole logarithm, magnetic flux and friction matrix of PMSM, respectively. $J_m \in R^{n\times n}$ denote the diagonal inertia matrix of PMSM. τ, $\tau_{mL} \in R^n$ are the vector of electromagnetic torque and motor load torque, respectively. ΔJ_m denote the modeling error of PMSM.

Assumption 1. *The input current of PMSM is strictly three-phase symmetrical.*

Assumption 2. *The core saturation of PMSM can be ignored.*

Property 1. *The manipulator system and the drive system are connected by the transmission with the reduction ratio of $\mu > 0$, that is $q = \mu\theta$ and $\tau_r = \mu^{-1}\tau_{mL}$.*

2.3. The Overall Model of the Manipulator Driven by PMSM

According to (1)–(7), combined with the model of the manipulator system and the PMSM system, the dynamic model of the manipulator driven by PMSM with unknown load and modeling errors can expressed as

$$\bar{M}(q)\ddot{q}+\bar{C}(q,\dot{q})\dot{q}+\bar{G}(q)=\tau-\mu(\tau_f+\tau_L+\Delta E)-\mu(\Delta M(q)\ddot{q}+\Delta C(q,\dot{q})\dot{q}+\Delta G(q))-\mu^{-1}\Delta J_m\ddot{q} \quad (8)$$

where $\bar{M}(q)=\mu M(q)+\mu^{-1}J_m$, $\bar{C}(q,\dot{q})=\mu C(q,\dot{q})+\mu^{-1}R_m$, $\bar{G}(q)=\mu G(q)$

Considering the input saturation of the drive motor in the project, the actual input of PMSM electromagnetic torque $\tau_s(t) = [\tau_{s1}(t), \cdots, \tau_{sn}(t)]^\mathrm{T}$ is defined as

$$\tau_{si}(t) = \begin{cases} \tau_{i\ max}, & \tau_i(t) \geq \tau_{i\ max} \\ \tau_i(t), & \tau_{i\ min} < \tau_i(t) < \tau_{i\ max} \\ \tau_{i\ min}, & \tau_i(t) \leq \tau_{i\ min} \end{cases} \tag{9}$$

where the subscript i denotes the i th element of the electromagnetic torque vector. $\tau_{i\ max}$, $\tau_{i\ min}$ are the upper and lower limit values of the electromagnetic torque input to the drive motor, respectively. The part of the control signal that cannot be executed by PMSM can be expressed as

$$\tau_{ni}(t) = \tau_i(t) - \tau_{si}(t) = \begin{cases} \tau_i(t) - \tau_{i\ max} & \tau_i(t) \geq \tau_{i\ max} \\ 0 & \tau_{i\ min} < \tau_i(t) < \tau_{i\ max} \\ \tau_i(t) - \tau_{i\ min} & \tau_i(t) \leq \tau_{i\ min} \end{cases} \tag{10}$$

Substituting (4) and (5), Equation (3) is rewritten as

$$\bar{M}(q)\ddot{q} + \bar{C}(q,\dot{q})\dot{q} + \bar{G}(q) = \tau_s - \mu\tau_f + f(\mu,\ \tau_n,\ \tau_L,\ q,\dot{q},\ddot{q}) \tag{11}$$

where $f(\mu,\ \tau_n,\ \tau_L,\ q,\dot{q},\ddot{q}) = \tau_n - \mu(\tau_L + \Delta E) - \mu(\Delta M(q)\ddot{q} + \Delta C(q,\dot{q})\dot{q} + \Delta G(q)) - \mu^{-1}\Delta J_m\ddot{q}$, represents unknown modeling error, load and unexecuted input. For the convenience of the following application we use f instead of $f(\mu,\ \tau_n,\ \tau_L,\ q,\dot{q},\ddot{q})$. Define the state vector as $x_1 = q$, $x_2 = \dot{q}$. The state equation of the manipulator system can be described as

$$\dot{x}_1 = x_2 \tag{12}$$
$$\dot{x}_2 = \bar{M}^{-1}(x_1)[-\bar{C}(x_1,x_2)x_2 - \bar{G}(x_1) + f + \tau_s - \mu\tau_f] \tag{13}$$
$$y = x_1 \tag{14}$$

In the actual servo tracking of the robot arm, it is necessary to ensure that the output of each joint is bounded and can normally track the desired position signal.

Assumption 3. *There exist time-varying output upper and lower bounds $y_{i\ max}(t)$ and $y_{i\ min}(t)$ ($i = 1, 2, \cdots, n$), such that $y_{i\ min}(t) \leq y_i(t) \leq y_{i\ max}(t)$, $\forall t > 0$.*

Assumption 4. *There are functions $y_{di\ min}(t)$ and $y_{di\ max}(t)$ that satisfy the inequality $y_{i\ min}(t) \leq y_{di\ min}(t)$ and $y_{di\ max}(t) \leq y_{i\ max}(t)$, $\forall t > 0$, so that the desired position satisfies $y_{di\ min}(t) \leq y_{di}(t) \leq y_{di\ max}(t)$.*

Lemma 1 ([34]). *For any $|\varsigma| < 1$, the inequality $\log\frac{1}{1-\varsigma^2} < \frac{\varsigma^2}{1-\varsigma^2}$ is satisfied.*

3. Design of Controller

In this section, the structure and approximation process of the adaptive RBF neural network are described. Then, the smooth-switching for backstepping gain method is designed. Finally, the BLF-ANBG controller is designed.

3.1. Design of Adaptive Neural Network Approximation

To estimate modeling error, external load torque and unexecuted input, an adaptive RBF neural network strategy composed of an input layer, middle layer and output layer is designed. The structure of the adaptive RBF neural network is shown in Figure 1.

The middle layer is composed of five neurons, the output of each neuron is

$$h_k = \exp\left(-\frac{\|z-c_k\|^2}{b_k^2}\right) \quad (k = 1, \cdots, 5) \tag{15}$$

where $z = [e_1, \dot{e}_1]^{\mathrm{T}}$ is the input vector. c_k, b_k represent the center point vector and width of the k th neuron, respectively. The output of the adaptive RBF neural network is

$$\hat{f}(\cdot) = \hat{w}^{\mathrm{T}} h(x) \tag{16}$$

where $h(x) = [h_1, \cdots, h_5]^{\mathrm{T}}$. $\hat{w}$ is the adaptive weight, and the adaptive law is described as

$$\dot{\hat{w}} = \psi h z^{\mathrm{T}} P B \tag{17}$$

where $\psi > 0$ is the constant gain. B represents the input matrix of the closed-loop system. P is the positive definite matrix, and there exists the matrix $Q \geq 0$ such that P satisfies the stability equation $PA + A^{\mathrm{T}}P = -Q$, where A is the state matrix of the closed-loop system and the equations are given in Section 3.3. The adaptive neural network proposed in this paper recalculates the adaptive weights through the position error, velocity error of the manipulator and the output of the hidden layer of the adaptive neural network in each iteration to achieve the training of the neural network.

Property 2. *Given a continuous function $f(\cdot)$, there is an ideal weight $\hat{w}^*$, and the adaptive neural network approximation error $\gamma = f(\cdot) - \hat{f}^*(\cdot)$ satisfies $\max\|\gamma\| \leq \gamma_0$. γ_0 is the upper bound of error, and satisfies $\gamma_0 \leq \varepsilon$, ε is a very small positive number.*

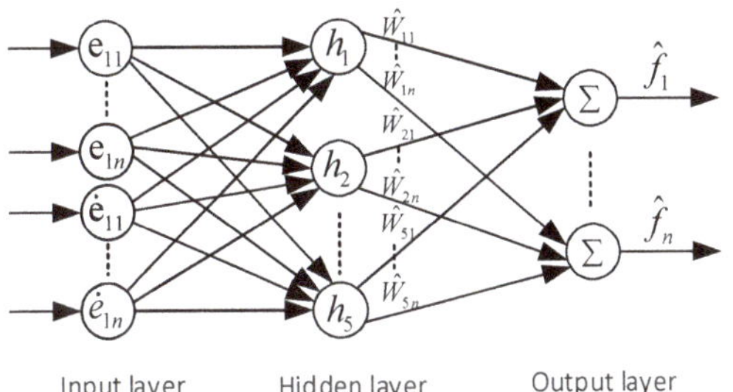

Figure 1. The structure of the Adaptive RBF Neural Network.

3.2. Design of Smooth-Switching for Backstepping Gain

The change in the control gain can produce the contradiction between the system dynamics characteristics and the steady-state characteristics. The larger the gain, the faster the response, the higher the accuracy, but the worse the stability [37,38]. The smaller the gain, the better the stability, but the response time will be longer and the accuracy will be lower. In addition, considering safety in engineering, the control gain is often within a certain range. In this part, a new smooth-switching for backstepping gain strategy is proposed, and the control gain is designed with the error surface and the change rate of the error surface, respectively.

3.2.1. The Variable Control Gain of the Error

Define the error surface as e_{ij}, where the subscript represents the j th component of the i th error surface. The variable control gain designed with error is expressed as

$$\Delta k_{ij}(e_{ij}) = \frac{2\alpha_{ij}}{\pi} \arctan\left(\frac{e_{ij}}{\beta_{ij}}\right)^2 + \delta_{ij} \tag{18}$$

where $\alpha_{ij} > 0$ is the magnification of the gain designed with the error. $\beta_{ij} > 0$ is the scale parameter of variable gain. δ_{ij} is a positive constant, ensure that $\Delta k_{ij}(e_{ij}) > 0, \forall e_{ij} \in R$.

3.2.2. The Variable Control Gain of the Change Rate of the Error

Take the derivative of the error surface e_{ij} to get $\dot{e}_{ij}$. The variable control gain designed with the change rate of error can be described as

$$\Delta k_{ij}(\dot{e}_{ij}) = \delta_{ij} - \frac{2\xi_{ij}}{\pi} \arctan\left(\frac{\dot{e}_{ij}}{\zeta_{ij}}\right)^2 \tag{19}$$

where $\xi_{ij} > 0$ is the magnification of the gain function designed based on the change rate of error. ζ_{ij} is the positive scale parameter. $\delta_{ij} > \xi_{ij}$, ensure that $\Delta k_{ij}(\dot{e}_{ij}) > 0$, $\forall \dot{e}_{ij} \in R$.

3.2.3. Design of Smooth-Switching for Backstepping Gain

To solve the smoothness of the gain switching transition process, a smooth-switching function based on the error surface is designed as

$$f(e_{ij}) = 1 - \exp\left(-\left(\frac{e_{ij}}{\sigma_{ij}}\right)^2\right) \tag{20}$$

where σ_{ij} is the positive scale constant. The smooth-switching function curve with different values of σ_{ij} are shown in Figure 2.

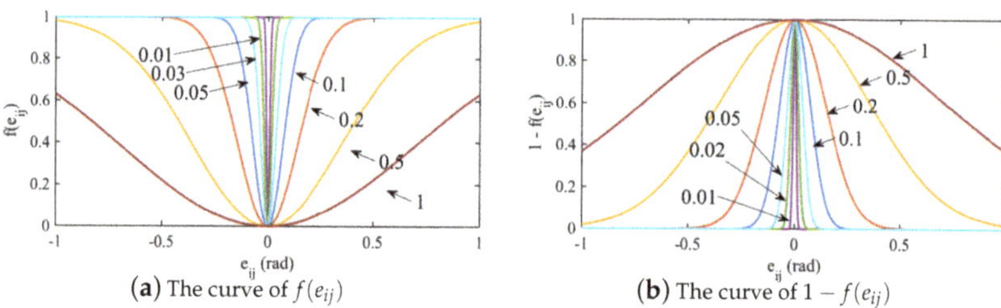

(a) The curve of $f(e_{ij})$ (b) The curve of $1 - f(e_{ij})$

Figure 2. The smooth-switching function curve.

According to (18)–(20), the smooth-switching for backstepping gain strategy is designed as

$$\Delta k_{ij} = f(e_{ij}) \cdot \Delta k_{ij}(e_{ij}) + [1 - f(e_{ij})] \cdot \Delta k_{ij}(\dot{e}_{ij}) \tag{21}$$

When the error is large, the control gain designed based on the error is mainly used. The greater the error, the greater the control gain, which ensures the rapid convergence of the system. When the error is small, the control gain designed by using the change rate of error is mainly used. The larger the change rate of error surface, the smaller the control gain. Thus, the jitter phenomenon caused by the large control gain of the manipulator system in the steady-state is reduced. When the error is in the middle range, the two gain functions transition in the form of smooth switching.

3.3. Design of BLF-ANBG Controller

To consider the time-varying asymmetric output limitation, the time-varying asymmetric BLF and adaptive neural network are used to design the smooth-switching for the backstepping gain controller. The block diagram of the manipulator system based on the BLF-ANBG controller is shown in Figure 3, and the design process is as follows

Step (1) Define the first tracking error surface vector as $e_1 = x_1 - x_d = [e_{11}, \cdots, e_{1n}]^{\mathrm{T}}$, x_d is the desired tracking position. The time-varying barrier of output is defined as

$$y_{ai}(t) = x_{di}(t) - y_{i\min}(t) \tag{22}$$

$$y_{bi}(t) = y_{i\max}(t) - x_{di}(t) \tag{23}$$

where $i = 1, \cdots, n$, which means the ith component of the vector. Define the time-varying asymmetric output constraints BLF as

$$V_1 = \sum_{i=1}^{n} V_{1i} = \sum_{i=1}^{n} \left[\frac{\lambda(e_{1i})}{2} \log \frac{1}{1 - \varsigma_{i+}^2(t)} + \frac{1 - \lambda(e_{1i})}{2} \log \frac{1}{1 - \varsigma_{i-}^2(t)} \right] \tag{24}$$

where

$$\lambda(e_{1i}) = \begin{cases} 1, & if\ e_{1i} > 0 \\ 0, & if\ e_{1i} \leq 0 \end{cases} \tag{25}$$

$$\varsigma_{i\ min}(t) = \frac{e_{1i}}{y_{ai}(t)},\ \varsigma_{i\ max}(t) = \frac{e_{1i}}{y_{bi}(t)} \tag{26}$$

Define the coordinate transformation as

$$\varsigma_i(t) = (1 - \lambda(e_{1i}))\varsigma_{i\ min}(t) + \lambda(e_{1i})\varsigma_{i\ max}(t) \tag{27}$$

Substituting (25) and (27) into (24), can obtain

$$V_1 = \sum_{i=1}^{n} V_{1i} = \sum_{i=1}^{n} \frac{1}{2}\log \frac{1}{1-\varsigma_i^2(t)},\ |\varsigma_i(t)| \leq 1 \tag{28}$$

It can be obtained from (28) that when $|\varsigma_i| \leq 1$, V_1 is positive definite. The differential of V_1 as

$$\dot{V}_1 = \sum_{i=1}^{n} \dot{V}_{1i} = \sum_{i=1}^{n} \left[\frac{\lambda(e_{1i})\varsigma_{i\ max}(t)}{y_{bi}(t)(1-\varsigma_{i\ max}^2(t))}\left(e_{2i} + x_{2di} - \dot{x}_{di} - e_{1i}\frac{\dot{y}_{bi}(t)}{y_{bi}(t)}\right) \right. \\ \left. + \frac{(1-\lambda(e_{1i}))\varsigma_{i\ min}(t)}{y_{ai}(t)(1-\varsigma_{i\ min}^2(t))}\left(e_{2i} + x_{2di} - \dot{x}_{di} - e_{1i}\frac{\dot{y}_{ai}(t)}{y_{ai}(t)}\right) \right] \tag{29}$$

where x_{2di} is the ith component of the virtual control vector x_{2d}, and the virtual control vector x_{2d} is designed by using the backstepping method as

$$x_{2d} = -(\Delta k_1 + \bar{k}_1(t))e_1 + \dot{x}_d \tag{30}$$

$$\Delta k_i = diag\{\Delta k_{i1}, \cdots, \Delta k_{in}\} \tag{31}$$

$$\bar{k}_1(t) = diag\{\bar{k}_{11}(t), \cdots, \bar{k}_{1n}(t)\} \tag{32}$$

where $\bar{k}_{1i}(t) = \sqrt{\left(\frac{\dot{y}_{ai}(t)}{y_{ai}(t)}\right)^2 + \left(\frac{\dot{y}_{bi}(t)}{y_{bi}(t)}\right)^2 + a}$, $a > 0$ is a constant to ensure that the derivative of x_{2di} is bounded. Substituting (26) and (27) and (30)–(32) into (29) can be rewritten as

$$\dot{V}_1 = \sum_{i=1}^{n} \dot{V}_{1i} \leq \sum_{i=1}^{n} \left[-\frac{\Delta k_{1i}\varsigma_i^2}{1-\varsigma_i^2} + \left(\frac{1-\lambda(e_{1i})}{y_{ai}^2(t)-e_{1i}^2} + \frac{\lambda(e_{1i})}{y_{bi}^2(t)-e_{1i}^2}\right)e_{1i}e_{2i} \right] \tag{33}$$

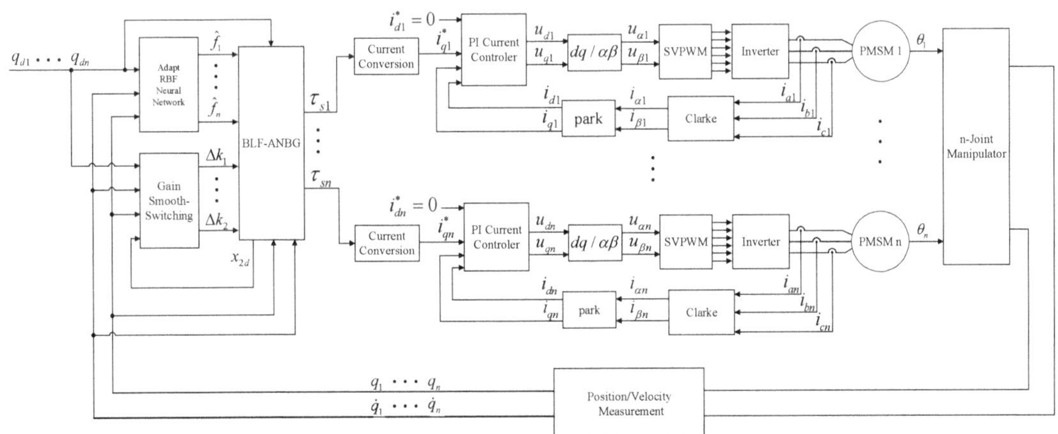

Figure 3. The Block Diagram of the Manipulator System Based on the BLF-ANBG Controller.

Step (2) The second error surface vector is defined as $e_2 = x_2 - x_{2d} = [e_{21}, \cdots, e_{2n}]^T$. Define the stability function as $V_2 = \sum\limits_{i=1}^{n} V_{2i} = \sum\limits_{i=1}^{n} \frac{1}{2} e_{2i}^2$. Substituting into (13), based on the backstepping control method, the actual control signal is selected as

$$\tau_s = \bar{M}(x_1)[-\Delta k_2 e_2 + \dot{x}_{2d} - \eta e_1] + \bar{C}(x_1, x_2) x_2 + \bar{G}(x_1) + \mu \tau_f - \hat{f} \tag{34}$$

where $\eta = diag\left\{ \left(\frac{1-\lambda(e_{11})}{y_{a1}^2(t) - e_{11}^2} + \frac{\lambda(e_{11})}{y_{b1}^2(t) - e_{11}^2} \right), \cdots, \left(\frac{1-\lambda(e_{1n})}{y_{an}^2(t) - e_{1n}^2} + \frac{\lambda(e_{1n})}{y_{bn}^2(t) - e_{1n}^2} \right) \right\}$, and $\hat{f}$ is the adaptive RBF neural network estimate of f. Substituting (30), Equation (34) can be rewritten as

$$\tau_s = \bar{M}(x_1)\left[\ddot{x}_d - \left(\Delta k_1 + \Delta k_2 + \bar{k}_1(t)\right)\dot{e}_1 - \left[\Delta k_2\left(\Delta k_1 + \bar{k}_1(t)\right) + \eta\right]e_1\right]$$
$$+\bar{C}(x_1, x_2) x_2 + \bar{G}(x_1) + \mu \tau_f - \hat{f} \tag{35}$$

Substitute (35) into (11), subtracting $\bar{M}(x_1)\dot{x}_2 + \bar{C}(x_1, x_2) x_2 + \bar{G}(x_1)$ from both sides, can get

$$\ddot{e}_1 + \left(\Delta k_1 + \Delta k_2 + \bar{k}_1(t)\right)\dot{e}_1 + \left[\Delta k_2\left(\Delta k_1 + \bar{k}_1(t)\right) + \eta\right]e_1 = \bar{M}^{-1}(x_1)\left(f - \hat{f}\right)$$
$$= \bar{M}^{-1}(x_1)\left(f - \hat{f}^* + \hat{f}^* - \hat{f}\right) = \bar{M}^{-1}(x_1)\left(\gamma - \tilde{w}^T h\right) \tag{36}$$

where $\tilde{w} = \hat{w} - \hat{w}^*$. Let $z = [e_1, \dot{e}_1]^T$, the closed-loop system equation is

$$\dot{z} = Az + B\left[\bar{M}^{-1}\left(\gamma - \tilde{w}^T h\right)\right] \tag{37}$$

where $A = \begin{bmatrix} 0_n & I_n \\ -\Delta k_2\left(\Delta k_1 + \bar{k}_1(t)\right) - \eta & -\Delta k_1 - \Delta k_2 - \bar{k}_1(t) \end{bmatrix}$, $B = \begin{bmatrix} 0 \\ I_n \end{bmatrix}$. 0_n and I_n represent n dimensional zero matrix and identity matrix, respectively.

4. The Analysis of BLF-ANBG Controller

In this section, the stability of the BLF-ANBG strategy is proved. Then, the manipulator joint output constraints are verified as not violated. Finally, the tracking error convergence is proved.

4.1. Proof of BLF-ANBG Controller Stability

The Lyapunov function of the BLF- ANBG controller is defined as

$$V = V_1 + V_2 + \frac{1}{2} z^T P z + \frac{1}{2\psi} \|\tilde{w}\|^2 = \sum\limits_{i=1}^{n} \left[\frac{\lambda(e_{1i})}{2} \log \frac{1}{1 - \varsigma_{i+}^2(t)} \right.$$
$$\left. + \frac{1-\lambda(e_{1i})}{2} \log \frac{1}{1 - \varsigma_{i-}^2(t)} \right] + \sum\limits_{i=1}^{n} \frac{1}{2} e_{2i}^2 + \frac{1}{2} z^T P z + \frac{1}{2\psi} \|\tilde{w}\|^2 \tag{38}$$

where $\|\tilde{w}\| = tr(\tilde{w}^T \tilde{w})$ is the Frobenius norm of $\tilde{w}$, and also represents the trace of matrix $\tilde{w}^T \tilde{w}$. From (38), it can be known that V is positive definite. Taking the derivative of $\tilde{w}$ and substituting (17) into $\dot{\tilde{w}}$ can obtain

$$\dot{\tilde{w}} = \dot{\hat{w}} - \dot{\hat{w}}^* = \psi h z^T P B \tag{39}$$

Taking the derivative of (38) can be known that

$$\dot{V} \le \sum_{i=1}^{n}\left[-\frac{\Delta k_{1i}\varsigma_i^2}{1-\varsigma_i^2} + \left(\frac{1-\lambda(e_{1i})}{y_{ai}^2(t)-e_{1i}^2} + \frac{\lambda(e_{1i})}{y_{bi}^2(t)-e_{1i}^2}\right)e_{1i}e_{2i}\right]$$
$$+ \sum_{i=1}^{n}\left[-\Delta k_{1i}e_{2i}^2 - \left(\frac{1-\lambda(e_{1i})}{y_{ai}^2(t)-e_{1i}^2} + \frac{\lambda(e_{1i})}{y_{bi}^2(t)-e_{1i}^2}\right)e_{1i}e_{2i}\right]$$
$$+ \frac{1}{2}\left[z^{\mathrm{T}}P\dot{z} + \dot{z}^{\mathrm{T}}Pz\right] + tr(B^{\mathrm{T}}Pzh^{\mathrm{T}}\tilde{w})$$
$$= \sum_{i=1}^{n}\left[-\frac{\Delta k_{1i}\varsigma_i^2}{1-\varsigma_i^2} - \Delta k_{1i}e_{2i}^2\right] + \frac{1}{2}\left[z^{\mathrm{T}}P\dot{z} + \dot{z}^{\mathrm{T}}Pz\right] + tr(B^{\mathrm{T}}Pzh^{\mathrm{T}}\tilde{w}) \tag{40}$$

Substituting (37) into (40), it becomes

$$\dot{V} \le \sum_{i=1}^{n}\left[-\frac{\Delta k_{1i}\varsigma_i^2}{1-\varsigma_i^2} - \Delta k_{1i}e_{2i}^2\right] + \frac{1}{2}\left[z^{\mathrm{T}}PAz + z^{\mathrm{T}}PB\gamma - z^{\mathrm{T}}PB\tilde{w}^Th\right.$$
$$\left. +z^{\mathrm{T}}A^{\mathrm{T}}Pz + \gamma^{\mathrm{T}}B^{\mathrm{T}}Pz - h^{\mathrm{T}}\tilde{w}B^{\mathrm{T}}Pz\right] + tr(B^{\mathrm{T}}Pzh^{\mathrm{T}}\tilde{w}) \tag{41}$$

Noting that $h^{\mathrm{T}}\tilde{w}B^{\mathrm{T}}Pz = z^{\mathrm{T}}PB\tilde{w}^Th = tr(B^{\mathrm{T}}Pzh^{\mathrm{T}}\tilde{w})$ and $\gamma^{\mathrm{T}}B^{\mathrm{T}}Pz = z^{\mathrm{T}}PB\gamma$, (41) is rewritten as

$$\dot{V} \le \sum_{i=1}^{n}\left[-\frac{\Delta k_{1i}\varsigma_i^2}{1-\varsigma_i^2} - \Delta k_{1i}e_{2i}^2\right] - \frac{1}{2}z^{\mathrm{T}}Qz + \gamma^{\mathrm{T}}B^{\mathrm{T}}Pz \tag{42}$$

Combining **Property 2** to get

$$\dot{V} \le \sum_{i=1}^{n}\left[-\frac{\Delta k_{1i}\varsigma_i^2}{1-\varsigma_i^2} - \Delta k_{1i}e_{2i}^2\right] - \|z\|\left[\frac{1}{2}\lambda_{\min}(Q)\|z\| - \|\gamma_0\|\lambda_{\max}(P)\right] \tag{43}$$

where $\lambda_{\min}(\cdot)$ and $\lambda_{\max}(\cdot)$ represent the upper and lower bounds of the eigenvalues, respectively. From (43), it can be seen that all signals in the system are uniformly ultimately bounded (UUB) [44,45]. The system is semi-global and practically stable [46].

Remark 1. *To satisfy $\dot{V} \le 0$, choose appropriate controller parameters such that $\|z\| \ge \frac{2\|\gamma_0\|\lambda_{\max}(P)}{\lambda_{\min}(Q)}$. In this paper, the appropriate eigenvalues of Q are selected to satisfy the above conditions to make the system asymptotically converge to the small neighborhood of the origin. When the approximation error γ tends to 0, the system asymptotically converges to the origin.*

4.2. Proof That the Manipulator Joint Outputs Constraints Are Not Violated

From (27), it can be obtained that when $e_{1i} < 0$, which yields $\varsigma_i(t) = \varsigma_{i\,\min}(t)$, and because $|\varsigma_i| \le 1$ and $y_{ai}(t) > 0$, according to (26) can be known that $-1 < \varsigma_{i\,\min}(t) \le 0$, then $-y_{ai}(t) < e_{1i}(t) \le 0$. Similarly, when $e_{1i} \ge 0$, $0 \le e_{1i}(t) < y_{bi}(t)$ can be obtained. In conclusion, that

$$-y_{ai}(t) < e_{1i}(t) < y_{bi}(t), \; |\varsigma_i(t)| < 1 \tag{44}$$

Adding $x_{di}(t)$ to each term in the inequality, (45) can be rewritten as

$$-y_{ai}(t) + x_{di}(t) < e_{1i}(t) + x_{di}(t) < y_{bi}(t) + x_{di}(t), \; |\varsigma_i(t)| < 1 \tag{45}$$

So $y_{i\,\min}(t) < y(t) < y_{i\,\max}(t)$, each joint is within the given constraints.

4.3. Proof of Tracking Error Convergence

According to **Lemma 1** and (38), $V(t) \le V(0)e^{-\rho t}$, $t \ge 0$ can be obtained [34,47], where $\rho = \min\{2\Delta k_{1j}, 2\Delta k_{2j}\}, j = 1, \cdots, n$. It can be obtained that

$$\frac{1}{2}\log\frac{1}{1-\varsigma_i^2(t)} \le V(0)e^{-\rho t} \tag{46}$$

Hence,

$$\varsigma_i^2(t) \leq 1 - e^{-2V(0)e^{-\rho t}} \tag{47}$$

When $e_{1i} \leq 0$, $\varsigma_i(t) = \varsigma_{i\,\min}(t) = \frac{e_{1i}}{y_{ai}(t)}$, $-y_{ai}(t)\sqrt{1-e^{-2V(0)e^{-\rho t}}} \leq e_{1i} \leq 0$. When $e_{1i} > 0$, $\varsigma_i(t) = \varsigma_{i\,\max}(t) = \frac{e_{1i}}{y_{bi}(t)}$, $0 < e_{1i} \leq y_{bi}(t)\sqrt{1-e^{-2V(0)e^{-\rho t}}}$. Combining both cases, can conclude that

$$-y_{ai}(t)\sqrt{1-e^{-2V(0)e^{-\rho t}}} \leq e_{1i} \leq y_{bi}(t)\sqrt{1-e^{-2V(0)e^{-\rho t}}} \tag{48}$$

where

$$\lim_{t\to\infty}\left(-y_{ai}(t)\sqrt{1-e^{-2V(0)e^{-\rho t}}}\right) = 0 \tag{49}$$

$$\lim_{t\to\infty}\left(y_{bi}(t)\sqrt{1-e^{-2V(0)e^{-\rho t}}}\right) = 0 \tag{50}$$

From (49) and (50), the upper and lower bounds of e_{1i} converge to 0, so e_{1i} converges to 0.

5. Simulink Results and Analysis

In this section, the BLF-ANBG control strategy is applied to a 2-DOF manipulator for simulation experiments to verify the feasibility of the strategy. The simulation experiment is divided into three parts. First, to verify the effectiveness of the smooth-switching for the backstepping control method, it is compared with the two variable gains functions without the smooth-switching strategy. Second, the BLF-ANBG strategy is used to control the manipulator to track the unit-step and periodic signals without modeling error and unknown load, and compare with other commonly used control strategies. Finally, the BLF-ANBG strategy is compared with the commonly used strategies in the presence of modeling errors and unknown loads. The parameters of the manipulator system in all simulation experiments in this section are shown in Table 1.

Table 1. The parameters of the manipulator system.

Parameters	Values	Parameters	Values
Weight (m_1, m_2)	0.5 kg	Length (l_1, l_2)	1 m
R_f	$diag\{5,5\}$ N	R_s	$diag\{2.875, 2.875\}\,\Omega$
n_p	$diag\{4,4\}$	L_d, L_q	$diag\{0.0085, 0.0085\}$ H
Φ	$diag\{0.175, 0.175\}$ Wb	J_m	$diag\{0.0025, 0.0025\}\,\mathrm{kg}\cdot\mathrm{m}^2$
η	$diag\{0.01, 0.01\}$	R_m	$diag\{6, 6\}$ N
F_c	$diag\{3,3\}$ N	Sampling Period	0.0001 s

The dynamics model of the 2-DOF manipulator system can be described as (8), where the inertia matrix $M(q)$, Coriolis force matrix $C(q, \dot{q})$ and gravity matrix $G(q)$ can be defined as

$$M(q) = \begin{bmatrix} M_{11} & M_{12} \\ M_{21} & M_{22} \end{bmatrix},\ C(q, \dot{q}) = \begin{bmatrix} C_{11} & C_{12} \\ C_{21} & C_{22} \end{bmatrix},\ G(q) = \begin{bmatrix} G_1 \\ G_2 \end{bmatrix}$$

where

$$\begin{cases} M_{11} = m_1 l_1^2 + m_2 l_1^2 + m_2 l_2^2 + 2m_2 l_1 l_2 \cos q_2 \\ M_{12} = M_{21} = m_2 l_2^2 + m_2 l_1 l_2 \cos q_2 \\ M_{22} = m_2 l_2^2 \\ C_{11} = -m_2 l_1 l_2 \dot{q}_2 \sin q_2 \\ C_{12} = -m_2 l_1 l_2 \dot{q}_1 \sin q_2 - -m_2 l_1 l_2 \dot{q}_2 \sin q_2 \\ C_{21} = m_2 l_1 l_2 \dot{q}_1 \sin q_2 \\ C_{22} = 0 \\ G_1 = m_1 l_1 g \cos q_1 + m_2 l_1 g \cos q_1 + m_2 l_2 g \cos(q_1 + q_2) \\ G_2 = m_2 l_2 g \cos(q_1 + q_2) \end{cases} \tag{51}$$

The parameters of the BLF-ANBG controller are described in Table 2.

Table 2. The parameters of controller.

Parameters	Values	Parameters	Values	Parameters	Values
β_{11}, β_{12}	0.5	ζ_{11}, ζ_{12}	0.5	β_{21}, β_{22}	0.05
ζ_{21}, ζ_{22}	0.05	σ_{11}, σ_{12}	0.5	σ_{21}, σ_{22}	10
δ_{11}, δ_{12}	2000	δ_{21}, δ_{22}	270	α_{11}, α_{12}	1000
α_{21}, α_{22}	5	$\check{\zeta}_{11}, \check{\zeta}_{12}$	1800	$\check{\zeta}_{21}, \check{\zeta}_{22}$	250
ε	0.1	ψ	10	Q	$diag\{60, 60, 60, 60\}$

From Table 2, $0 < \beta_{ij} < 1$, $0 < \zeta_{ij} < 1$, the values of β_{ij} and ζ_{ij} are obtained by trial and error within an appropriate range. When β_{ij} and ζ_{ij} are larger, the dynamic response of the system is faster, and when β_{ij} and ζ_{ij} are smaller, the transition process is smoother. If α_{ij} is larger, the upper bound of the variable gain function of the error is larger, and the systems response is faster, which can be appropriately increased on the premise of ensuring the safety of the system. δ_{ij} is the lower and upper bounds of the two gain functions, and $\delta_{ij} > \check{\zeta}_{ij}$ ensures that the system is stable. Q can be appropriately increased to enhance the system convergence effect.

5.1. Simulation Comparison Results of Smooth-Switching for Backstepping Gain Strategy

In this part, the manipulator system uses the smooth-switching for backstepping gain strategy to track the unit-step signal, and compares it with two variable gain functions without the smooth-switching strategy. The initial position of each joint of the manipulator system is $q(0) = [0, 0]^T$, the execution saturation is ± 20 N·m, and the output constraints of each joint are $y_{i\min} = -0.3 + 0.1 \sin(0.8t + \frac{\pi}{3})$, $y_{i\max} = 1.3 + 0.2 \sin(0.5t + \frac{\pi}{2})$. The tracking and error curves for the unit-step signal are provided in Figure 4.

As can be seen from Figure 4, using only the gain function based on the error has a faster response speed and can reach the steady-state faster, but there will be overshoot, and large jitter will occur after reaching the steady state. Using only the gain function based on the change rate of error can improve steady-state performance but has a long response time. The smooth-switching for backstepping gain strategy combines the advantages of two gain functions and can have good dynamic and steady-state performances at the same time.

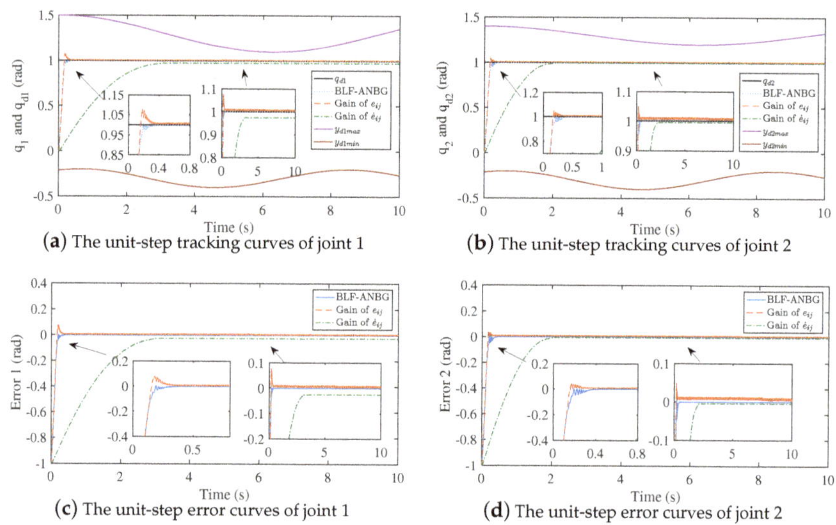

(a) The unit-step tracking curves of joint 1

(b) The unit-step tracking curves of joint 2

(c) The unit-step error curves of joint 1

(d) The unit-step error curves of joint 2

Figure 4. The unit-step tracking curves compared to two variable gain functions.

5.2. Without Modeling Error and Unknown Load

In this part, the manipulator tracks the unit-step signal and the expected periodic signal $q_d = \left[0.8sin(0.5t + \frac{\pi}{2}),\ 0.9sin(0.8t + \frac{\pi}{3})\right]^{\mathrm{T}}$ without modeling error and load, respectively. The input torque saturation of each joint is ± 20 N·m. The joint constraints that track the unit-step signal are $y_{i\,\min} = -0.3 + 0.1\sin(0.8t + \frac{\pi}{3})$, $y_{i\,\max} = 1.3 + 0.2\sin(0.5t + \frac{\pi}{2})$. The joint constraints that track the periodic signal signal are $y_{i\,\min} = -1.1 + 0.1\sin(0.6t + \frac{\pi}{3})$, $y_{i\,\max} = 1.2 + 0.2\sin(0.5t)$. The proposed BLF-ANBG strategy is compared with the conventional backstepping strategy under large gain and small gain. The tracking and error curves for the unit-step signal and periodic signal are as shown in Figures 5 and 6. The dynamic performance and steady-state performance of the tracking two signals are as shown in Table 3.

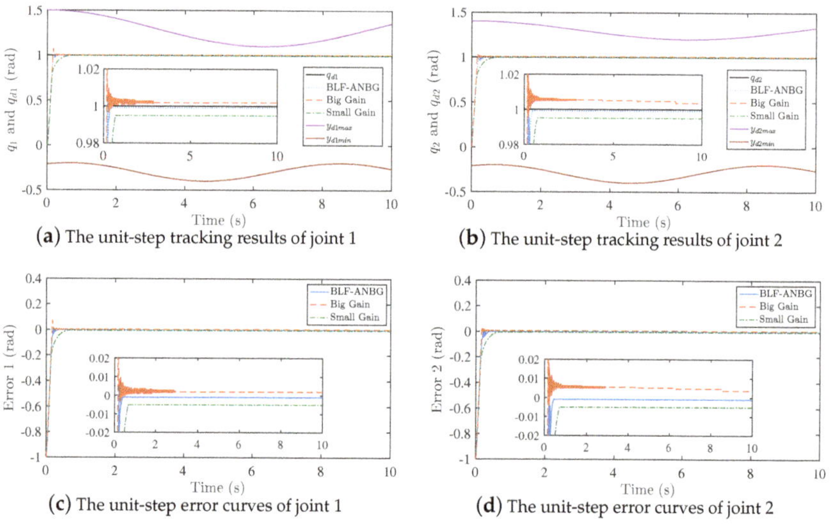

(a) The unit-step tracking results of joint 1

(b) The unit-step tracking results of joint 2

(c) The unit-step error curves of joint 1

(d) The unit-step error curves of joint 2

Figure 5. The unit-step tracking curves compared to traditional backstepping.

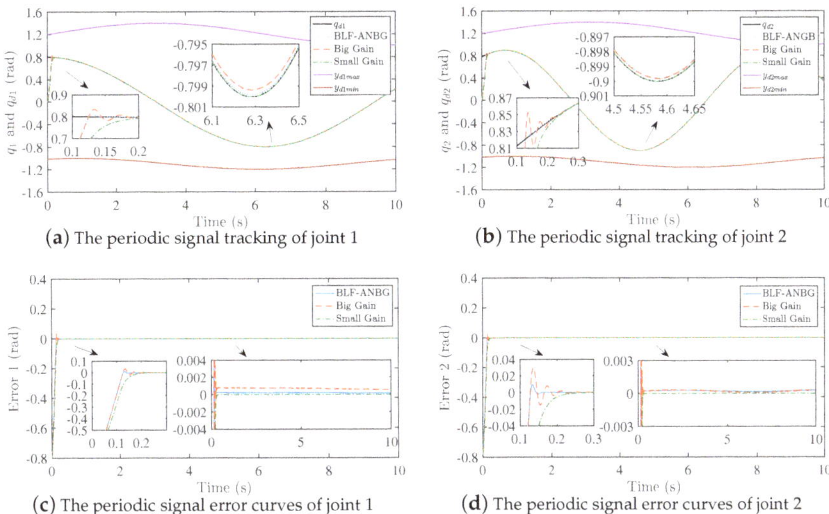

(**a**) The periodic signal tracking of joint 1

(**b**) The periodic signal tracking of joint 2

(**c**) The periodic signal error curves of joint 1

(**d**) The periodic signal error curves of joint 2

Figure 6. The periodic signal tracking curves compared to traditional backstepping.

Table 3. The rise time and error range compared to traditional backstepping.

Signal	Joint	Description	BLF-ANGB	Big Gain	Small Gain
Unit-step	joint 1	Rise Times (s)	0.1865	0.1525	0.6653
		Tracking Error (rad)	±0.0013	±0.0095	±0.0059
		Jitter Range (rad)	0.0005	0.032	0.0002
	joint 2	Rise Times (s)	0.1894	0.1503	0.6712
		Tracking Error (rad)	±0.0011	±0.0065	±0.0004
		Jitter Range (rad)	0.00054	0.0035	0.0001
Period	joint 1	Rise Times (s)	0.1255	0.1244	0.2312
		Tracking Error (rad)	±0.0013	±0.0342	±0.0011
		Jitter Range (rad)	0.0007	0.0022	0.0001
	joint 2	Rise Times (s)	0.1269	0.1255	0.2716
		Tracking Error (rad)	±0.0009	±0.0296	±0.0005
		Jitter Range (rad)	0.00011	0.00025	0.00001

From these figures and tables, it is easy to see that the joint outputs are all within the given constraints. The fixed gain of the traditional backstepping strategy without smooth-switching for gain can affect the performance of the system tracking. When the gain is large, although the system reaches the steady state in 0.1525 s and the accuracy is high, the system has a large jitter in the steady state, and the jitter range is 0.032 rad. When the gain is small, the jitter range is 0.0002 rad, which is smaller than that when the gain is large, but the system reaches the steady state at 0.6712 s and the accuracy is low. Therefore, the value of the backstepping gain will cause the contradiction between the dynamic characteristics and the steady-state characteristics of the system. When the BLF- ANBG strategy is applied, this contradiction can be effectively solved, the system can quickly reach the steady state and have better steady-state performance.

Secondly, the feasibility of this strategy is verified by comparing two commonly used SMC strategies and PD gravity compensation strategies. The tracking and error curves of the unit-step signal and periodic signal of the three strategies are presented in Figures 7 and 8, respectively. The tracking performances are shown in Table 4.

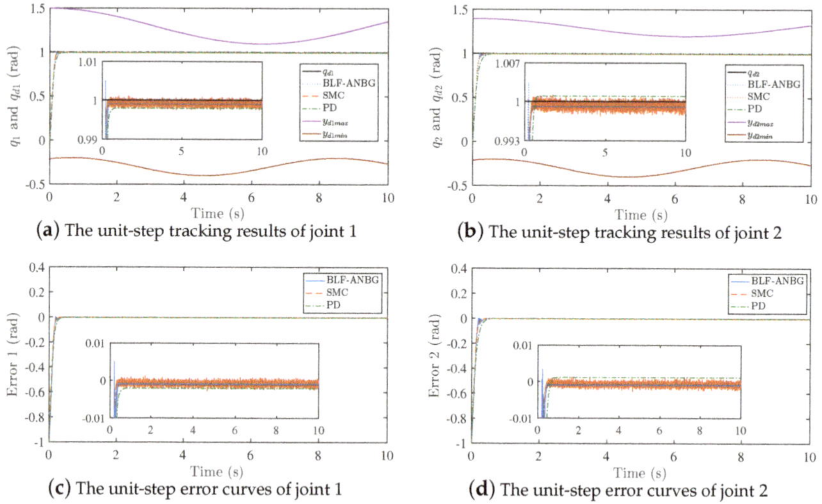

(a) The unit-step tracking results of joint 1

(b) The unit-step tracking results of joint 2

(c) The unit-step error curves of joint 1

(d) The unit-step error curves of joint 2

Figure 7. The unit-step signal tracking and error curves of the three strategies.

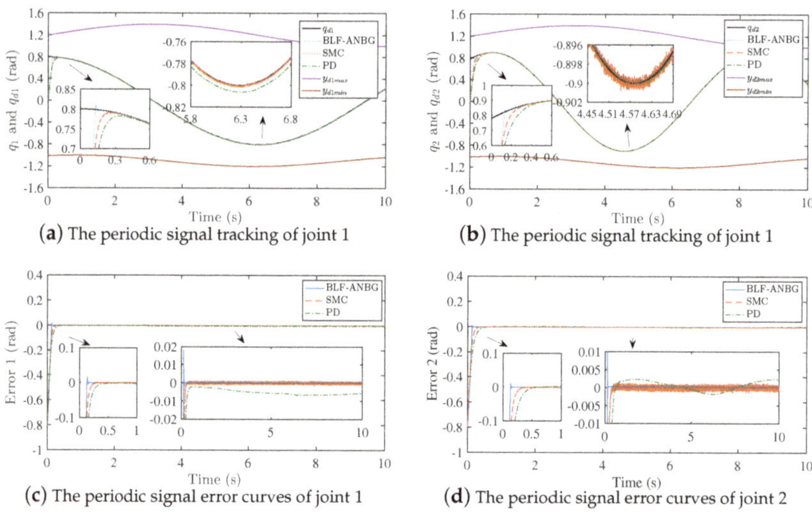

(a) The periodic signal tracking of joint 1

(b) The periodic signal tracking of joint 1

(c) The periodic signal error curves of joint 1

(d) The periodic signal error curves of joint 2

Figure 8. The expected periodic signal tracking and error curves of the three strategies.

Table 4. The rise time and error range without modeling error and load.

Signal	Joint	Description	BLF-ANGB	SMC	PD
Unit-step	joint 1	Rise Times (s)	0.1865	0.267	0.5041
		Tracking Error (rad)	± 0.0013	± 0.002	± 0.0021
	joint 2	Rise Times (s)	0.1894	0.296	0.6201
		Tracking Error (rad)	± 0.0011	± 0.002	± 0.0011
Period	joint 1	Rise Times (s)	0.1255	0.1956	0.3913
		Tracking Error (rad)	± 0.0013	± 0.0023	± 0.0067
	joint 2	Rise Times (s)	0.1269	0.2154	0.4541
		Tracking Error (rad)	± 0.0009	± 0.0021	± 0.0024

It can be seen from Figures 7 and 8 and Table 4 that the SMC can quickly reach the steady-state when tracking the unit-step signal and the periodic signal, but the chattering phenomenon occurs, and the jitter range of each joint is 0.004 rad and 0.0046 rad. The jitter of the PD gravity compensation strategy is obviously reduced, yet the tracking accuracy is poor, and the tracking error is ±0.0067 rad. Compared with these two strategies, BLF-ANBG has better tracking accuracy and stability, and can reach steady state faster.

5.3. With Modeling Error and Unknown Load

In this part, time-varying modeling errors and time-varying unknown loads are added to simulate the tracking effect of the manipulator in practical engineering applications. The controller parameters are the same as in Table 2. The parameters of modeling error and unknown load are described in Table 5.

Table 5. The parameters of modeling error and unknown load.

Parameters	Values	Parameters	Values
$\Delta M(q)$	$0.15M(q)$	$\Delta C(q,\dot{q})$	$0.25C(q,\dot{q})$
$\Delta G(q)$	$0.2M(q)$	ΔJ_m	$0.1J_m$
τ_L	$\begin{bmatrix} 10\sin(0.5t + \pi/2) \\ 10\sin(0.5t + \pi/2) \end{bmatrix}$	ΔE	$\begin{bmatrix} 0.1\cos 0.5t \\ 0.1\cos 0.5t \end{bmatrix}$

To verify the performance of the BLF-ANBG strategy, the BLF-GSS strategy, the SMC strategy and the PD gravity compensation strategy are used for comparison. The initial position of the joint is 0. The approximation curves of the adaptive neural network strategy are shown in Figure 9.

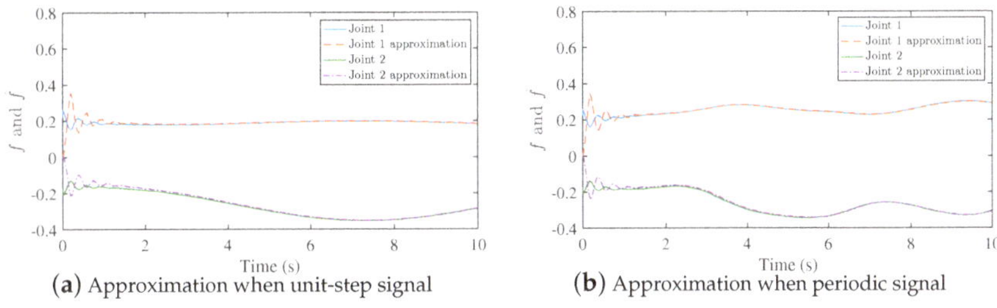

Figure 9. The approximation curves for the unit-step signal and expected periodic signal.

Figure 9 clearly shows that although the adaptive neural network has large approximation error in the initial stage, it can effectively approximate the modeling error and unknown load within 2 s. The tracking result and error curves of the unit-step signal are given in Figure 10, the tracking result and error curves of the expected periodic signal are provided in Figure 11. The controller performance is shown in Table 6.

According to Figures 10 and 11 and Table 6, it can be known that modeling errors and unknown loads have an impact on the tracking accuracy of the manipulator. In the absence of adaptive neural network compensation, the BLF-GSS strategy, the SMC strategy and the PD gravity compensation strategy all generate large tracking errors; the maximum tracking errors of the three strategies are ±0.0007 rad, ±0.0022 rad, ±0.0054 rad. The BLF-ANBG strategy can effectively reduce the modeling error and the influence of unknown loads, and can cause each joint to reach the steady-state within 0.156 s; the steady-state accuracy is also high. At the same time, the joints are all within the given constraints.

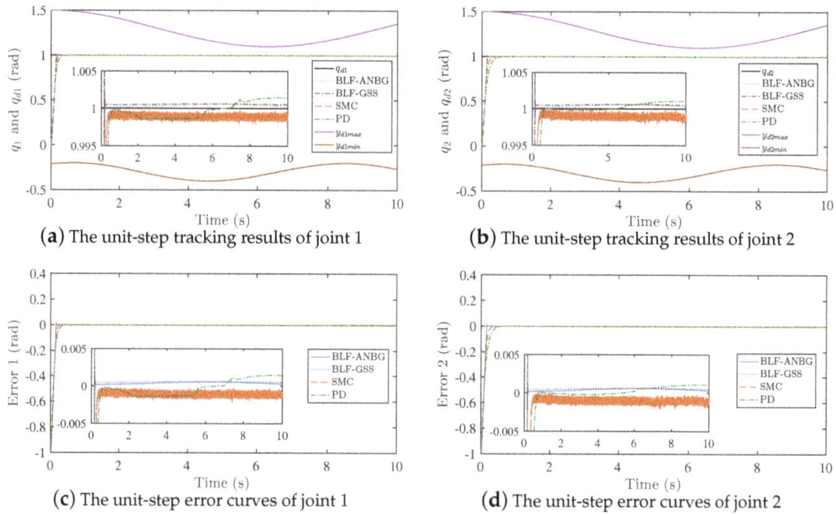

(**a**) The unit-step tracking results of joint 1

(**b**) The unit-step tracking results of joint 2

(**c**) The unit-step error curves of joint 1

(**d**) The unit-step error curves of joint 2

Figure 10. The unit-step signal tracking and error curves of the four strategies.

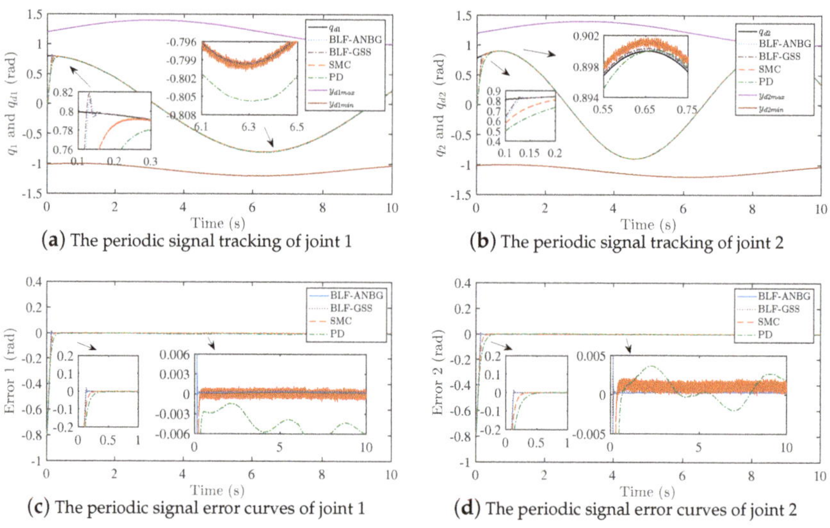

(**a**) The periodic signal tracking of joint 1

(**b**) The periodic signal tracking of joint 2

(**c**) The periodic signal error curves of joint 1

(**d**) The periodic signal error curves of joint 2

Figure 11. The expected periodic signal tracking and error curves of the four strategies.

Table 6. The tracking performance with modeling error and load.

Signal	Joint	Description	BLF-ANGB	BLF-GSS	SMC	PD
Unit-step	joint 1	Rise Times (s)	0.1562	0.1506	0.2817	0.4634
		Tracking Error (rad)	±0.0005	±0.0006	±0.0016	±0.0014
	joint 2	Rise Times (s)	0.1552	0.1496	0.2142	0.5652
		Tracking Error (rad)	±0.0005	±0.0007	±0.002	±0.0012
Period	joint 1	Rise Times (s)	0.1247	0.1235	0.2721	0.4254
		Tracking Error (rad)	±0.0002	±0.0004	±0.0017	±0.0054
	joint 2	Rise Times (s)	0.1223	0.1269	0.3256	0.4481
		Tracking Error (rad)	±0.0003	±0.0006	±0.0022	±0.0047

Summarizing the above results, the strategy proposed in this article can track the desired position signal quickly and stably in the presence of modeling errors and unknown loads. The contradiction between dynamic and steady-state caused by backstepping method gain is significantly improved.

6. Conclusions

This article presents a new tracking control strategy to solve the contradiction between the dynamic and steady-state characteristic caused by the control gain of the manipulator when there are uncertain interference terms. In this work, an overall model of the manipulator driven by PMSM with multiple constraints, modeling errors, and unknown loads is established. The proposed BLF-ANBG control strategy is applied to the 2-DOF manipulator. The simulation comparison shows that the proposed control strategy can effectively improve the contradiction between the dynamic and the steady-state performances of the system, and make the system take into account both excellent dynamic and steady-state characteristics. Additionally, the strategy effectively compensates for model errors, external disturbances and actuator saturation, while limiting the output of the system within the time-varying asymmetric constraint, which is more suitable for practical engineering needs. In actual engineering, the state of the speed and acceleration of the manipulator will also be constrained. In the future work, we will work hard to solve the problem of the full-state constraints of the manipulator system.

Author Contributions: Conceptualization, H.Y. (Haisheng Yu) and Q.Y.; methodology, Q.Y.; software, Q.Y.; validation, X.M., W.Y. and H.Y. (Huan Yang); formal analysis, Q.Y.; investigation, Q.Y.; resources, H.Y. (Haisheng Yu); data curation, X.M.; writing—original draft preparation, Q.Y.; writing—review and editing, X.M.; visualization, W.Y. and H.Y. (Huan Yang); supervision, H.Y. (Huan Yang); project administration, H.Y. (Haisheng Yu); funding acquisition, H.Y. (Haisheng Yu). All authors have read and agreed to the published version of the manuscript.

Funding: This work was supported by the National Natural Science Foundation of China with grant number 61573203 and the Shandong Province Natural Science Foundation with grant number ZR2021MF005.

Institutional Review Board Statement: Not applicable.

Informed Consent Statement: Not applicable.

Data Availability Statement: Not applicable.

Conflicts of Interest: The authors declare no conflict of interest.

Abbreviations

The following abbreviations are used in this manuscript:

BLF	Barrier Lyapunov Function
ANBG	smooth-switching for backstepping gain based on adaptive neural network
DOF	degree of freedom
PMSM	permanent magnet synchronous motor
PID	proportional integral derivative
SMC	sliding mode control
GSS	smooth-switching for backstepping gain
UUB	uniformly ultimately bounded

References

1. Saab, S.S.; Ghanem, P. A Multivariable Stochastic Tracking Controller for Robot Manipulators without Joint Velocities. *IEEE Trans. Autom. Control* **2018**, *63*, 2481–2495. [CrossRef]
2. Nguyen, V.C.; Le, P.N.; Kang, H.J. An Active Fault-Tolerant Control for Robotic Manipulators Using Adaptive Non-Singular Fast Terminal Sliding Mode Control and Disturbance Observer. *Actuators* **2021**, *10*, 332. [CrossRef]
3. Park, K.M.; Kim, J.; Park, J.; Park, F.C. Learning-Based Real-Time Detection of Robot Collisions without Joint Torque Sensors. *IEEE Robot. Autom. Lett.* **2021**, *6*, 103–110. [CrossRef]

4. Li, L.; Xiao, J.; Zhao, Y.; Liu, K.; Li, K. Robust position anti-interference control for PMSM servo system with uncertain disturbance. *CES Trans. Elecerical Mach. Syst.* **2020**, *4*, 10. [CrossRef]

5. Yu, Y.; Cong, L.; Tian, X.; Mi, Z.; Li, Y.; Fan, Z.; Fan, H. A stator current vector orientation based multi-objective integrative suppressions of flexible load vibration and torque ripple for PMSM considering electrical loss. *CES Trans. Elecerical Mach. Syst.* **2020**, *4*, 161–171. [CrossRef]

6. Hong, D.K.; Hwang, W.; Lee, J.Y.; Woo, B.C. Design, Analysis, and Experimental Validation of a Permanent Magnet Synchronous Motor for Articulated Robot Applications. *IEEE Trans. Magn.* **2018**, *54*, 1–4. [CrossRef]

7. Wen, S.; Qin, G.; Zhang, B.; Lam, H.; Zhao, Y.; Wang, H. The study of model predictive control algorithm based on the force/position control scheme of the 5-DOF redundant actuation parallel robot. *Robot. Auton. Syst.* **2016**, *79*, 12–25. [CrossRef]

8. Pradhan, S.K.; Subudhi, B. Position control of a flexible manipulator using a new nonlinear self-tuning PID controller. *IEEE/CAA J. Autom. Sin.* **2020**, *7*, 136–149. [CrossRef]

9. Shojaei, K.; Kazemy, A.; Chatraei, A. An Observer-Based Neural Adaptive PID^2 Controller for Robot Manipulators Including Motor Dynamics with a Prescribed Performance. *IEEE/ASME Trans. Mechatron.* **2021**, *26*, 1689–1699. [CrossRef]

10. Kim, M.J.; Chung, W.K. Disturbance-Observer-Based PD Control of Flexible Joint Robots for Asymptotic Convergence. *IEEE Trans. Robot.* **2015**, *31*, 1508–1516. [CrossRef]

11. Feng, Y.; Yu, X.; Man, Z. Non-singular terminal sliding mode control of rigid manipulators. *Automatica* **2002**, *38*, 2159–2167. [CrossRef]

12. Yeh, Y.L. Output Feedback Tracking Sliding Mode Control for Systems with State- and Input-Dependent Disturbances. *Actuators* **2021**, *10*, 117. [CrossRef]

13. Huang, A.C.; Chen, Y.C. Adaptive sliding control for single-link flexible-joint robot with mismatched uncertainties. *IEEE Trans. Control Syst. Technol.* **2004**, *12*, 770–775. [CrossRef]

14. Buondonno, G.; De Luca, A. Efficient Computation of Inverse Dynamics and Feedback Linearization for VSA-Based Robots. *IEEE Robot. Autom. Lett.* **2016**, *1*, 908–915. [CrossRef]

15. Meng, X.; Yu, H.; Zhang, J.; Xu, T.; Wu, H.; Yan, K. Disturbance Observer-Based Feedback Linearization Control for a Quadruple-Tank Liquid Level System. *ISA Trans.* **2021**, *122*, 146–162. [CrossRef]

16. Bagheri, M.; Karafyllis, I.; Naseradinmousavi, P.; Krstic, M. Adaptive control of a two-link robot using batch least-square identifier. *IEEE/CAA J. Autom. Sin.* **2021**, *8*, 86–93. [CrossRef]

17. Yoo, B.K.; Ham, W.C. Adaptive control of robot manipulator using fuzzy compensator. *IEEE Trans. Fuzzy Syst.* **2000**, *8*, 186–199.

18. Wang, H. Adaptive Control of Robot Manipulators with Uncertain Kinematics and Dynamics. *IEEE Trans. Autom. Control* **2017**, *62*, 948–954. [CrossRef]

19. Kanellakopoulos, I.; Kokotovic, P.; Morse, A. Systematic design of adaptive controllers for feedback linearizable systems. *IEEE Trans. Autom. Control* **1991**, *36*, 1241–1253. [CrossRef]

20. Cheng, X.; Zhang, Y.; Liu, H.; Wollherr, D.; Buss, M. Adaptive neural backstepping control for flexible-joint robot manipulator with bounded torque inputs. *Neurocomputing* **2021**, *458*, 70–86. [CrossRef]

21. Farrell, J.A.; Polycarpou, M.; Sharma, M.; Dong, W. Command Filtered Backstepping. *IEEE Trans. Autom. Control* **2009**, *54*, 1391–1395. [CrossRef]

22. Chang, W.; Li, Y.; Tong, S. Adaptive Fuzzy Backstepping Tracking Control for Flexible Robotic Manipulator. *IEEE/CAA J. Autom. Sin.* **2021**, *8*, 1923–1930. [CrossRef]

23. Yang, C.; Jiang, Y.; Na, J.; Li, Z.; Cheng, L.; Su, C.Y. Finite-Time Convergence Adaptive Fuzzy Control for Dual-Arm Robot with Unknown Kinematics and Dynamics. *IEEE Trans. Fuzzy Syst.* **2019**, *27*, 574–588. [CrossRef]

24. Ling, S.; Wang, H.; Liu, P.X. Adaptive Fuzzy Tracking Control of Flexible-Joint Robots Based on Command Filtering. *IEEE Trans. Ind. Electron.* **2020**, *67*, 4046–4055. [CrossRef]

25. Wai, R.J.; Yang, Z.W. Adaptive Fuzzy Neural Network Control Design via a T–S Fuzzy Model for a Robot Manipulator Including Actuator Dynamics. *IEEE Trans. Syst. Man Cybern. Part B* **2008**, *38*, 1326–1346.

26. Liu, Q.; Li, D.; Ge, S.S.; Ji, R.; Ouyang, Z.; Tee, K.P. Adaptive bias RBF neural network control for a robotic manipulator. *Neurocomputing* **2021**, *447*, 213–223. [CrossRef]

27. Narayanan, V.; Jagannathan, S.; Ramkumar, K. Event-Sampled Output Feedback Control of Robot Manipulators Using Neural Networks. *IEEE Trans. Neural Netw. Learn. Syst.* **2019**, *30*, 1651–1658. [CrossRef]

28. Zhang, Z.; Li, Z.; Zhang, Y.; Luo, Y.; Li, Y. Neural-Dynamic-Method-Based Dual-Arm CMG Scheme with Time-Varying Constraints Applied to Humanoid Robots. *IEEE Trans. Neural Netw. Learn. Syst.* **2015**, *26*, 3251–3262. [CrossRef]

29. Singh, A.P.; Deb, D.; Agarwal, H. On selection of improved fractional model and control of different systems with experimental validation. *Commun. Nonlinear Sci. Numer. Simul.* **2019**, *79*, 104902. [CrossRef]

30. Singh, A.P.; Deb, D.; Agarwal, H.; Bingi, K.; Ozana, S. Modeling and Control of Robotic Manipulators: A Fractional Calculus Point of View. *Arab. J. Sci. Eng.* **2021**, *46*, 9541–9552. [CrossRef]

31. Singh, A.P.; Deb, D.; Agrawal, H.; Balas, V.E. Fractional Modeling of Robotic Systems. In *Fractional Modeling and Controller Design of Robotic Manipulators: With Hardware Validation*; Springer International Publishing: Berlin/Heidelberg, Germany, 2021; pp. 19–43.

32. Meng, X.; Yu, H.; Zhang, J.; Yan, K. Optimized control strategy based on EPCH and DBMP algorithms for quadruple-tank liquid level system. *J. Process. Control* **2022**, *110*, 121–132. [CrossRef]

33. Liu, A.; Yu, H. Smooth-Switching Control of Robot-Based Permanent-Magnet Synchronous Motors via Port-Controlled Hamiltonian and Feedback Linearization. *Energies* **2020**, *13*, 5731. [CrossRef]
34. Tee, K.P.; Ren, B.; Ge, S.S. Control of nonlinear systems with time-varying output constraints. *Automatica* **2011**, *47*, 2511–2516. [CrossRef]
35. Liu, A.; Li, H. Stabilization of Delayed Boolean Control Networks with State Constraints: A Barrier Lyapunov Function Method. *IEEE Trans. Circ. Syst. II Express Briefs* **2021**, *68*, 2553–2557. [CrossRef]
36. Fuentes-Aguilar, R.Q.; Chairez, I. Adaptive Tracking Control of State Constraint Systems Based on Differential Neural Networks: A Barrier Lyapunov Function Approach. *IEEE Trans. Neural Netw. Learn. Syst.* **2020**, *31*, 5390–5401. [CrossRef]
37. Yoo, S.J.; Park, J.B.; Choi, Y.H. Adaptive Output Feedback Control of Flexible-Joint Robots Using Neural Networks: Dynamic Surface Design Approach. *IEEE Trans. Neural Netw.* **2008**, *19*, 1712–1726.
38. Cheng, X.; Liu, H.; Lu, W. Chattering-Suppressed Sliding Mode Control for Flexible-Joint Robot Manipulators. *Actuators* **2021**, *10*, 288. [CrossRef]
39. Yueneng, Y.; Ye, Y. Backstepping sliding mode control for uncertain strict-feedback nonlinear systems using neural-network-based adaptive gain scheduling. *J. Syst. Eng. Electron.* **2018**, *29*, 580–586.
40. Zhou, Q.; Zhao, S.; Li, H.; Lu, R.; Wu, C. Adaptive Neural Network Tracking Control for Robotic Manipulators with Dead Zone. *IEEE Trans. Neural Netw. Learn. Syst.* **2019**, *30*, 3611–3620. [CrossRef]
41. Li, B.; Lin, H.; Xing, H. Adaptive adjustment of iterative learning control gain matrix in Harsh noise environment. *J. Syst. Eng. Electron.* **2013**, *24*, 128–134. [CrossRef]
42. Li, H.; Liu, Q.; Feng, G.; Zhang, X. Leader–follower consensus of nonlinear time-delay multiagent systems: A time-varying gain approach. *Automatica* **2021**, *126*, 109444. [CrossRef]
43. Fromion, V.; Monaco, S.; Normand-Cyrot, D. Asymptotic properties of incrementally stable systems. *IEEE Trans. Autom. Control* **1996**, *41*, 721–723. [CrossRef]
44. Feng, G. A compensating scheme for robot tracking based on neural networks. *Robot. Auton. Syst.* **1995**, *15*, 199–206. [CrossRef]
45. Wang, M.; Huang, L.; Yang, C. NN-Based Adaptive Tracking Control of Discrete-Time Nonlinear Systems with Actuator Saturation and Event-Triggering Protocol. *IEEE Trans. Syst. Man Cybern. Syst.* **2021**, *51*, 7613–7621. [CrossRef]
46. Yu, Z.; Yang, Y.; Li, S.; Sun, J. Observer-Based Adaptive Finite-Time Quantized Tracking Control of Nonstrict-Feedback Nonlinear Systems With Asymmetric Actuator Saturation. *IEEE Trans. Syst. Man Cybern. Syst.* **2020**, *50*, 4545–4556. [CrossRef]
47. Li, G.; Yu, J.; Chen, X. Adaptive Fuzzy Neural Network Command Filtered Impedance Control of Constrained Robotic Manipulators with Disturbance Observer. *IEEE Trans. Neural Netw. Learn. Syst.* **2021**, 1–10. [CrossRef]

Article

Structural Design and Experiments of a Dynamically Balanced Inverted Four-Bar Linkage as Manipulator Arm for High Acceleration Applications

Matthijs J. J. Zomerdijk and Volkert van der Wijk *

Mechatronic System Design, Department of Precision and Microsystems Engineering, Faculty of Mechanical, Maritime and Materials Engineering, Delft University of Technology, Mekelweg 2, 2628 CD Delft, The Netherlands; mjjzomerdijk@outlook.com
* Correspondence: v.vanderwijk@tudelft.nl

Abstract: Industrial robotic manipulators in pick-and-place applications require short settling times to achieve high productivity. The fluctuating reaction forces and moments on the base of a dynamically unbalanced manipulator, however, cause base vibrations, leading to increased settling times. These base vibrations can be eliminated with dynamic balancing, which is achieved, in general, with the addition of counter-masses and counter-inertias. Adding these elements, however, comes at the cost of increased moving mass and inertia, resulting in lower natural frequencies and again higher settling times. For a minimal settling time it is therefore essential that a balanced mechanism has high natural frequencies with an optimal mass distribution. A dynamically balanced inverted four-bar linkage architecture is therefore favoured over architectures which depend on counter-masses and counter-rotating flywheels. The goal of this paper is to present and experimentally verify a structural design of a manipulator arm with high natural frequencies that is based on a dynamically balanced inverted four-bar linkage. The dynamical properties and the robustness to manufacturing tolerances are both verified with simulations and experiments. Experiments for 5.2 G tip accelerations show, when fully balanced, a reduction of 99.3% in reaction forces and 97.8% in reaction moments as compared to the unbalanced mechanism. The manipulator reached 21 G tip accelerations and a first natural frequency of 212 Hz was measured, which is significantly high and more than adequate for implementation in high acceleration applications.

Keywords: shaking force and shaking moment balancing; low settling time; natural frequencies; inverted four-bar mechanism; robotic manipulators

Citation: Zomerdijk, M.J.J.; van der Wijk, V. Structural Design and Experiments of a Dynamically Balanced Inverted Four-Bar Linkage as Manipulator Arm for High Acceleration Applications. *Actuators* **2022**, *11*, 131. https://doi.org/10.3390/act11050131

Academic Editors: Marco Carricato and Edoardo Idà

Received: 10 January 2022
Accepted: 14 April 2022
Published: 5 May 2022

Publisher's Note: MDPI stays neutral with regard to jurisdictional claims in published maps and institutional affiliations.

1. Introduction

Staying competitive in industry requires to reduce production costs and increase production rates. For robotic manipulators in pick-and-place applications in the semiconductor and packaging industry, for example, this requires the minimization of the cycle times. Reducing cycle times, in general, requires higher speeds and accelerations, which on the other hand may worsen vibration phenomena and lead to increased settling times and a reduced precision. The settling time is defined as the time to reach and stay within a certain error band of the final position after a motion is initiated. A longer settling time means that additional waiting time is added to the cycle for vibrations to die out. Multiple design approaches exist to achieve optimal settling time in multiple degree-of-freedom (DoF) manipulators [1–3]. These approaches focus on the manipulator solely and assume the base to be rigid. In reality, however, the base is elastic and therefore vibrations in the base can also affect the precision and the settling time significantly [4], not only of itself but also of surrounding machinery.

Base vibrations can be minimized with dynamic balancing, by which its source, the fluctuating reaction forces and moments exerted by the manipulator on its base, are eliminated. In fact, dynamic balancing leads to a dynamic decoupling between the mechanism and its base, for which there is no need for force frames and vibration isolation [5]. A manipulator or mechanism is considered dynamically balanced when both the sum of linear momentum and the sum of angular momentum stay constant during motions. A disadvantage of dynamic balance solutions often is that significant mass and inertia needs to be added to the mechanism, which degenerates the dynamical properties and the natural frequencies [6,7] and increases the settling time. It is therefore key to find designs which are dynamically balanced with low additional mass and inertia and have significantly high natural frequencies in order to achieve low settling times.

The literature on dynamic balancing is predominantly theoretical and the majority of the published experiments does not take the dynamic properties into account [8–10]. The experimentally verified high speed dynamically balanced manipulators which can be found are the DUAL-V and Hummingbird manipulator. The DUAL-V manipulator [11] relies on actuation redundancy and is based on a duplicate pantograph architecture. Accelerations over 10 G were reached during experiments with a 17 cm motion distance. The Hummingbird [12] is a force-balanced manipulator with a separately controlled reaction wheel for moment balancing, an approach which is known as active balancing. Accelerations up to 50 G were achieved for 5 mm motion distance and the first natural frequency of the mechanism showed 1.3 kHz. The active moment balancing in the Hummingbird resulted in a 90% reduction of the reaction moment, which is relatively low due to non-ideal actuators and limited control performance. For a much larger manipulator comparable to the Hummingbird, accelerations up to 10 G were reported with 25 cm motion distance; however, no further experimental information was published [13,14].

The goal of this paper is to present and to experimentally verify a structural design of a manipulator arm with high natural frequencies that is based on a dynamically balanced inverted four-bar linkage. The dynamically balanced inverted four-bar linkage was found by Ricard and Gosselin in 2000 [15] and can be considered a 1-DoF inherently dynamically balanced mechanism as it needs—besides a specific mass distribution—no additional counter rotating inertia (counter-inertia) for moment balance. If one link of the four-bar linkage is regarded as the manipulator arm, with the other elements solely for balancing, then the design can be used as a building block in the design of multi-DoF dynamically balanced manipulators [16].

In Section 2 the dynamically balanced inverted four-bar linkage is introduced. Section 3 presents the design of the manipulator arm with the investigation of the natural frequencies. In Section 4 the robustness to manufacturing tolerances of the designed mechanism is evaluated. The experimental setup and the experimental results are presented in Sections 5 and 6, respectively, with a discussion in Section 7.

2. Mechanism of the Balanced Manipulator Arm

2.1. Architecture of the Balanced Manipulator Arm

Achieving dynamic balance with counter rotating flywheel based architectures, in general, requires a rotary transmission such as a pair of gears or a belt drive. Such a transmission adds mass, compliance, and backlash to the mechanism, which affect the dynamic properties and the settling times negatively. In contrast, inherently balanced architectures, such as the dynamically balanced inverted four-bar linkage, only require a specific distribution of masses and inertias of its links without the need for additional elements. The balanced inverted four-bar linkage is a closed loop mechanism which has the highest concentration of its masses and inertias located close to the base pivots or on the coupler link, which is beneficial for high natural frequencies as will become clear from this paper.

The single rotatable link shown in Figure 1a can be regarded as the starting point of the structural design and can be considered as the manipulator arm that has been integrated into the inverted four-bar linkage to achieve the dynamically balanced mechanism in Figure 1b. The unbalanced rotatable link has a length L, a tip mass m_p, and a tip inertia I_p. In both Figure 1a,b each link has a mass m_i and an inertia I_i with the link center of mass (CoM) at a distance r_i as illustrated, with i the link number. The angle of the link with respect to the horizontal axis is denoted with θ_1. The full length of the first link in Figure 1b is the sum of length L of the rotatable link and extension L_1 and the length of the third link is the sum of L_3 and r_3. The distance between the two base pivots is denoted with L_4, while the coupler link has a length L_2. The mechanism is dynamically balanced when Equations (1) and (2) [15] are satisfied and the CoM of each link is located on the centre line of the link as shown in Figure 1b. As compared to counter rotating flywheel based architectures, in the balanced inverted four-bar linkage link 3 can be regarded a counterrotation with coupler link 2 as the transmission that is stiff and without backlash, a significant advantage of this solution.

Force balance conditions:

$$r_2 = L_2(1 - \frac{m_1^* r_1^*}{m_2 L_1}), \quad r_3 = \frac{m_2 r_2 L_3}{m_3 L_2} \tag{1}$$

Moment balance conditions:

$$L_1 = L_3, \quad L_2 = L_4,$$
$$I_2 = m_2(L_2 r_2 - r_2^2) - I_{c1},$$
$$I_3 = -m_3(L_3 r_3 + r_3^2) + I_{c1}, \tag{2}$$
$$I_{c1} = I_1^* + m_1(r_1^{*2} + r_1^* L_1)$$

With:

$$m_1^* = m_1 + m_p,$$
$$r_1^* = \frac{m_1 r_1 + m_p L}{m_1 + m_p}, \tag{3}$$
$$I_1^* = I_1 + I_p + m_1 r_1^2 + m_p L^2$$

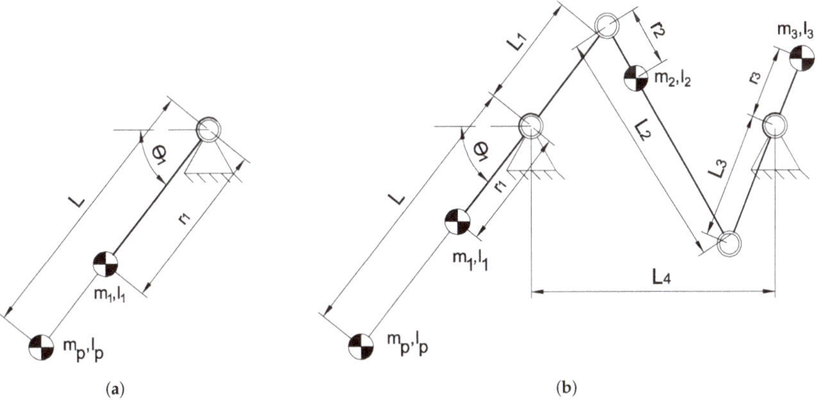

Figure 1. (**a**) Unbalanced rotatable link; (**b**) Rotatable link as manipulator arm within a dynamically balanced inverted four-bar linkage (adapted from [15]).

2.2. Mode Shapes of the Balanced Manipulator Arm

The natural frequencies indicate at which frequencies elements of a mechanism are prone to vibrate, whereas mode shapes show how the mechanism deforms at each natural frequency. The natural frequencies of a design can be increased by adding stiffness or by reducing the mass for which the mode shapes show the locations with significant deformation requiring additional stiffness. The first three distinctive in-plane mode shapes of the mechanism in Figure 1 are found as shown in Figure 2. For the modal analysis the actuation is located in the base pivot of the first link and is assumed rigid as indicated in red.

The three mode-shapes in Figure 2 were determined with the help of Spacar [17], which is a numerical flexible multi-body software package based on the Euler Bernoulli beam theory. The first mode shown in Figure 2a is predominantly caused by the tip mass m_p in combination with the relatively low lateral stiffness of the first link. The second mode shape in Figure 2b is mainly the result of the elasticity in link 1 and 3, combined with the as balancing mass acting mass of link 3. The third mode shape in Figure 2c is mainly depending on the elasticity of link 2 combined with its mass distribution. To increase the stiffness of these links, without changing the material, link lengths can be reduced or the second moment of area could be increased. The cross-section and link length therefore need to be taken into account in order to optimize natural frequencies. It needs to be noted that although reducing the length of the links increases stiffness, it comes at the cost of additional mass to satisfy the balance equations. The natural frequency associated with the second in-plane mode shape depends also on the position of the mechanism. When θ_1 is larger than 90 deg the excitation of the second mode shape loads link 1 and link 3 predominantly longitudinally. For smaller angles these links are loaded predominantly transversely, for which they have a lower stiffness resulting in lower natural frequencies.

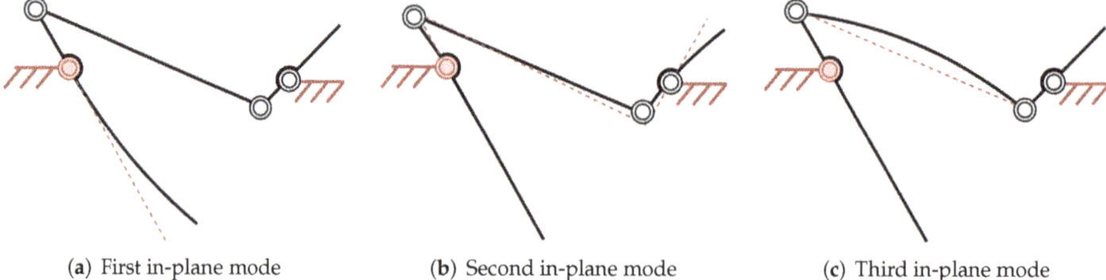

(**a**) First in-plane mode   (**b**) Second in-plane mode   (**c**) Third in-plane mode

Figure 2. In-plane mode shapes of the balanced inverted four-bar mechanism assuming an infinitely stiff actuator in the first base pivot, shown in red.

3. Structural Design

The structural design of the prototype dynamically balanced manipulator arm is presented in Figure 3. All revolute joints consist of plain bearings, except for the ETEL RTMBi140-100 direct drive motor which has two ball bearings. Since the balls in a ball bearing have an angular velocity different from the links and therefore a deviating angular momentum, they could affect the moment balance quality, which is not the case for plain bearings. The plain bearings were assembled with two bearing tensioners, one on each side of the link. These bearing tensioners are an additional set of plain bearings which are tensioned in order to remove play, therefore preventing another source of vibrations. Since the friction forces in the joints result solely in internal forces, they do not affect the overall balance quality.

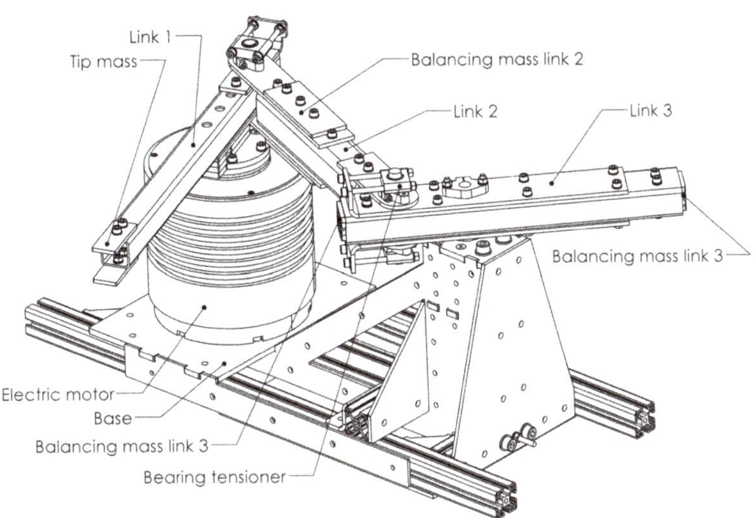

Figure 3. Final design of the dynamically balanced manipulator arm link 1 mounted on the electric motor and on the base.

3.1. Design Details

All links were made of stainless steel tubes of which the dimensions are listed in Table 1. Additional balancing masses m_2^* and twice $m_3^*/2$ for tuning the mass distribution and also the bearing mountings were made from stainless steel sheet metal. The axles of link 2, the coupler link, and the axle mounts were all made from aluminium. The axle of the base pivot of link 3 was made from steel for a higher strength.

Table 1. Dimensions of the tubes of the links.

	Link 1	Link 2	Link 3
Width (mm)	30	40	50
Height (mm)	30	40	30
Wall thickness (mm)	2	4	2

A top view of the balanced mechanism with its geometric design parameters is shown in Figure 4 and the inertial and geometric parameter values have been listed in Table 2. In this table link parameters m_i, r_i, and I_i include all masses and inertias of the link i and the tip mass, including the values for m_2^* and m_3^* and for the actuator inertia. The parameters related to the individual balancing masses m_2^* and m_3^* are denoted with an asterisk and the inertia of each link is taken at the link CoM. The admissible range of motion of the mechanism is from $\theta_1 = 60$ deg to $\theta_1 = 110$ deg.

Table 2. Parameter values of the balanced manipulator.

[mm]	[g]	[mm]	[kgm^2]
$L = 300$	$m_p = 112.12$	$r_p = 300.00$	$I_p = 0.0000377$
$L_1 = 70$	$m_1 = 2140.93$	$r_1 = 36.26$	$I_1 = 0.0295905$
$L_2 = 320$	$m_2 = 2139.95$	$r_2 = 154.16$	$I_2 = 0.0168973$
$L_3 = 70$	$m_3 = 2539.59$	$r_3 = 28.41$	$I_3 = 0.0307103$
$cL_4 = 320$	$m_2^* = 208.04$	$r_2^* = 124.00$	$I_2^* = 0.0001351$
	$m_3^* = 506.11$	$r_3^* = 46.00$	$I_3^* = 0.0124768$

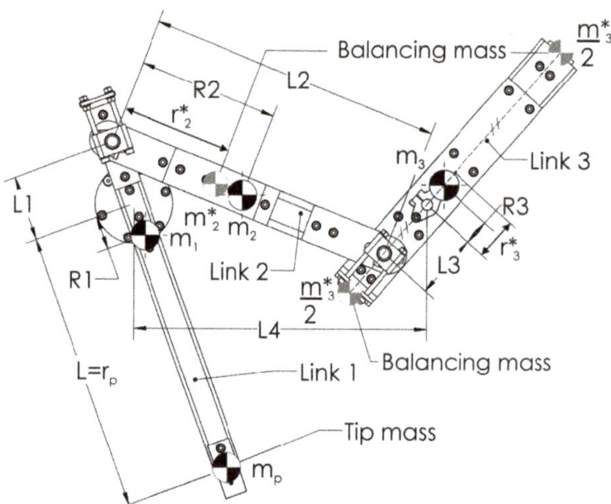

Figure 4. Top view of the dynamically balanced manipulator arm link 1 with the main parameters.

The values in Table 2 have been found by an iterative design process based on three steps: (1) Initial parameter optimization with Spacar, (2) detailed computer aided design (CAD), and (3) verification of natural frequencies.

The starting point in the design was the initial unbalanced mechanism in Figure 1a with a tip mass (m_p) of 112.12 g connected to the actuator with a link length $L = 0.3$ m. The initial lengths of L_1, L_2, L_3, and L_4 were determined before the detailed design by a genetic algorithm in combination with Spacar. Since incorporating the unbalanced manipulator arm in the balanced four-bar mechanism introduces additional natural frequencies, for optimal dynamic performance these frequencies need to be higher than the first natural frequency of the manipulator arm itself. Since for analysis the base pivot of link 1 is fixed, the first natural frequency of the balanced manipulator arm is equal to the unbalanced arm. The goal of this optimization therefore is to maximize the second natural frequency of the balanced mechanism. Constraints were considered for the cross-section of the links to keep the tubes rectangular and the lengths of the tubes were constrained to keep the design manufacturable. The mass and inertia of the balancing masses and of the joints were added to the model, while no constraints were applied to the dimensions of the balancing masses. The optimization resulted in an initial length for L_1, L_2, L_3, and L_4 of 78 mm, 310 mm, 78 mm, and 310 mm, respectively, for which the second natural frequency of the mechanism resulted in 565 Hz and is associated with the second in-plane mode shape as shown in Figure 2b. This frequency can be assumed to the theoretical optimum of the design.

The CAD design requires to take dimensional constraints into account. Therefore the final parameter values in Table 2 are a trade off between natural frequencies, balance conditions, dimensional constraints, and mass. By satisfying the balance conditions of the second link and the resulting dependency among mass, inertia, and CoM, it was required to shorten the length of L_1 to 70 mm.

In order to reduce the mass of the third link (m_3), it follows from the balance conditions that the inertia (I_3) of the link needs to be increased. Therefore link 3 includes the two balance masses $m_3^*/2$ which are located at the extremities of the link with their common CoM at a distance r_3^* from the base pivot. This construction reduces the total mass m_3 of the third link by 36% as compared to having a single balance mass. In addition, the second in-plane natural frequency increases by 5%.

3.2. Finite Elements Simulation

The natural frequencies of the CAD model were analysed with Comsol [18]. The analysis did not take the stiffness of the bolted connections into account in order to reduce the computational effort. The axles and the bearing tensioners were added in the simulation as point masses. To omit internal collisions of the links during frequency sweeps, it was chosen to perform the analysis around the centre of the admissible motion range at $\theta_1 = 88.2$ deg. The three identified mode shapes of Figure 2 are clearly recognizable in the simulation results shown in Figure 5.

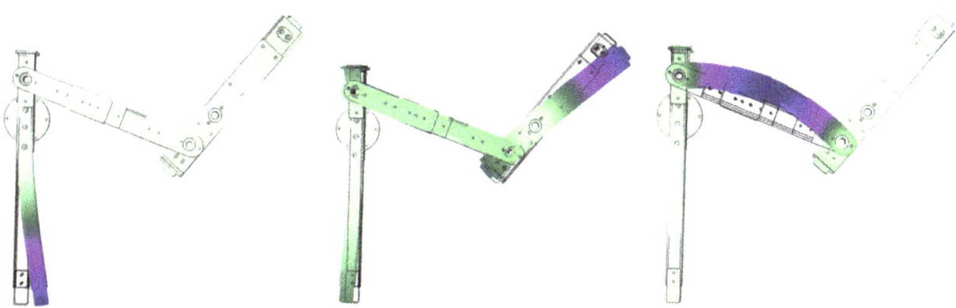

(**a**) First in-plane mode of 312 Hz

(**b**) Second in-plane mode of 353 Hz

(**c**) Third in-plane mode of 826 Hz

Figure 5. First three in-plane mode shapes with their natural frequency from finite element analysis at $\theta_1 = 88.2$ deg. With dark blue the largest displacements of a specific mode shape are shown. The displacements are strongly amplified.

The first natural frequency of 312 Hz in Figure 5a is located in the manipulator arm and therefore equal to the unbalanced rotatable arm in Figure 1a. Of the second and third natural frequencies in Figure 5b,c, respectively, the latter is significantly higher. The natural frequency of the second in-plane mode shape depends significantly on the stiffness along the length L_1. For a construction based on tubes, a high stiffness, however, is challenging to obtain.

In addition to the in-plane modes shapes, three out-of-plane modes were observed. The first out-of-plane mode shown in Figure 6a affects link 1 and is equal for both the unbalanced rotatable link in Figure 1a as for the same link in the balanced design in Figure 1b. The second mode shown in Figure 6b is a result of the balancing mass added at the end of the third link. The excitation of this mode is limited, since the fluctuating forces in this direction are insignificant during motion. In contrast, the third out-of-plane mode shape shown in Figure 6c is easily excited during motion, as this link then is loaded longitudinally. Still, the natural frequency associated with this mode shape is significantly higher than that of the second in-plane mode shape.

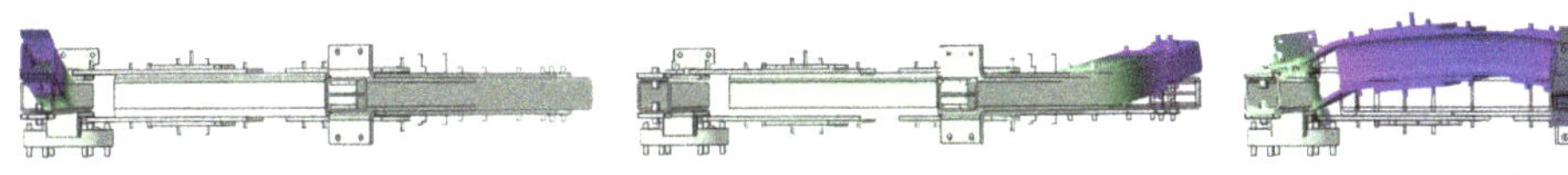

(**a**) First out-of-plane mode of 324 Hz

(**b**) Second out-of-plane mode of 696 Hz

(**c**) Third out-of-plane mode of 721 Hz

Figure 6. The first three out-of-plane mode shapes from finite element analysis at $\theta_1 = 88.2$ deg. With dark blue the largest displacements due to the excitation of a specific mode shape are shown. The displacements are strongly amplified.

By altering the angle θ_1 it was confirmed that the natural frequency associated with the second in-plane mode shape is significantly affected by the position of the mechanism. At an angle $\theta_1 = 60$ deg the frequency is significantly lower with 346 Hz, due to the increased transverse loading in link 1 when this mode is excited. Increasing angle θ to 115 degrees results in a natural frequency of 390 Hz, which then is due to an increased longitudinal loading. The natural frequencies of the other mode shapes in Figures 5 and 6 are marginally affected by the position of the mechanism.

3.3. Bearing Loads and Actuation Torques

The actuation torque of the dynamically balanced manipulator arm about the base pivot of link 1 is 2.2 times higher than that for the unbalanced rotatable link 1, which is due to the increased reduced inertia of the mechanism. The peak loads in the bearings of the base pivots are 4.2 times higher for the balanced manipulator arm.

4. Robustness and Manufacturing Tolerances

During manufacturing, errors in mass and dimensions occur due to manufacturing and material tolerances. To minimize mass errors all elements were weighed with an accuracy of 0.01 g and physical measurements were incorporated in the CAD model before calculation of the separate balancing masses. The influence on the balance quality of the remaining mass errors in counter-masses, dimensional tolerances in links, and tip mass deviations were studied by simulations in Simulink [19]. It was assumed that the mass distribution and radius of gyration were not affected by the mass errors, which makes mass and inertia errors linearly dependent. The balance quality is defined as a percentage where 100% corresponds to a perfectly dynamically balanced mechanism and 0% corresponds to the unbalanced single rotatable link 1. The parameter values of the unbalanced mechanism, including the actuator, are shown in Table 3. For this analysis the mechanism was assumed rigid and an ODE45 solver was used. The force balance quality was determined by the in-plane peak reaction force, calculated with $\max(\sqrt{F_x^2 + F_y^2})$. The moment balance quality was determined by $\max(M_z)$. The reference trajectory was based on the S-curve motion profile shown in Figure 7.

Table 3. Parameter values of the unbalanced rotatable manipulator including the inertia of the actuator.

[g]	[mm]	[kgm²]
$m_u = 1719.16$	$r_u = 64.75$	$I_u = 0.0222$

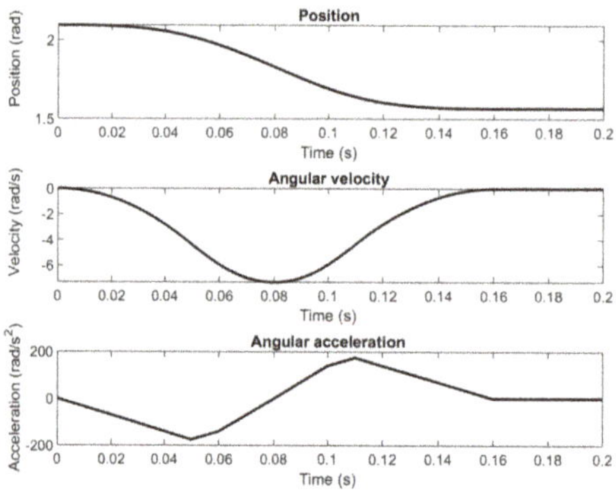

Figure 7. Reference trajectory based on the S-curve profile in which the mechanism is moved from $\theta_1 = 98.2$ deg to $\theta_1 = 68.2$ deg with a peak rotational acceleration of 174.5 rad/s².

4.1. Mass Errors

The influence of mass errors in the balance masses on both the force and moment balance quality is shown in Figure 8. It reveals that the performance is significantly more sensitive to mass errors in the third link than of mass errors in the second link. The impact of mass errors in the balance masses of the second link could still be further reduced by taking these errors into account during the calculation of the balance masses of the third link. This is possible since the force balance condition of the third link, Equation (1), can be addressed after link 2 has been completed, which was not done here.

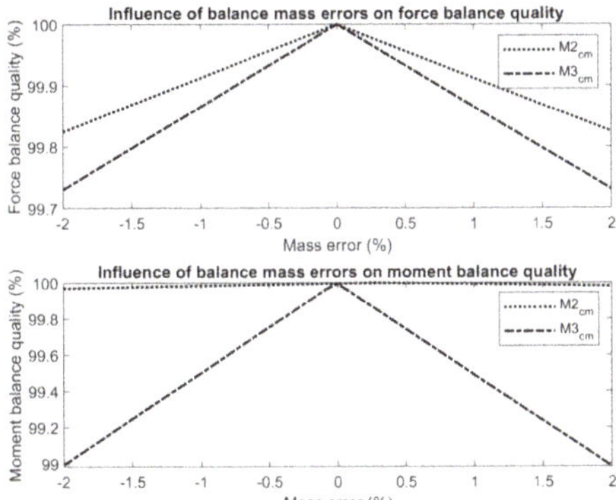

Figure 8. The influence of the mass errors in the balancing masses on the balance quality shows to be relatively small.

For moment balance the angular momentum of the complete mechanism must be constant [6], which depends on the inertia and the angular velocity of each link [20]. Due to the higher angular velocity and inertia of the third link as compared to the second link and therefore the larger contribution of link 3 to the moment balance, mass errors in the third link have a significantly larger impact on the moment balance quality.

The balance masses were produced by laser cutting where multiple test pieces averaged a raw mass error of 0.61%. With the help of post processing these errors could possibly be further reduced.

4.2. Geometry Errors

Figure 9 shows the influence of link length deviations on the balance quality, revealing that the force balance quality is most sensitive to deviations of L_1, while the moment balance quality is most sensitive to deviations of L_4. Deviations in length L_4 not only affect the moment balance, but also alter the torque arm between the two base joints, with significant impact on the moment balance quality. Sensitivity to deviations of L_2 and L_4 are significantly lower when those lengths remain equal to each other.

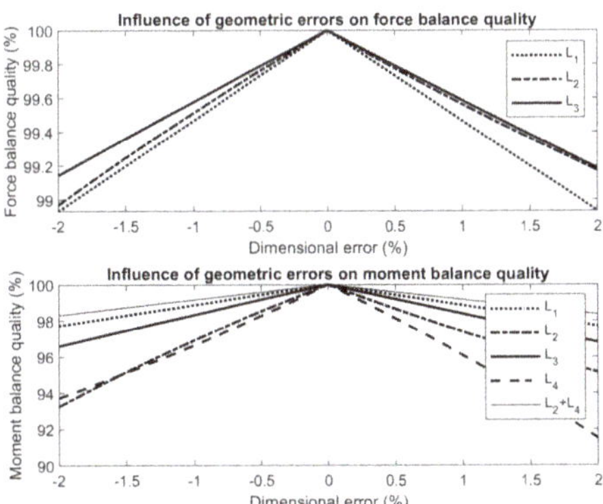

Figure 9. The influence of the geometric deviations on the balance quality shows that geometric error have a significantly larger impact on the moment balance quality than on the force balance quality.

The length deviations due to the production tolerances were estimated to be below 0.1 mm. By the tensioning devices removing play in the plain bearings, deviations of up to 0.05 mm in each bearing can occur. Therefore in total the dimensional deviations could be up to 0.2 mm. Within these tolerances it was calculated that the force balance quality remains above 99% and the moment balance quality remains above 98%.

4.3. Fluctuating Tip Mass

When the tip mass m_p would be regarded a payload, for example in pick-and-place applications, then this mass would vary significantly for each cycle. Figure 10 shows that when the tip mass is completely removed, the force balance quality has reduced to 75% while the moment balance quality has reduced to 65%. This means that deviation of the tip mass has a significant influence, which, however, remains relatively low for small deviations.

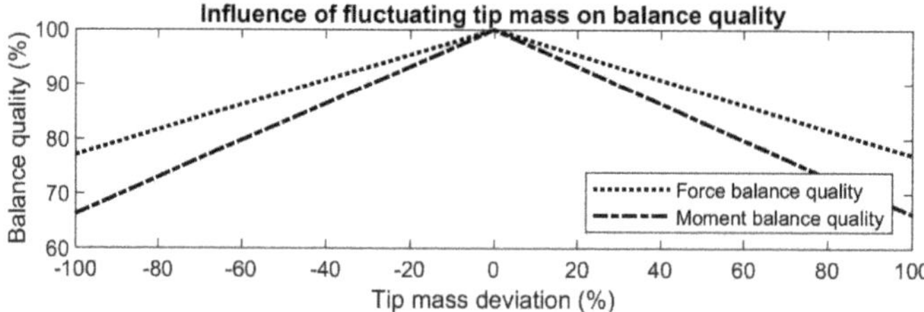

Figure 10. The influence of the tip mass on the balance quality shows to be significant.

5. Experimental Setup

A prototype of the dynamically balanced manipulator arm was fabricated in an experimental setup, which is shown in Figure 11. The base consists of aluminium extrusion profiles with stainless steel plates. To allow for small movements of the base for measurements within the horizontal plane, the base is suspended by four chains from an external aluminium frame. For measuring the reaction forces and the reaction moment within the horizontal plane, three single point loads-cells were installed as indicated with numerical balloons in Figure 12. Each of the load cells could measure a maximum force of 49 N with a precision of 0.05%. Load cell 1 measured the forces along the X-axis, while load cells 2 and 3 measured the forces along the Y-axis. From load cells 2 and 3 the in-plane shaking moment about the Z-axis was calculated by multiplying the measured force at load cell 3 with the distance between load cells 2 and 3. To minimize any parasitic forces in transverse directions, the connections between the load cells and the base were made of thin flexible rods of steel as shown in detail A in Figure 12. The rod connected to load cell 1 has a diameter of 1 mm, while the rods connected to load cells 2 and 3 have a diameter of 0.5 mm. Each load cell was electronically connected to a Penko SGM 720 load cell transducer, which sampled each load cell at a frequency of 1 kHz. In order to determine the balance quality and to compare the performance with the theoretical model, the reaction forces and the reaction moment were measured for the mechanism moving with the trajectory of Figure 7.

The manipulator arm was actuated by an ETEL RTMBi140-100 direct drive motor, which can deliver a peak torque of 131 Nm. The trajectory was achieved by feedback control only, in which the PID control was automatically tuned by the ETEL motion controller. The natural frequencies of the manipulator arm were determined with a frequency sweep conducted by the same ETEL motion controller. Videos of the manipulator in action are available as Supplementary Material.

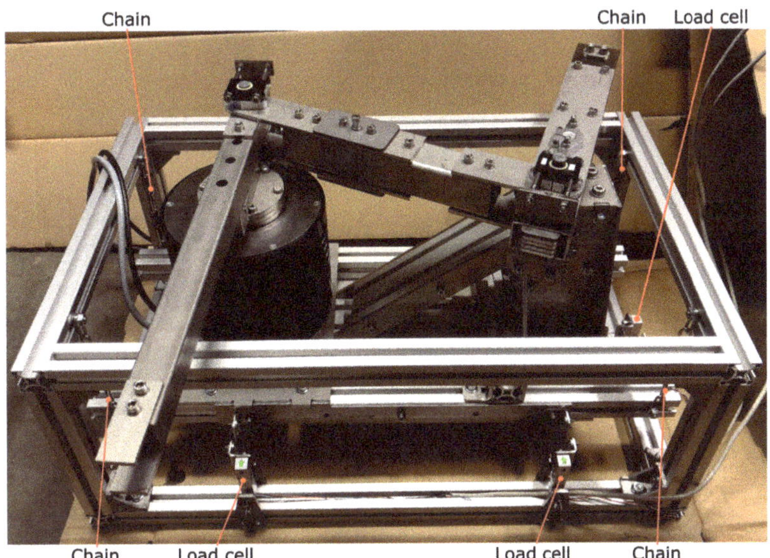

Figure 11. Experimental setup of the balanced manipulator arm with its base suspended by chains and mounted to the load cells for measurements of the in-plane shaking forces and shaking moment.

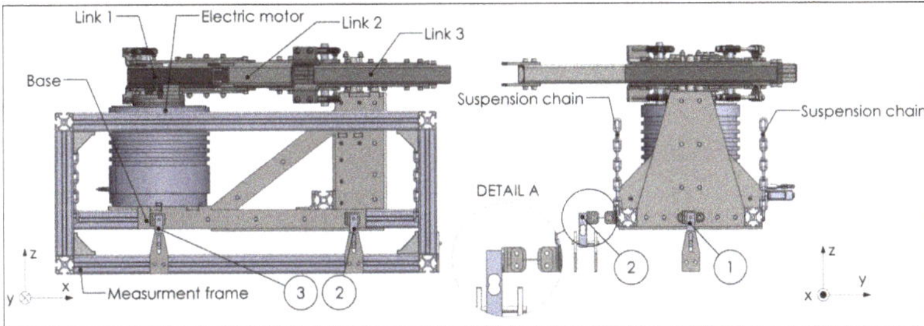

Figure 12. CAD model of the experimental setup with the manipulator base suspended by chains to allow for small in-plane motions of the base. The numbers denote the positions of the load cells. In the right figure the measurement frame is hidden. Detail A shows the connection of a thin rod between a load cell and the manipulator base.

6. Experiments and Experimental Results

Four experimental evaluations were performed with the experimental setup: (1) The balance quality when dynamically balanced, (2) the balance quality with reduced tip mass (i.e., when partly balanced), (3) identification of the natural frequencies, and (4) the maximal acceleration capabilities of the prototype. The balance quality of the mechanism was determined similarly as in Section 4. The simulated unbalanced rotatable link 1 was chosen as the reference to keep results between the simulations and the experiments comparable. The shaking forces and shaking moment exerted on the base by the simulated unbalanced manipulator arm for the trajectory in Figure 7 are shown in Figure 13. The reaction force in this figure is the magnitude of the reaction force vector in the horizontal plane. In these simulations the mass and inertia of all components of the real prototype were taken into account. The motion consisted of a rotation of 30 degrees within 160 ms. During this motion a peak rotational acceleration was reached of 174.5 rad/s^2, meaning that the point of the tip mass reached a transverse acceleration of 51.1 m/s^2 or 5.2 G.

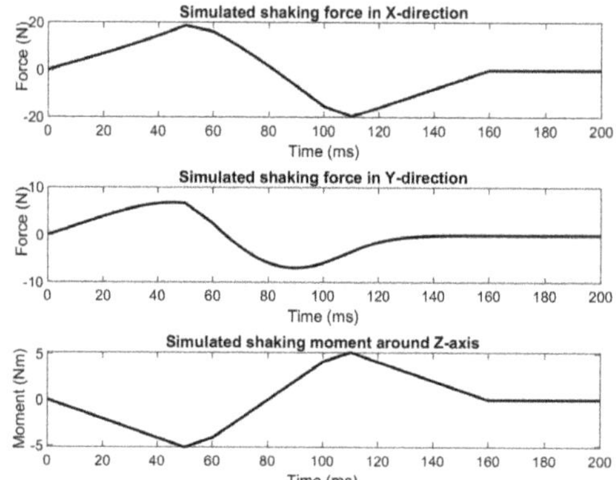

Figure 13. Simulated shaking forces and shaking moment of the prototype unbalanced manipulator arm for the motion in Figure 7. The peak shaking force is 25.54 N and the peak shaking moment is 5.16 Nm.

6.1. Balance Quality When Dynamically Balanced

Figure 14 shows the shaking forces and the shaking moment exerted to the base of the prototype from simulations (including the manufacturing errors) and from measurements with the experimental setup. Comparing the simulations with the measurements reveals that the measured shaking forces and shaking moment are significantly higher than from the simulations. The measured results are smoother than the results from simulation due to damping. Based on the measurements, the force balance and moment balance quality during motion results in 99.3% and 97.8%, respectively. The moment balance quality is slightly lower than expected, which could have been caused by a deviation in the reported motor inertia of the manufacturer and by the third link which turned out not being perfectly straight but slightly curved and without perfect rectangular cross-section, which was not taking into account in the simulations.

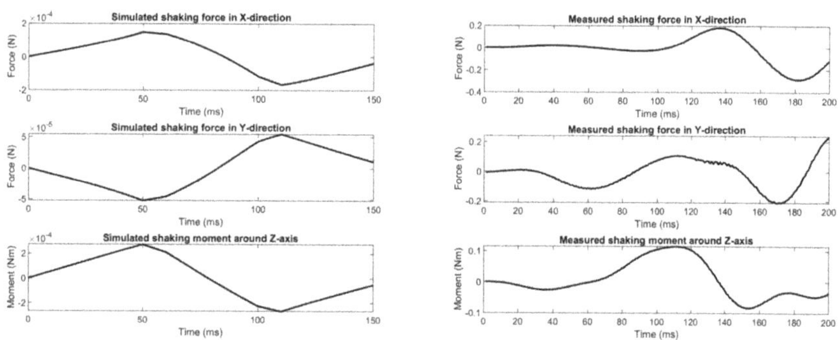

Figure 14. Comparison between simulated and measured shaking forces and shaking moment of the balanced mechanism, with a maximal tip acceleration of 5.2 G.

6.2. Balance Quality with Reduced Tip Mass

The balance quality for a reduced tip mass was measured with a 50% lower mass m_p of 56.06 g of which the results are shown Figure 15. The motion trajectory was the same as in the previous experiment, however with 90% lower accelerations and a peak tip acceleration of 0.52 G since for peak accelerations of 5.2 G the deformations of the load cells and the connecting rods due to the shaking forces and shaking moment caused unreliable measurements.

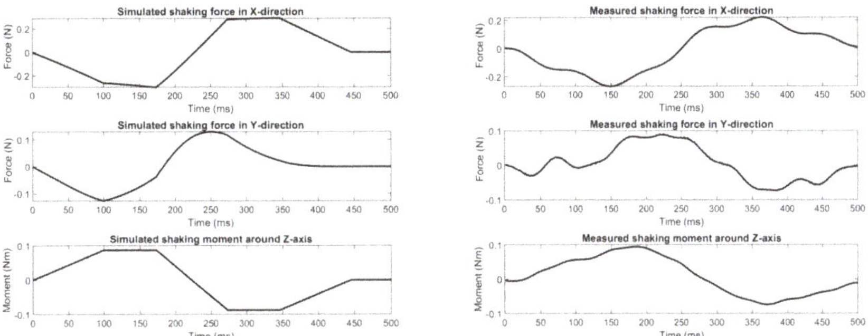

Figure 15. Comparison of the simulated and the measured shaking forces and shaking moment of the balanced manipulator arm with 50% of the tip mass with a maximal tip acceleration of 0.52 G.

It is observed in Figure 15 that the measurements are close to the simulated values. The shaking forces are slightly lower than simulated, which is likely due to a combination of the transverse stiffness of the connection rods and vibrations caused by the unbalance in the measurement setup. By comparing the measurements of Figure 15 with Figure 14, then it can be observed that the shaking forces and shaking moment have clearly increased. With half of the tip mass, the measured force and moment balance quality are 90.3% and 81.9%, respectively, which is higher than the predicted values of 85.3% and 83.12%, respectively.

6.3. Identification of the Natural Frequencies

A frequency sweep was conducted to determine the natural frequencies of the prototype in the experimental setup. The sweep was done in the position $\theta_1 = 88.2$ deg, which is approximately in the centre of the range of motion. Table 4 shows the results of the frequency sweep and of the FEA simulations. A bode plot of the frequency sweep can be found in Figure A1.

Table 4. Calculated and measured natural frequencies of the balanced manipulator arm for $\theta_1 = 88.2$ deg.

	f_1	f_2	f_3
FEA	312 Hz	721 Hz	826 Hz
Measurements	212 Hz	443 Hz	637 Hz

The results in Table 4 show that the first measured frequency is 32% lower than obtained from the FEA. The main reason for this deviation is caused by the relatively high elasticity of the base, which was assumed rigid in the FEA simulation. This is supported by earlier experiments that were carried out with a base design with fewer structural members, resulting in even more elasticity with a 55% lower first natural frequency that was measured. This means that the stiffness between the two base pivots have a significant influence on the natural frequencies of the mechanism. Moreover, the bolted connection between the actuator and first link, which was modelled rigidly in the simulations, is expected to have resulted in the lower measured frequencies.

6.4. Maximal Acceleration Capabilities of the Prototype

Finally, the maximal tip accelerations of the experimental setup were investigated, resulting in tip accelerations of over 21 G. Since the ETEL motion controller could not directly measure the angular accelerations during motion, they were derived from the measured angular velocity by central finite difference. The motor torque was calculated from the measured motor current with a motor torque constant of 4.264 Nm/A. Both the angular accelerations and the motor torque are shown for comparison with the simulations in Figure 16.

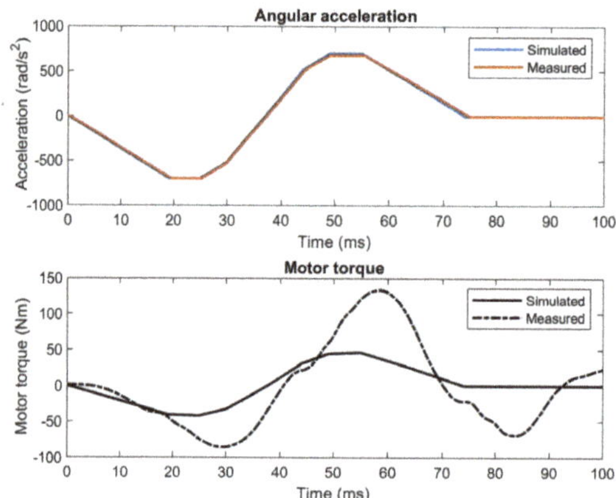

Figure 16. Comparison of the angular acceleration and the motor torque of the measurements and the simulations for a motion with a peak transverse tip acceleration of just over 21 G.

As can be seen in Figure 16, the measured angular acceleration is about equal to the simulation, confirming that the tip has successfully reached a transverse tip acceleration of over 21 G (21.35 G precisely). While the simulations predicted a peak actuation torque of 46.7 Nm, a peak of 133.2 Nm was measured during experiments. This is expected to be caused by the poor PID controller and also by joint friction, which were not considered in the simulations. This means that the prototype still has potential to move significantly faster after optimizing the PID-controller and after implementing feed-forward control. The actuation torque shows to not be instantly zero after the motion is finished, which is caused by the PID controller that is still controlling the damping to reach the final position.

Figure 17 shows the measured shaking forces and shaking moment for 21 G tip accelerations, which turn out to increase significantly after the motion was finished after 75 ms. This increase could have been caused by rotational motion about the z-axis of the base within the experimental setup as a result of a higher shaking moment. The high actuation torques could also have resulted in elastic deformations of the manipulator links for which the manipulator becomes slightly unbalanced, increasing the shaking moment. The measured values show a force balance quality and a moment balance quality of 97.2% and 96.9%, respectively. These values are slightly lower than the measured values for accelerations up to 5 G.

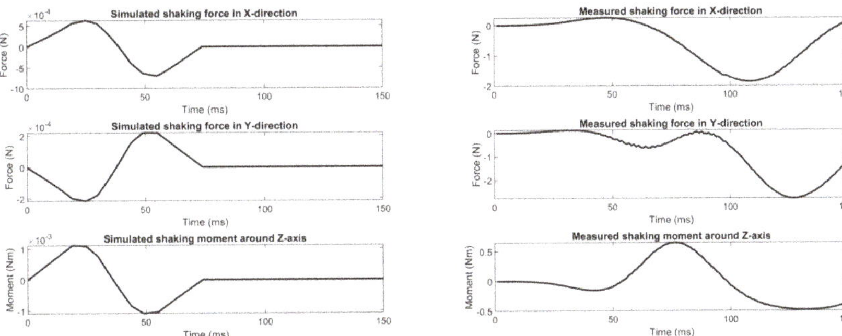

Figure 17. Comparison of the simulated and the measured shaking forces and shaking moment of the prototype for a maximal tip acceleration of just over 21 G.

7. Discussion

This paper presented the structural design and the experimental verification of a dynamically balanced inverted four-bar mechanism, aimed at high acceleration robotic manipulator applications. The main focus in the design was on optimizing the natural frequencies with the help of the mode shapes. The significance of each mode shape, however, was not taken into account. Analysis including the mode participation factors could gain insight in the contribution of each mode to the dynamic response when actuated in a particular direction [21]. The bode plot of the manipulator arm in Figure A1 reveals that the second resonance peak (at 443 Hz) is higher than the first peak (at 212 Hz), meaning that it is the second peak that is limiting the controller performance here. By moving the amplitude of the second resonance peak below the amplitude of the first resonance peak would increase the controller performance significantly.

The experimental setup showed a first natural frequency of 212 Hz and it successfully performed a motion with a peak transverse tip acceleration of about 21 G. While significant improvement of the prototype manipulator is still possible, cycle times of around 200 ms are within reach of the current design for this motion, resulting in cycle rates of around five cycles/second with a motion distance of 7.8 cm. The measured natural frequencies of the prototype therefore are already sufficiently high for realistic implementation in high acceleration applications.

The use of stainless steel tubes and sheet metal limited the design and the dimensional tolerances of the prototype and resulted in a reduction of the balance quality. Perhaps producing the links more precisely by CNC milling of a (topology) optimized design could result in significantly higher natural frequencies [22]. The objective of the topology optimization then should be the maximization of the natural frequencies with the balance conditions added as constraints of the optimization.

In Figures 14, 15 and 17 a vibration with a frequency of 18.18 Hz is noticeable in the shaking moment measurements. This vibration was caused by the low transverse stiffness of the connecting rods between the load cells and the base in combination with the high inertia of the base, mechanism and actuator. Adding stiffness to these connecting rods would lower the vibrational amplitude, but would also cause measurement errors due to higher transverse loading of the load cells. Comparable issues have been reported with a measurement setup based on a multi DoF force measurement sensor [11].

The tensioning devices removed the play in the plain bearings; however, it was challenging to tune them well. Tensioning the bearings too much caused a large integral term in the PID controller due to high friction and both the integral term and the friction caused significantly higher actuation torques and settle times. Reducing the tension in the bearings improved the PID controller performance, but it increased the backlash in the joints. A solution for finding an optimum between the two would be by extending the PID controller with feedforward control. With the help of feedforward control the PID can be tuned for error rejection, which would result in fewer actuator saturation. The calculation of the feedforward torques then also would require a real time operating system which communicates with the ETEL control system.

Pick-and-place applications often require multi-DoF manipulators, whereas this study focussed on a single DoF manipulator only. Future work therefore is needed to investigate how the theory of the balanced inverted four-bar linkage can be effectively extended to multi-DoF applications to obtain designs of dynamically balanced multi-DoF manipulators with short settling time.

8. Conclusions

In this paper the structural design of a manipulator arm with high natural frequencies that is based on a dynamically balanced inverted four-bar linkage was presented and experimentally verified. The dynamic properties of the manipulator arm were evaluated by analysing the first three in-plane natural frequencies, showing that the transverse stiffness of the first link, the manipulator arm, and the second link, the coupler link, have most influence on the natural frequencies. For the robustness to manufacturing tolerances it was shown that the manipulator is more prone to geometric deviations than to deviations of mass.

A prototype manipulator in an experimental setup was built to verify both the dynamic balance and the dynamic properties. The prototype successfully performed high acceleration motions with minimal shaking forces and shaking moment. When compared to the unbalanced case, a reduction of 99.3% in shaking forces and 97.8% in shaking moment was measured for end-effector accelerations of 5.2 G. For the partly balanced case with only half of the required end-effector mass, a force and moment balance quality was achieved of 90.3% and 81.9%, respectively. While these values are lower than for the fully balanced case, the shaking forces and shaking moment are still significantly lower than for the unbalanced case. A first natural frequency of 212 Hz was measured, which is significantly lower than the 312 Hz obtained from simulations, which is primarily caused by the lower stiffness of the base design. While significant improvement of the prototype manipulator is still possible, the natural frequencies are already sufficiently high to achieve short settling times and short cycle times during high-speed motions.

The experimental setup achieved end-effector accelerations of 21 G. Since the prototype was fabricated with relatively basic production methods, it shows that it is relatively simple to manufacture a dynamically balanced manipulator suitable for high acceleration applications.

Supplementary Materials: The following supporting information can be downloaded at: https://www.mdpi.com/article/10.3390/act11050131/s1, Video S1: Unbalanced and Balanced Motion of a robotic manipulator arm.

Author Contributions: Conceptualization, M.J.J.Z. and V.v.d.W.; methodology, M.J.J.Z.; software, M.J.J.Z.; validation, M.J.J.Z. and V.v.d.W.; formal analysis, M.J.J.Z.; investigation, M.J.J.Z.; resources, M.J.J.Z. and V.v.d.W.; data curation, M.J.J.Z.; writing—original draft preparation, M.J.J.Z.; writing—review and editing, M.J.J.Z. and V.v.d.W.; visualization, M.J.J.Z.; supervision, V.v.d.W.; project administration, V.v.d.W. All authors have read and agreed to the published version of the manuscript.

Funding: This research received no external funding.

Institutional Review Board Statement: Not applicable.

Informed Consent Statement: Not applicable.

Data Availability Statement: Not applicable.

Conflicts of Interest: The authors declare no conflict of interest.

Appendix A. Frequency Response of the Prototype

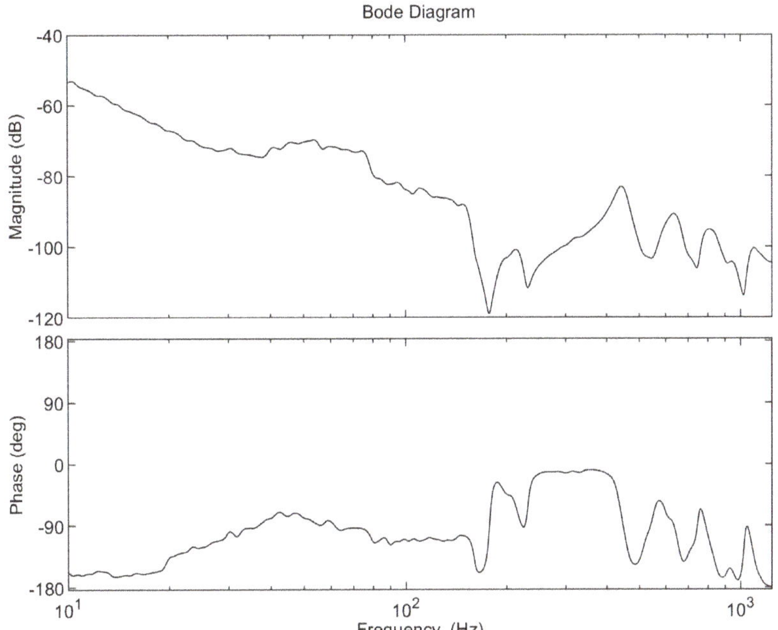

Figure A1. Bode plot of the balanced prototype with motor torque as input and rotary position as output.

References

1. Germain, C.; Caro, S.; Briot, S.; Wenger, P. Optimal design of the IRSBot-2 based on an optimized test trajectory. In Proceedings of the ASME 2013 International Design Engineering Technical Conferences and Computers and Information in Engineering Conference, Portland, OR, USA, 4–7 August 2013.
2. Briot, S.; Pashkevich, A.; Chablat, D. On the optimal design of parallel robots taking into account their deformations and natural frequencies. In Proceedings of the ASME 2009 International Design Engineering Technical Conferences and Computers and Information in Engineering Conference, San Diego, CA, USA, 3 August–2 September 2009.
3. Pierrot, F.; Baradat, C.; Nabat, V.; Company, O.; Krut, S.; Gouttefarde, M. Above 40g acceleration for pick-and-place with a new 2-dof PKM. *IEEE Int. Conf. Robot. Autom.* **2009**, *2*, 1794–1800. [CrossRef]
4. Arakelian, V. Inertia forces and moments balancing in robot manipulators: A review. *Adv. Robot.* **2017**, *31*, 717–726. [CrossRef]
5. van der Wijk, V. Methodology for Analysis and Synthesis of Inherently Force and Moment-Balanced Mechanisms: Theory and Applications. Ph.D. Thesis, University of Twente, Enschede, The Netherlands, 2014. [CrossRef]
6. van der Wijk, V.; Herder, J.L.; Demeulenaere, B. Comparison of various dynamic balancing principles regarding additional mass and additional inertia. *J. Mech. Robot.* **2009**, *1*, 1–9. [CrossRef]
7. De Jong, J.J.; Schaars, B.E.M.; Brouwer, D.M. The influence of flexibility on the force balance quality: A frequency domain approach. In Proceedings of the 19th International Conference of the European Society for Precision Engineering and Nanotechnology, Bilbao, Spain, 3–7 June 2019; Volume 19, pp. 2–5.
8. Foucault, S.; Gosselin, C.M. Synthesis, design, and prototyping of a planar three degree-of-freedom reactionless parallel mechanism. *J. Mech. Des. Trans. ASME* **2004**, *126*, 992–999. [CrossRef]
9. Laliberté, T.; Gosselin, C. Synthesis, optimization and experimental validation of reactionless two-DOF parallel mechanisms using counter-mechanisms. *Meccanica* **2016**, *51*, 3211–3225. [CrossRef]
10. de Jong, J.; van Dijk, J.; Herder, J. A screw based methodology for instantaneous dynamic balance. *Mech. Mach. Theory* **2019**, *141*, 267–282. [CrossRef]
11. van der Wijk, V.; Krut, S.; Pierrot, F.; Herder, J.L. Design and experimental evaluation of a dynamically balanced redundant planar 4-RRR parallel manipulator. *Int. J. Robot. Res.* **2013**, *32*, 744–759. [CrossRef]
12. Karidis, J.; McVicker, G.; Pawletko, J.; Zai, L.; Goldowsky, M.; Brown, R.; Comulada, R. The Hummingbird minipositioner-providing three-axis motion at 50 g's with low reactions. In Proceedings of the 1992 IEEE International Conference on Robotics and Automation, Nice, France, 12–14 May 1992; Volume 1, pp. 685–692. [CrossRef]

13. Menschaar, H.F.; Ariens, A.B.; Herder, J.L.; Bakker, B.M. Five-Bar Mechanism with Dynamic Balancing Means and Method for Dynamically Balancing a Five-Bar Mechanism. WIPO WO2006080846A1, 3 August 2006.
14. Raaijmakers, R. Besi zoekt snelheidslimiet pakken en plaatsen op (besi attacks the speedlimit for pick and place motion). *Mechatronica Nieuws* **2007**, 26–31. (In Dutch)
15. Ricard, R.; Gosselin, C.M. On the Development of Reactionless Parallel Manipulators. In Proceedings of the ASME 2000 Design Engineering Technical Conferences and Computers and Information in Engineering Conference, Baltimore, MD, USA, 10–13 September 2000.
16. Wu, Y.; Gosselin, C.M. Synthesis of reactionless spatial 3-DoF and 6-DoF mechanisms without separate counter-rotations. *Int. J. Robot. Res.* **2004**, *23*, 625–642. [CrossRef]
17. Jonker, J.B.; Meijaard, J.P. SPACAR — Computer Program for Dynamic Analysis of Flexible Spatial Mechanisms and Manipulators. In *Multibody Systems Handbook*; Springer: Berlin/Heidelberg, Germany, 1990; pp. 123–143. [CrossRef]
18. COMSOL AB. COMSOL Multiphysics® 5.4. Available online: https://www.comsol.com/ (accessed on 13 April 2022).
19. Mathworks. Matlab® R2019a Update 9. Available online: https://www.mathworks.com/ (accessed on 13 April 2022).
20. Berkof, R.S.; Lowen, G.G. A New Method for Completely Force Balancing Simple Linkages. *J. Eng. Ind.* **1969**, *91*, 21–26. [CrossRef]
21. Park, J.H.; Asada, H. Concurrent design optimization of mechanical structure and control for high speed robots. *J. Dyn. Syst. Meas. Control Trans. ASME* **1994**, *116*, 344–356. [CrossRef]
22. Briot, S.; Goldsztejn, A. Topology optimization of a reactionless four-bar linkage. *Mech. Mach. Sci.* **2018**, *50*, 414–421. [CrossRef]

Communication

A Finite-Time Trajectory-Tracking Method for State-Constrained Flexible Manipulators Based on Improved Back-Stepping Control

Yiwei Zhang [1,*], **Min Zhang** [2], **Caixia Fan** [1] and **Fuqiang Li** [1]

[1] College of Sciences, Henan Agricultural University, Zhengzhou 450001, China; fancaixia@henau.edu.cn (C.F.); fqlihenau@163.com (F.L.)

[2] State Grid Henan Skills Training Center, Zhengzhou 450001, China; zmmin_zhang@163.com

[*] Correspondence: zhangyiwei@henau.edu.cn; Tel.: +86-150-9317-7903

Abstract: In order to solve the trajectory-tracking-control problem of the state-constrained flexible manipulator systems, a finite-time back-stepping control method based on command filtering is presented in this paper. Considering that the virtual signal requires integration in each step, which will lead to high computational complexity in the traditional back-stepping, the finite-time command filter is used to filter the virtual signal and to obtain the intermediate signal in finite time, to thus reduce the computational complexity. The compensation mechanism is used to eliminate the error generated by the command filter. Furthermore, the adaptive estimation method is introduced to approach the uncertainty of the state-constrained flexible manipulator system. Then, the Lyapunov function is used to prove that the tracking error of the system can be stabilized in a sufficiently small origin neighborhood within a finite time. The simulation of a single rod flexible manipulator system demonstrates the effect of the proposed approach.

Keywords: flexible manipulator; state-constrained; finite-time control; back-stepping control

Citation: Zhang, Y.; Zhang, M.; Fan, C.; Li, F. A Finite-Time Trajectory-Tracking Method for State-Constrained Flexible Manipulators Based on Improved Back-Stepping Control. *Actuators* **2022**, *11*, 139. https://doi.org/10.3390/act11050139

Academic Editors: Marco Carricato and Edoardo Idà

Received: 24 April 2022
Accepted: 16 May 2022
Published: 19 May 2022

Publisher's Note: MDPI stays neutral with regard to jurisdictional claims in published maps and institutional affiliations.

1. Introduction

Due to the advantages of light weight, low power consumption, low cost and large payloads, flexible manipulators have been widely used in important fields, such as intelligent manufacturing, microsurgery and space operations [1–4].

A dynamic model of a flexible-joint manipulator system has the characteristics of nonlinearity, strong coupling and time variability. The design of its controller has always been a challenging problem. In recent years, many control methods have been applied to control manipulator systems, such as PID control, adaptive control, robust control, vibration control, fuzzy control and collaborative control [5–10]. Reference [5] proposed an adaptive sliding-mode control for uncertain single link flexible manipulator system; however, this controller does not deal with the chattering problem well.

Compared with sliding-mode control, the back-stepping control method overcomes this disadvantage, and thus the back-stepping controller design method is widely used in high-order nonlinear flexible manipulator system. In reference [7], the back-stepping method is applied to the controller design of a flexible manipulator system; however, the virtual control signal needs integration in each step of operation, which increases the computational complexity of the system. The dynamic surface control is introduced in reference [8–10] to reduce the computational complexity of the system by using a first-order filter but does not consider the filtering error caused by the introduction of filter, which reduces the tracking effect of a closed-loop system.

In addition, considering that the system is often subject to various restrictions in the actual operation process (such as saturation, physical restrictions, etc.) if the system state exceeds the given limit range, it will reduce the control effect and even lead to system instability. Therefore, how to design a controller to control the input and output of the system within the desired range should be considered. In order to control the system state within the desired range, the literature [11,12] has applied the barrier Lyapunov function to the nonlinear system; however, they did not consider the application of this control method in the manipulator system.

References [13–18] applied the barrier Lyapunov function to a manipulator system with limited output, and reference [19] further applied it to an n-order rigid manipulator system with full state constraints; however, the references [13–19] did not consider the flexibility in the actual manipulator system, and the design of the controller is based on the traditional back-stepping method, which requires a large amount of calculation. It was also indicated that the convergence speed of the system has not been considered in the literature [13–19], which will limit the control effect of the actual system.

At the same time, fast convergence, rapid response and good robustness are very important for the manipulator control system, and the finite-time control is very effective to improve these performances. Reference [20] applied finite-time control to the tracking control of a spacecraft system. Reference [21] applied finite-time control to the manipulator system with terminal sliding mode but ignores the flexibility of the manipulator system. Reference [22] studied the application of finite-time control based on a neural network with a flexible manipulator but does not consider the condition of limited state. Reference [23] applied adaptive command filtering control to nonlinear systems with full state constraints, without considering finite-time control.

At present, the problem of finite-time trajectory-tracking control of flexible manipulators with limited states has not been well solved. Aiming at this problem, a trajectory-tracking control algorithm based on the barrier Lyapunov function and the command filter back-stepping control method is proposed in this paper, which can achieve finite-time convergence and solves the trajectory-tracking problem of the state-constrained flexible manipulator system.

Compared with dynamic surface control and traditional back-stepping control, the method designed in this paper not only eliminates the computational complexity by designing the finite-time command filter in the process of establishing the finite-time virtual control function but also designs the finite-time error-compensation mechanism to eliminate the error in the filtering process, and verifies the finite-time convergence of the closed-loop system by Lyapunov function.

This paper is divided into five sections. The next section introduces the dynamic model of flexible manipulators and the problem statement. Designs of the command filter back-stepping controller for the flexible manipulator system are given in Section 3. Simulation results is provided in Section 4, followed by a brief conclusion in Section 5.

2. Preliminaries

2.1. Dynamic Model of Flexible Manipulator

The flexible-joint model of a manipulator is shown in Figure 1.

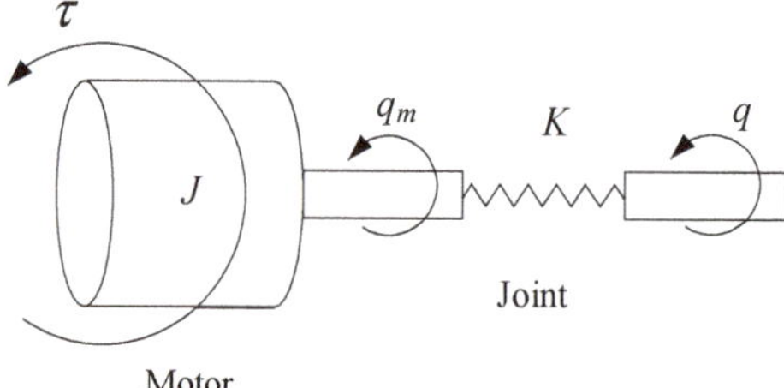

Figure 1. The flexible-joint model of a manipulator.

The dynamic model of flexible manipulators studied in this paper is as follows.

$$\begin{cases} M(q)\ddot{q} + C(q,\dot{q})\dot{q} + F(\dot{q}) = K(q_m - q) \\ J\ddot{q}_m + B\dot{q}_m + K(q_m - q) = \tau \end{cases} \tag{1}$$

where $q \in \mathbb{R}^n, \dot{q} \in \mathbb{R}^n, \ddot{q} \in \mathbb{R}^n$ are the joint position, velocity and acceleration, respectively. $M(q) \in \mathbb{R}^{n \times n}$ is a symmetric positive definite inertia matrix, and $C(q,\dot{q}) \in \mathbb{R}^{n \times n}$ is the Coriolis matrix. $F \in \mathbb{R}^{n \times n}$ is the joint friction coefficient matrix. $q_m \in \mathbb{R}^n, \dot{q}_m \in \mathbb{R}^n, \ddot{q}_m \in \mathbb{R}^n$ are the position, velocity and acceleration of the motor rotor angle, respectively. $K \in \mathbb{R}^{n \times n}$ represents the flexibility of the model joints. $J \in \mathbb{R}^{n \times n}$ represents the inertia term of the model, $B \in \mathbb{R}^{n \times n}$ represents the damping term of the model joints, $\tau \in \mathbb{R}^n$ is the control input vector of the system. The dynamic model satisfies the following properties [24].

Property 1. $M(q)$ is a symmetric positive definite matrix, $M^{-1}(q)$ is bounded, and $M_h \leq \|M(q)\|_2 \leq M_H$, where M_h and M_H are normal numbers.

Property 2. $M^{-1}(q)K$ is bounded, and $\|M^{-1}(q)K\|_2 \leq \rho$, where ρ is a constant and satisfies $\rho > 0$.

Let $q = x_1, \dot{q} = x_2, q_m = x_3, \dot{q}_m = x_4$, and then system (1) is equivalently transformed into

$$\begin{cases} \dot{x}_1 = x_2 \\ \dot{x}_2 = M^{-1}(x_1)[-C(x_1, x_2)x_2 - F(x_2) - K(x_1 - x_3)] \\ \dot{x}_3 = x_4 \\ \dot{x}_4 = J^{-1}[-Bx_4 - K(x_3 - x_1) + \tau] \end{cases} \tag{2}$$

where $x_i = [x_{i,1}, x_{i,2}, \cdots, x_{i,n}], (i = 1, 2, 3, 4)$ is the state variable of the system, and τ is the control input of the system. The state variable of the system satisfies the following assumption

$$|x_i| \leq k_i, k_i > 0, i = (1, 2, 3, 4) \tag{3}$$

Furthermore, we define the desired trajectory $x_d = [x_{d,1}, x_{d,2}, \cdots, x_{d,n}]^T \in \mathbb{R}^n$, where x_d and $\dot{x}_d$ are continuous and bounded.

2.2. Problem Statement

Design a nonlinear trajectory-tracking control strategy for the above state-constrained flexible manipulator system (1) to ensure that the joint position of the flexible manipulator q tracks the desired trajectory q_d. The tracking error can converge to a neighborhood near zero, that is, $\forall \beta > 0, \exists t_0 \geq 0$, when $t > t_0$, there exists $|q - q_d| < \beta$, and all state variables of the closed-loop system are continuously stable and bounded.

3. Controller Design

3.1. Error Compensation

According to the trajectory-tracking task in the problem description, the tracking error signal in the virtual control signal is defined as

$$\begin{cases} z_1 = x_1 - x_d \\ z_2 = x_2 - \phi_2 \\ z_3 = x_3 - \phi_3 \\ z_4 = x_4 - \phi_4 \end{cases} \tag{4}$$

where $\phi_{i+1} = [\phi_{i,1,1}, \phi_{i,2,1}, \cdots, \phi_{i,n,1}]^T, i = (1, 2, 3)$, which is output by the following command filter

$$\begin{cases} \dot{\phi}_{i,s,1} = l_{i,s,1} \\ l_{i,s,1} = -\lambda_{i,s,1}|\phi_{i,s,1} - \alpha_{i,s}|^{1/2} sign(\phi_{i,s,1} - \alpha_{i,s}) + \phi_{i,s,2} \\ \dot{\phi}_{i,s,2} = -\lambda_{i,s,2} sign\left(\phi_{i,s,2} - \dot{\phi}_{i,s,1}\right) \end{cases} \tag{5}$$

where $\phi_{i,n,1}$ is the state of the nth command filter in ith step, $\lambda_{i,s,1}$ and $\lambda_{i,s,2}$ are all positive numbers, and the virtual control function $\alpha_{i,s}$ is the input of the command filter.

Note that, in order to ensure that the system state converges in a finite time, the filtering error of the command filter satisfies $\lim\limits_{t \to T_1} \|\phi_2 - \alpha_1\| \le \omega_1$, $\lim\limits_{t \to T_2} \|\phi_3 - \alpha_2\| \le \omega_2$, $\lim\limits_{t \to T_3} \|\phi_4 - \alpha_3\| \le \omega_3$, where ω_1, ω_2 and ω_3 are all positive numbers, T_1, T_2 and T_3 are the convergence times of the command filter used for the first step, the second step and the third step, respectively.

In order to reduce the error existing between the virtual control signal and the finite-time command filter output signal, the error-compensation mechanism is constructed as follows

$$
\begin{cases}
\dot{v}_1 = -p_1 v_1 + v_2 + (\phi_2 - \alpha_1) - [q_{1,1}sign(v_{1,1}), \cdots, q_{1,n}sign(v_{1,n})]^T \\
\dot{v}_2 = -p_2 v_2 + (\phi_3 - \alpha_2) - v_1 + v_3 - [q_{2,1}sign(v_{2,1}), \cdots, q_{2,n}sign(v_{2,n})]^T \\
\dot{v}_3 = -p_3 v_3 + v_4 + (\phi_4 - \alpha_3) - [q_{3,1}sign(v_{3,1}), \cdots, q_{3,n}sign(v_{3,n})]^T \\
\dot{v}_4 = -p_4 v_4 - v_3 - [q_{4,1}sign(v_{4,1}), \cdots, q_{4,n}sign(v_{4,n})]^T
\end{cases}
\tag{6}
$$

where v represents the error compensation vector, and $p_i > 0$ and $q_i > 0$ are the tuning parameters that need to be designed.

Then, the compensated tracking error signal is designed as

$$
\begin{cases}
v_1 = z_1 - v_1 \\
v_2 = z_2 - v_2 \\
v_3 = z_3 - v_3 \\
v_4 = z_4 - v_4
\end{cases}
\tag{7}
$$

Ultimately, the virtual control signals in the controller are

$$
\begin{cases}
\alpha_1 = -p_1 z_1 + \dot{x}_d - \left[a_{1,1}v^r_{1,1}, \cdots, a_{1,n}v^r_{1,n}\right]^T \\
\alpha_2 = -p_2 z_2 - z_1 + \dot{\phi}_2 - \frac{1}{2}v_2 - \frac{1}{2h^2}\left[v_{2,1}\hat{\theta}_2 S^T_{2,1}S_{2,1}, \cdots, v_{2,n}\hat{\theta}_2 S^T_{2,n}S_{2,n}\right]^T - \left[a_{2,1}v^r_{2,1}, \cdots, a_{2,n}v^r_{2,n}\right]^T \\
\alpha_3 = -p_3 z_3 - z_2 + \dot{\phi}_3 - \left[a_{3,1}v^r_{3,1}, \cdots, a_{3,n}v^r_{3,n}\right]^T \\
\tau = -p_4 z_4 - z_3 + \dot{\phi}_4 - \frac{1}{2}v_4 - \frac{1}{2h^2}\left[v_{4,1}\hat{\theta}_4 S^T_{4,1}S_{4,1}, \cdots, v_{4,n}\hat{\theta}_4 S^T_{4,n}S_{4,n}\right]^T - \left[a_{4,1}v^r_{4,1}, \cdots, a_{4,n}v^r_{4,n}\right]^T
\end{cases}
\tag{8}
$$

where a and h are all positive numbers, and $0 < r < 1$. $\hat{\theta}_2$ and $\hat{\theta}_4$ are the estimated variables obtained by the adaptive update law in (21).

Remark 1. *In the process of designing the tracking controller of the flexible manipulator system using the traditional back-stepping method, each step needs to design a virtual control signal as (4) to ensure that each subsystem has the desired performance. However, using virtual controlled derivatives results in increased computational complexity. In this paper, the output of the finite-time command filter (5) is used to approximate the virtual signal and the derivative of the virtual signal to replace the calculation of the derivative of the virtual signal in the traditional back-stepping process; thereby, the computational complexity is eliminated. However, filter errors will persist until the finite-time command filter stabilizes, which will affect the control quality. Therefore, this paper proposes a finite-time error-compensation mechanism of (6) to quickly eliminate filtering errors.*

3.2. Stability Analysis

This section investigates the command filtering back-stepping control strategy with state constraints and error compensation. Some necessary and sufficient conditions are derived for the main results.

Lemma 1 [25]. *If $x_i \in R, i = (1, 2, \cdots, n)$ and $0 < p \leq 1$, then the following inequality holds*

$$\left(\sum_{i=1}^{n} |x_i| \right)^p \leq \sum_{i=1}^{n} |x_i|^p \leq n^{1-p} \left(\sum_{i=1}^{n} |x_i| \right)^p \tag{9}$$

Lemma 2 [26]. *If there exists a real number $\lambda_1 > 0$, $\lambda_2 > 0$ and $0 < \gamma < 1$, the finite-time stable extended Lyapunov condition can be obtained by $\dot{V}(x) + \lambda_1 V(x) + \lambda_2 V^\gamma(x) \leq 0$, and the convergence time $T_r \leq t_0 + [1/(\lambda_1(1-\gamma))] \ln\left[(\lambda_1 V^{1-\gamma}(t_0) + \lambda_2)/\lambda_2 \right]$.*

Theorem 1. *For the flexible-joint manipulator (1), using the error-compensation mechanism in (6) and the virtual control signal in (8), the joint position track the desired joint position in a finite time, and all system states in the closed-loop system are bounded in a finite time.*

Proof of Theorem 1. The stability of the closed-loop system is proven by the following four steps.

Step 1. Select Lyapunov function $V_1 = \frac{1}{2} v_1^T v_1$, taking the time derivative of V_1 yields

$$\begin{aligned} \dot{V}_1 &= v_1^T \dot{v}_1 = v_1^T (\dot{z}_1 - \dot{v}_1) = v_1^T (\dot{x}_1 - \dot{x}_d - \dot{v}_1) \\ &= v_1^T [\alpha_1 + z_2 + (\phi_2 - \alpha_1) - \dot{x}_d - \dot{v}_1] \end{aligned} \tag{10}$$

Substitute α_1 and v_1 into (20) to find

$$\dot{V}_1 = -k_1 v_1^T v_1 + v_2^T v_2 - \left[a_{1,1} v_{1,1}^r, \cdots, a_{1,n} v_{1,n}^r \right]^T + \left[v_{1,1} q_{1,1} sign(v_{1,1}), \cdots, v_{1,n} q_{1,n} sign(v_{1,n}) \right]^T \tag{11}$$

Step 2. Select Lyapunov function $V_2 = V_1 + \frac{1}{2} v_2^T v_2$, and taking the derivative of V_2 yields

$$\begin{aligned} \dot{V}_2 &= \dot{V}_1 + v_2^T \dot{v}_2 = \dot{V}_1 + v_2^T (\dot{z}_2 - \dot{v}_2) \\ &= \dot{V}_1 + v_2^T \left(\dot{x}_2 - \dot{\phi}_2 - \dot{v}_2 \right) \\ &= \dot{V}_1 + v_2^T \left(f_2 + g_2 x_2 - \dot{\phi}_2 - \dot{v}_2 \right) \end{aligned} \tag{12}$$

where $f_2 = -M^{-1}(x_1)[C(x_1, x_2)x_2 + Fx_2 + Kx_1]$, $g_2 = M^{-1}(x_1)K$.

Since the function f_2 contains uncertainty, it is approximated by a neural network [27], then $f_2 = [f_{2,1}, f_{2,2}, \cdots, f_{2,n}]^T$ can be approximately expressed as

$$f_{2,i} = W_{2,i}^T S_{2,i} + \zeta_{2,i}, i = (1, 2, \cdots, n) \tag{13}$$

where $W_{2,i}$ is the weight matrix, $S_{2,i}$ is the basis function vector, and $\zeta_{2,i}$ is the approximation error and satisfies $\|\zeta_{2,i}\| \leq \varepsilon_2, \varepsilon_2 > 0$.

According to the Young inequality, we can find

$$v_2^T f_2 \leq \sum_{i=1}^{n} \left(\frac{v_{2,i}^2 \|W_{2,i}\|^2 S_{2,i}^T S_{2,i}}{2h^2} + \frac{h^2 + v_{2,i}^2 + \varepsilon_2^2}{2} \right) \tag{14}$$

where h is a positive number. Substitute α_2, $\dot{v}_2$ and $v_2^T f_2$ into (14), we obtain

$$\begin{aligned} \dot{V}_2 \leq &-\sum_{i=1}^{2} \left(p_i v_i^T v_i \right) + g_2 v_2^T v_3 + \sum_{i=1}^{n} \left(\frac{v_{2,i}^2 \left(\|W_{2,i}\|^2 - \hat{\theta}_2 \right) S_{2,i}^T S_{2,i}}{2h^2} + \frac{h^2 + \varepsilon_2^2}{2} \right) \\ &- \sum_{i=1}^{n} \left(a_{1,i} v_{1,i}^{\gamma+1} + a_{2,i} v_{2,i}^{\gamma+1} \right) + \sum_{i=1}^{n} \left(v_{1,i} q_{1,i} sign(v_{1,i}) + v_{2,i} q_{2,i} sign(v_{2,i}) \right) \end{aligned} \tag{15}$$

Step 3. Select Lyapunov function $V_3 = V_2 + \frac{1}{2}v_3^T v_3$, and taking the derivative of V_3 yields

$$
\begin{aligned}
\dot{V}_3 &= \dot{V}_2 + v_3^T \dot{v}_3 \\
&= \dot{V}_2 + v_3^T \left(\alpha_3 + z_4 + (\phi_4 - \alpha_3) - \dot{\phi}_3 - \dot{v}_3 \right)
\end{aligned}
\tag{16}
$$

Substitute α_3 and $\dot{v}_3$ into (20) to obtain

$$
\begin{aligned}
\dot{V}_3 &= \dot{V}_2 - p_3 v_3^T v_3 - g_2 v_3^T v_2 + v_3^T v_4 \\
&\quad - \left[a_{3,1} v_{3,1}^{r+1}, \cdots, a_{3,n} v_{3,n}^{r+1} \right]^T + \left[v_{3,1} q_{3,1} sign(v_{3,1}), \cdots, v_{3,n} q_{3,n} sign(v_{3,n}) \right]^T
\end{aligned}
\tag{17}
$$

Step 4. Select Lyapunov function $V_4 = V_3 + \frac{1}{2}v_4^T v_4$, and take the derivative of V_4 yields

$$
\begin{aligned}
\dot{V}_4 &= \dot{V}_3 + v_4^T \dot{v}_4 = \dot{V}_3 + v_4^T \left(\dot{z}_4 - \dot{v}_4 \right) \\
&= \dot{V}_3 + v_4^T \left(\dot{x}_4 - \dot{\phi}_4 - \dot{v}_4 \right) \\
&= \dot{V}_3 + v_4^T \left(f_4 + g_4 \tau - \dot{\phi}_4 - \dot{v}_4 \right)
\end{aligned}
\tag{18}
$$

where $f_4 = -J^{-1}[Bx_4 + K(x_3 - x_1)]$, $g_4 = J^{-1}$. Similar to (13), f_4 can be written as

$$
f_{4,i} = W_{4,i}^T S_{4,i} + \zeta_{4,i}, i = (1, 2, \cdots, n)
\tag{19}
$$

where $W_{4,i}$ is weight matrix, $S_{4,i}$ is the basis function vector, and $\zeta_{4,i}$ is the approximation error and satisfies $\|\zeta_{4,i}\| \leq \varepsilon_4, \varepsilon_4 > 0$.

In view of (18) and the Young inequality, we have

$$
\begin{aligned}
\dot{V}_4 &\leq -\sum_{i=1}^{4} \left(p_i v_i^T v_i \right) + \sum_{i=1}^{n} \left(\frac{v_{2,i}^2 \left(\|W_{2,i}\|^2 - \hat{\theta}_2 \right) S_{2,i}^T S_{2,i}}{2h^2} + \frac{h^2 + \varepsilon_2^2}{2} \right) \\
&\quad + \sum_{i=1}^{n} \left(\frac{v_{4,i}^2 \left(\|W_{4,i}\|^2 - \hat{\theta}_4 \right) S_{4,i}^T S_{4,i}}{2h^2} + \frac{h^2 + \varepsilon_4^2}{2} \right) - \sum_{i=1}^{4} \sum_{j=1}^{n} \left(a_{i,j} v_{i,j}^{\gamma+1} \right) + \sum_{i=1}^{4} \sum_{j=1}^{n} \left(v_{i,j} q_{i,j} sign(v_{i,j}) \right) \\
&\leq -\sum_{i=1}^{4} \left(p_i - \frac{q_i}{2} \right) v_i^T v_i + \sum_{i=1}^{n} \left(\frac{v_{2,i}^2 \left(\|W_{2,i}\|^2 - \hat{\theta}_2 \right) S_{2,i}^T S_{2,i}}{2h^2} + \frac{h^2 + \varepsilon_2^2}{2} \right) \\
&\quad + \sum_{i=1}^{n} \left(\frac{v_{4,i}^2 \left(\|W_{4,i}\|^2 - \hat{\theta}_4 \right) S_{4,i}^T S_{4,i}}{2h^2} + \frac{h^2 + \varepsilon_4^2}{2} \right) - \sum_{i=1}^{4} \sum_{j=1}^{n} \left(a_{i,j} v_{i,j}^{\gamma+1} \right) + \sum_{i=1}^{4} \sum_{j=1}^{n} \left(\frac{q_{i,j}}{2} \right)
\end{aligned}
\tag{20}
$$

Let $\theta_2 = \max\left(\|W_{2,i}\|^2 \right)$, $\theta_4 = \max\left(\|W_{4,i}\|^2 \right)$, $i = (1, 2, \cdots, n)$, then the estimate $\hat{\theta}_2$ and $\hat{\theta}_4$ of θ_2 and θ_4 can be obtained by the following adaptive update law

$$
\begin{cases}
\dot{\hat{\theta}}_2 = -m_2 l_2 \hat{\theta}_2 + \frac{1}{2h^2} m_2 \sum_{i=1}^{n} v_{2,i}^2 S_{2,i}^T S_{2,i} \\
\dot{\hat{\theta}}_4 = -m_4 l_4 \hat{\theta}_2 + \frac{1}{2h^2} m_4 \sum_{i=1}^{n} v_{4,i}^2 S_{4,i}^T S_{4,i}
\end{cases}
\tag{21}
$$

where $m_2 > 0$, $m_4 > 0$, $l_2 > 0$, $l_4 > 0$. Let $\tilde{\theta}_2 = \theta_2 - \hat{\theta}_2$, $\tilde{\theta}_4 = \theta_4 - \hat{\theta}_4$; furthermore, we construct the following Lyapunov function $\tilde{V}_4 = V_4 + \frac{1}{2m_2}\tilde{\theta}_2^2 + \frac{1}{2m_4}\tilde{\theta}_4^2$, and take the derivative of $\tilde{V}_4$, we have

$$
\begin{aligned}
\dot{\widetilde{V}}_4 &= \dot{V}_4 + \frac{1}{2m_2}\widetilde{\theta}_2\dot{\widetilde{\theta}}_2 + \frac{1}{2m_4}\widetilde{\theta}_4\dot{\widetilde{\theta}}_4 \\[4pt]
&\leq -\sum_{i=1}^{4}\left(p_i v_i^T v_i - \frac{q_i}{2} v_i^T v_i\right) + \frac{1}{m_2}\widetilde{\theta}_2\left(\dot{\theta}_2 - \dot{\hat{\theta}}_2\right) + \frac{1}{m_4}\widetilde{\theta}_4\left(\dot{\theta}_4 - \dot{\hat{\theta}}_4\right) - \sum_{i=1}^{4}\sum_{j=1}^{n}\left(a_{i,j}v_{i,j}^{\gamma+1}\right) \\[4pt]
&\quad + \sum_{i=1}^{n}\left(\frac{v_{2,i}^2\widetilde{\theta}_2 S_{2,i}^T S_{2,i} + v_{4,i}^2\widetilde{\theta}_4 S_{4,i}^T S_{4,i}}{2h^2} + \frac{\varepsilon_2^2 + \varepsilon_4^2}{2} + h^2\right) \\[4pt]
&\leq -\sum_{i=1}^{4}\left(p_i v_i^T v_i - \frac{q_i}{2} v_i^T v_i\right) + \frac{n}{2}\left(\varepsilon_2^2 + \varepsilon_4^2\right) + nh^2 - \sum_{i=1}^{4}\sum_{j=1}^{n}\left(a_{i,j}v_{i,j}^{\gamma+1}\right) \\[4pt]
&\quad + l_2\widetilde{\theta}_2\hat{\theta}_2 + l_4\widetilde{\theta}_4\hat{\theta}_4 \\[4pt]
&\leq -\sum_{i=1}^{4}\left(p_i v_i^T v_i - \frac{q_i}{2} v_i^T v_i\right) + \frac{n}{2}\left(\varepsilon_2^2 + \varepsilon_4^2\right) + nh^2 - \sum_{i=1}^{4}\sum_{j=1}^{n}\left(a_{i,j}v_{i,j}^{\gamma+1}\right) \\[4pt]
&\quad + l_2 d_2\theta_2^2 + l_4 d_4\theta_4^2 - \frac{\chi_2}{p_2}\widetilde{\theta}_2^2 - \frac{\chi_4}{p_4}\widetilde{\theta}_4^2
\end{aligned}
\tag{22}
$$

where $\chi_2 = (p_2 l_2(2d_2 - 1))/2d_2$, $\chi_4 = (p_4 l_4(2d_4 - 1))/2d_4$, $d_2 > 0.5$, $d_4 > 0.5$. By Lemma 1, we obtain

$$
\dot{\widetilde{V}}_4 \leq -\rho\widetilde{V}_4 - \sigma\widetilde{V}_4^{\frac{\gamma+1}{2}} + c
\tag{23}
$$

in which

$$
\rho = \min(2p_i - q_i, 2\chi_2, 2\chi_4),
$$

$$
\sigma = \min\left(a_{i,j}2^{\frac{\gamma+1}{2}}, 2\chi_2^{\frac{\gamma+1}{2}}, 2\chi_2^{\frac{\gamma+1}{2}}\right),
$$

$$
c = \frac{n}{2}\left(\varepsilon_2^2 + \varepsilon_4^2\right) + nh^2 + n\sum_{i=1}^{4}\frac{q_i}{2} + l_2 d_2\theta_2^2 + l_4 d_4\theta_4^2.
$$

It is clear that we can find a positive number $0 < \delta < 1$ such that

$$
\dot{\widetilde{V}}_4 \leq -\delta\rho\widetilde{V}_4 - (1-\delta)\rho\widetilde{V}_4 - \sigma\widetilde{V}_4^{\frac{\gamma+1}{2}} + c
\tag{24}
$$

Furthermore, it can be seen from Lemma 2 that, if $\widetilde{V}_4 > \frac{c}{(1-\delta)\rho}$, then we have

$$
\dot{\widetilde{V}}_4 \leq -\delta\rho\widetilde{V}_4 - \sigma\widetilde{V}_4^{\frac{\gamma+1}{2}}
\tag{25}
$$

which indicates that $v_i, \widetilde{\theta}_2, \widetilde{\theta}_4$ will converge to the following region in finite time

$$
\left(v_i, \widetilde{\theta}_2, \widetilde{\theta}_4\right) \in \left\{\widetilde{V}_4 \leq \frac{c}{(1-\delta)\rho}\right\}
\tag{26}
$$

The time T_r required to reach the region in (26) is

$$
T_r \leq \frac{1}{\delta\rho\left(1 - \frac{\gamma+1}{2}\right)}\ln\left(\left(\delta\rho\widetilde{V}_4^{1-\frac{\gamma+1}{2}}(0) + \sigma\right)/\sigma\right),
\tag{27}
$$

which means that the state of the system converges in the desired neighborhood near the origin in finite time, and all control signals in the closed-loop system are bounded in finite time. $\square$

Remark 2. *In view of the definition of convergence region, it can be seen that we can obtain a smaller convergence area by increasing the parameters* $p_i, a_i, q_i, \chi_2, \chi_4, i = (1,2,3,4)$.

4. Simulation

In this section, a simulation example of a single joint flexible manipulator is provided to illustrate the theoretical result. The parameter of system (1) is shown in Table 1.

Table 1. Parameters of the flexible manipulator.

Parameters	Value
M	0.2/kg
L	0.3/m
g	9.8/m/s^2
K	6.47/N·m/rad
B	0.01
J	0.21/ m/s^2

The structure diagram of the control system is shown in Figure 2.

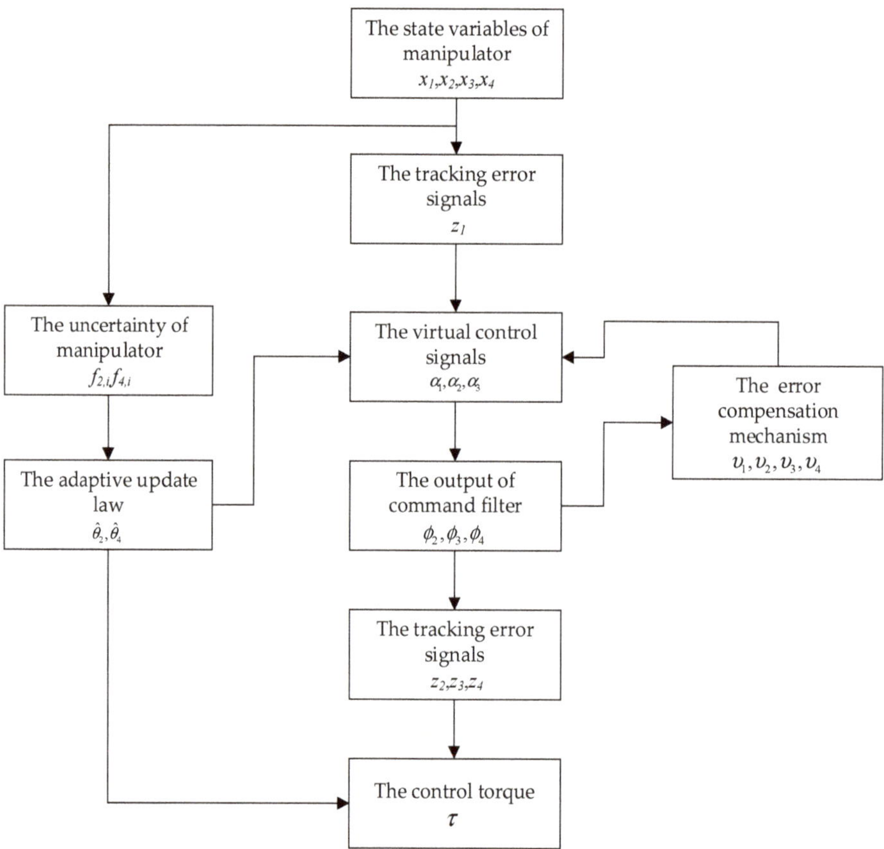

Figure 2. The structure diagram of the control system.

Suppose that the desired tracking trajectory is set as $q_d = \sin(t)$, then the initial state is $x_1(0) = q(0) = 0.2$, $x_2(0) = \dot{q}(0) = 0$. The term of joint friction is set as $F(\dot{q}) = 0.01\cos\dot{q}$. The constraints of the system state are $\|x_1\| \leq 1.8$, $\|x_2\| \leq 2$, $\|x_3\| \leq 3$, $\|x_4\| \leq 10$, respectively.

The parameters of the command filter (5) are set as $\lambda_{i,s,1} = \lambda_{i,s,2} = 30$. The parameters of the error-compensation mechanism (6) are selected as $p_1 = 1$, $p_2 = 15$, $p_3 = p_4 = 20$, and $q_1 = q_2 = q_3 = q_4 = 1$. The parameters of virtual control signals (8) are set as $a_1 = 1, a_2 = a_3 = 5, a_4 = 7, h = 1, r = 3/5$. The parameters of the adaptive update law (21) are set as $m_2 = m_4 = 1, l_2 = l_4 = 1$. The other parameters used in (22) are selected as $\gamma = 1/2, \varepsilon_2 = \varepsilon_4 = 0.1, d_2 = d_4 = 1$.

First, in order to verify the effectiveness of the error-compensation mechanism for the flexible manipulator system (1), the position and velocity tracking curves are shown in Figure 3 in the case that the error-compensation mechanism is not used in the controller, and the position tracking error curves without error compensation are shown in Figure 4. Then, the position and velocity tracking curves are shown in Figure 5 by using the error-compensation mechanism, and the position tracking error curves with error compensation are shown in Figure 6.

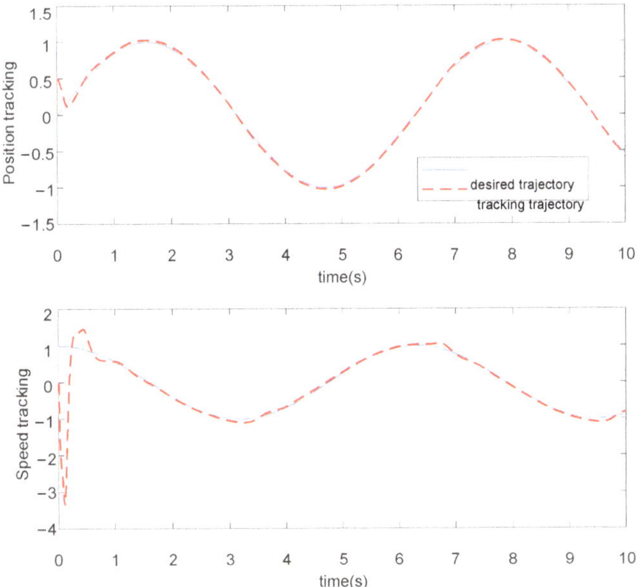

Figure 3. Position and velocity tracking curves without error compensation.

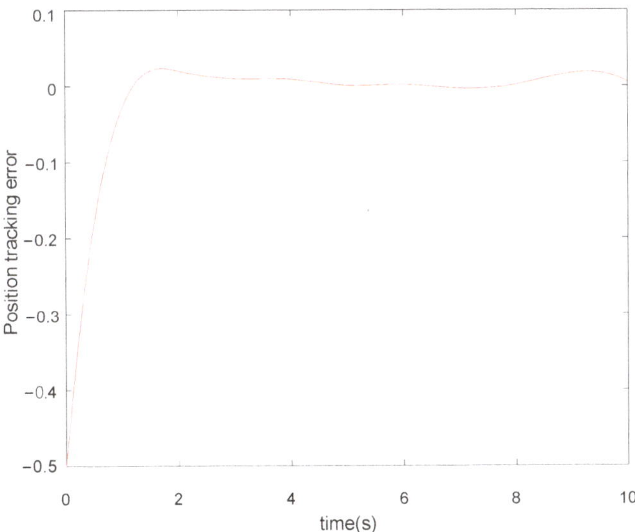

Figure 4. Position tracking error curves without error compensation.

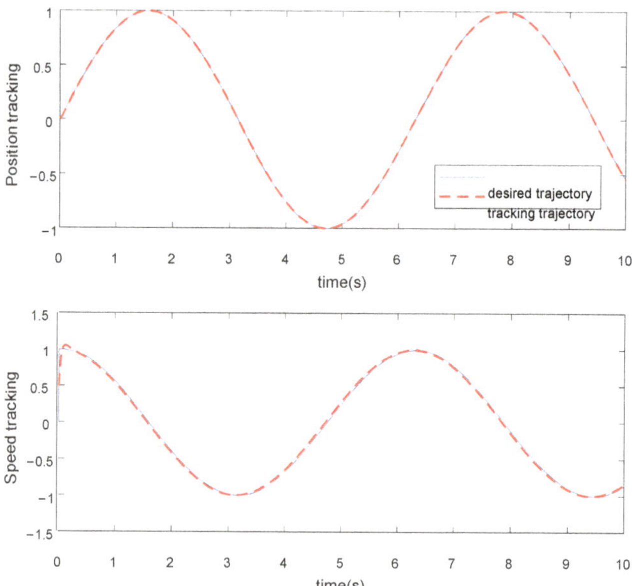

Figure 5. Position and velocity tracking curves with error compensation.

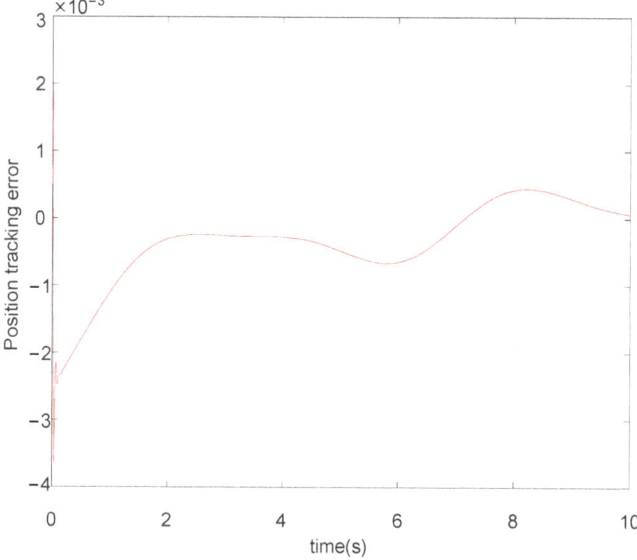

Figure 6. Position tracking error curves with error compensation.

As can be seen from Figure 3 to Figure 6, the use of the error-compensation mechanism in the controller can improve the accuracy of the position and velocity tracking effectively. In addition, after using the error-compensation mechanism, the system state can satisfy the state constraints at the same time. Then, the effectiveness of the error-compensation mechanism designed in this paper is verified.

Secondly, the finite-time control effect of the control algorithm proposed in this paper is verified by adjusting parameters in the following two cases. The parameters is set as

$p_i = 5, a_i = 2, q_i = 1, i = (1, 2, 3, 4)$ in the first case. The parameters is set as $p_i = 10, a_i = 5$, $q_i = 1, i = (1, 2, 3, 4)$ in the second case.

Figures 7 and 8 show the joint position tracking curves under two different control parameters. It can be seen from Figure 7 that, in the first case, the joint trajectory can track the desired trajectory in 2.3 s, while by increasing the adjustment parameters, the joint can track the desired trajectory in 0.2 s, as shown in Figure 8. As can be seen from Figures 7 and 8, we can improve the convergence efficiency by increasing the gain of the control parameters.

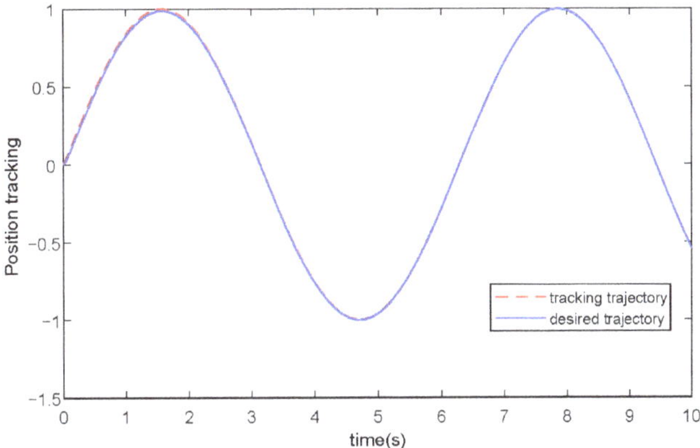

Figure 7. Position tracking curves under the first case.

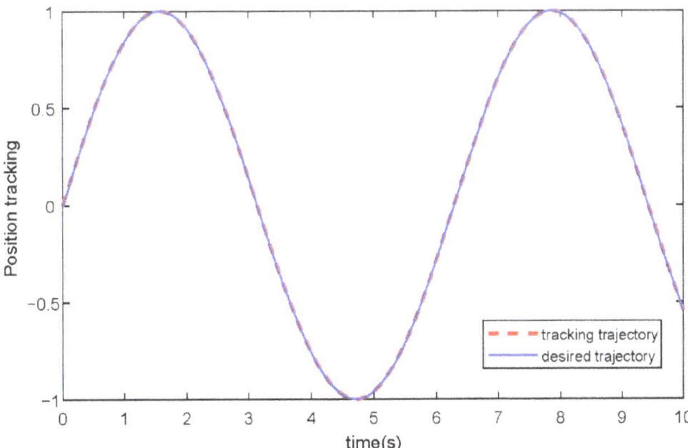

Figure 8. Position tracking curves under the second case.

5. Conclusions

Aiming at the finite-time tracking-control problem of state-constrained flexible manipulator systems, this paper proposes an adaptive neural network command filtering back-stepping control method. Compared with the traditional back-stepping control, the computational complexity is eliminated by designing a finite-time command filter in the process of establishing a finite-time virtual control function, and a finite-time error-compensation mechanism is designed to eliminate errors in the filtering process. The Lyapunov function is used to verify the finite-time convergence of the closed-loop system. In future work, the method designed in this paper will be extended to a multi-joint flexible manipulator system, and an adaptive robust control method will be considered to overcome the influence of unknown external disturbances.

Author Contributions: Conceptualization, Y.Z. and M.Z.; methodology, Y.Z.; software, M.Z.; validation, Y.Z., C.F. and F.L.; formal analysis, C.F.; investigation, F.L.; resources, Y.Z.; data curation, Y.Z.; writing—original draft preparation, Y.Z.; writing—review and editing, C.F.; visualization, C.F.; supervision, Y.Z.; project administration, Y.Z.; funding acquisition, Y.Z. All authors have read and agreed to the published version of the manuscript.

Funding: This research was funded by the National Natural Science Foundation of China, grant number 61703146; the Scientific and Technological Project of Henan Province, grant number 202102110126; the Backbone teacher project of Henan Province grant number, 2020GGJS048; the key scientific research projects of colleges and universities in Henan Province, grant number 19B413002.

Institutional Review Board Statement: Not applicable.

Informed Consent Statement: Not applicable.

Data Availability Statement: Not applicable.

Conflicts of Interest: The authors declare no conflict of interest. The funders had no role in the design of the study; in the collection, analyses, or interpretation of data; in the writing of the manuscript, or in the decision to publish the results.

References

1. Huang, D.; Huang, X. Neural Network Compensation Control for Model Uncertainty of Flexible Space Manipulator Based on Hybrid Trajectory. *J. Eng. Sci. Technol. Rev.* **2021**, *14*, 86–94.
2. Shin, H.C.; Choi, S.B. Position control of a two-link flexible manipulator featuring piezoelectric actuators and sensors. *Mechatronics* **2001**, *11*, 707–729. [CrossRef]
3. Mahmood, I.A.; Moheimani, S.O.; Bhikkaji, B. Precise tip positioning of a flexible manipulator using resonant control. *IEEE/ASME Trans. Mechatron.* **2008**, *13*, 180–186. [CrossRef]
4. Chang, W.; Li, Y.; Tong, S. Adaptive fuzzy backstepping tracking control for flexible robotic manipulator. *IEEE/CAA J. Autom. Sin.* **2018**, *8*, 1923–1930. [CrossRef]
5. Sun, C.; Wei, H.; Jie, H. Neural Network Control of a Flexible Robotic Manipulator Using the Lumped Spring-Mass Model. *IEEE Trans. Syst. Man Cybern. Syst.* **2017**, *47*, 1863–1874. [CrossRef]
6. Reddy, M.; Jacob, J. Vibration Control of Flexible Link Manipulator Using SDRE Controller and Kalman Filtering. *Stud. Inform. Control.* **2017**, *26*, 143–150.
7. Belherazem, A.; Chenafa, M. Passivity Based Adaptive Control of a Single-Link Flexible Manipulator. *Autom. Control Comput. Sci.* **2021**, *55*, 1–14. [CrossRef]
8. Cao, F.; Liu, J. An adaptive iterative learning algorithm for boundary control of a coupled ODE–PDE two-link rigid–flexible manipulator. *J. Frankl. Inst.* **2016**, *354*, 277–297. [CrossRef]
9. Yao, W.; Guo, Y.; Wu, Y.F. Robust Adaptive Dynamic Surface Control of Multi-link Flexible Joint Manipulator with Input Saturation. *Int. J. Control Autom. Syst.* **2022**, *20*, 577–588. [CrossRef]
10. Kivila, A.; Book, W.; Singhose, W. Modeling spatial multi-link flexible manipulator arms based on system modes. *Int. J. Intell. Robot. Appl.* **2021**, *5*, 300–312. [CrossRef]
11. Ji, N.; Liu, J. Vibration control for a flexible satellite with input constraint based on Nussbaum function via backstepping method. *Aerosp. Sci. Technol.* **2018**, *77*, 563–572. [CrossRef]
12. Cao, F.; Liu, J. Three-dimensional modeling and input saturation control for a two-link flexible manipulator based on infinite dimensional model. *J. Frankl. Inst.* **2020**, *357*, 1026–1042. [CrossRef]
13. Zhang, S.J.; Cao, Y. Cooperative Localization Approach for Multi-Robot Systems Based on State Estimation Error Compensation. *Sensors* **2019**, *19*, 3842. [CrossRef] [PubMed]
14. Zhu, X.; Shen, X.; Wang, L. Tip Tracking Control of a Linear-Motor-Driven Flexible Manipulator with Controllable Damping. *IFAC-Pap.* **2020**, *53*, 9163–9168. [CrossRef]
15. Xing, X.; Liu, J. PDE model-based state-feedback control of constrained moving vehicle-mounted flexible manipulator with prescribed performance. *J. Sound Vib.* **2019**, *441*, 126–151. [CrossRef]
16. Ma, X.; Wang, P.; Ye, M. Shared Autonomy of a Flexible Manipulator in Constrained Endoluminal Surgical Tasks. *IEEE Robot. Autom. Lett.* **2019**, *4*, 3106–3112. [CrossRef]
17. Cheng, X.; Liu, H.S. Bounded decoupling control for flexible-joint robot manipulators with state estimation. *IET Control Theory Appl.* **2020**, *14*, 2348–2358. [CrossRef]
18. Zhang, S.; Cao, Y. Consensus in networked multi-robot systems via local state feedback robust control. *Int. J. Adv. Robot. Syst.* **2019**, *16*, 1729881419893549. [CrossRef]
19. Tang, Y.; Xing, X.; Karimi, H.R. Tracking control of networked multi-agent systems under new characterizations of impulses and its applications in robotic systems. *IEEE Trans. Ind. Electron.* **2016**, *63*, 1299–1307. [CrossRef]

20. Xia, Y.; Zhang, J.; Lu, K.; Zhou, N. Finite-Time Tracking Control of Rigid Spacecraft Under Actuator Saturations and Faults. *IEEE Trans. Autom. Sci. Eng.* **2016**, *18*, 368–381.
21. Anjum, Z.; Guo, Y.; Yao, W. Fault tolerant control for robotic manipulator using fractional-order backstepping fast terminal sliding mode control. *Trans. Inst. Meas. Control* **2021**, *43*, 3244–3254. [CrossRef]
22. Zhao, Z.; Liu, Z. Finite-Time Convergence Disturbance Rejection Control for a Flexible Timoshenko Manipulator. *IEEE/CAA J. Autom. Sin.* **2021**, *8*, 161–172. [CrossRef]
23. Qiu, J.; Sun, K.; Rudas, I.J.; Gao, H. Command Filter-Based Adaptive NN Control for MIMO Nonlinear Systems with Full-State Constraints and Actuator Hysteresis. *IEEE Trans. Cybern.* **2020**, *50*, 2905–2915. [CrossRef] [PubMed]
24. Huang, A.C.; Chen, Y.C. Adaptive sliding control for single-link flexible-joint robot with mismatched uncertainties. *IEEE Trans. Control Syst. Technol.* **2004**, *12*, 770–775. [CrossRef]
25. Abdollahi, F.; Talebi, H.A.; Patel, R.V. A stable neural network based observer with application to flexible-joint manipulators. *IEEE Trans. Neural Netw.* **2006**, *17*, 118–129. [CrossRef]
26. Yu, S.H.; Yu, X.H.; Shirinzadeh, B. Continuous finite-time control for robotic manipulators with terminal sliding mode. *Automatica* **2005**, *41*, 1957–1964. [CrossRef]
27. Wang, D.H.; Zhang, S.J. Improved neural network-based adaptive tracking control for manipulators with uncertain dynamics. *Int. J. Adv. Robot. Syst.* **2020**, *17*, 1823–1838. [CrossRef]

An Active Fault-Tolerant Control Based on Synchronous Fast Terminal Sliding Mode for a Robot Manipulator

Quang Dan Le and Hee-Jun Kang *

Department of Electrical, Electronic and Computer Engineering, University of Ulsan, Ulsan 44610, Korea; ledantm@gmail.com
* Correspondence: hjkang@ulsan.ac.kr

Abstract: To maintain the safe operation and acceptable performance of robot manipulators when faults occur inside the system, fault-tolerant control must deal differently uncertainties and disturbances, especially with the occurrence of loss-effective faults. Therefore, in this paper, an active fault-tolerant control for robot manipulators based on the combination of a novel finite-time synchronous fast terminal sliding mode control and extended state observer is proposed. Due to the internal constraints of the synchronization technique, the position error at each actuator simultaneously approaches zero and tends to be equal. Therefore, the proposed controller can suppress the effects of faults and guarantee the acceptable performance of robot manipulators when faults occur. First, an extended state observer is designed to estimate the lumped uncertainties, disturbance and faults. Then, the information from the observer is used to combine with the main novel synchronous fast terminal sliding mode controller as a compensator. By combining the merits of the observer compensation, sliding mode and synchronization technique, the proposed fault-tolerant controller is able to deal with uncertainties and disturbances in normal operation mode and reduce the effects of faults in case faults occur, especially in the occurrence of loss-effective faults. Finally, the enhanced safety, reality and effectiveness of the proposed fault-tolerant control are evaluated through the control of a 3-DOF robot manipulator in both a simulated environment and experiment.

Keywords: active fault-tolerant control; synchronous fast terminal sliding mode control; active fault-tolerant control; robot manipulator control; extended state observer; fault

Citation: Le, Q.D.; Kang, H.-J. An Active Fault-Tolerant Control Based on Synchronous Fast Terminal Sliding Mode for a Robot Manipulator. *Actuators* **2022**, *11*, 195. https://doi.org/10.3390/act11070195

Academic Editors: Marco Carricato and Edoardo Idà

Received: 15 June 2022
Accepted: 13 July 2022
Published: 17 July 2022

Publisher's Note: MDPI stays neutral with regard to jurisdictional claims in published maps and institutional affiliations.

1. Introduction

At present, robot manipulators play an important role in the manufacturing industry and in daily life. In parallel with the development of hardware and advantageous control technology, robot manipulators are required to enhance safety in all situations. Safety is a key role when robots share the workspace with humans or cooperate with humans. Therefore, fault-tolerant control (FTC) was introduced to increase safety and acceptable performance when faults occur in the system. Generally, AFTC can be divided into two main types: (1) active FTC (AFTC) [1–3] and (2) passive FTC (PFTC) [4–6]. Each FTC strategy has advantages and disadvantages depending on the characteristics of the system, the knowledge about the system, and the types of faults. FTC became very important in the systems with human application systems such as aircraft [2], human robot interaction [7], and wearable systems [8] which require high safety and reliability.

In robot manipulator systems, PFTC has a simple architecture with one controller for both normal and fault conditions. PFTC responds quickly to faults due to the elimination of a fault diagnosis process. Some techniques have been classified into PFTC such as sliding mode control (SMC) [9,10], adaptive control [11–13], and so on. For instance, the SMC [10] is widely known due to its robustness and ability to deal with uncertainties and disturbances. In order to guarantee the stability of systems, the design of the SMC required the upper-bound knowledge of faults which is difficult to gain in a real system.

Therefore, PFTC with SMC is less flexible and limited to tolerating the fault's capability in real systems. The second PFTC technique is adaptive control which does not need exact knowledge of faults, but the excessive adaptation rate is a problem with this kind of controller. From the above analyses, it can be seen that PFTC has the advantage in the case of the systems with a good knowledge about the upper bound of uncertainties, disturbances and faults. In contrast, to guarantee the stability and acceptable performance of a system, AFTC uses fault detection and fault diagnosis (FDD) to compensate for the normal controller. In AFTC, the FDD process is a very important process which leads to increasing the ability of tolerating faults. Therefore, FDD [14–17] has been developed with several techniques. The advantage of the AFTC strategy is it has a strong ability to deal with high-magnitude faults and multiple faults. In addition, the upper bound of faults does not require being exactly known in the same way as PFTC does. However, this strategy slowly responds to faults because the FDD process needs time to feed back the information of faults as well as the time delay problem in a real system. Hence, the performance of AFTC is affected. In a real system, the degradation performance of AFTC can be obtained by the occurrence of the picking phenomenon due to the slow response of AFTC and high magnitude of uncertainties and disturbances. In robot manipulator systems, the FDD has been widely replaced by the fault estimation process (FE) which uses fault detection, fault isolation and fault diagnosis within one step. The structure controller is combined with the estimator and is known as disturbance observer base (DOB). The difference between DOB and AFTC with the estimator is that in AFTC with an estimator, the output value is used to compare the threshold to detect faults and isolation faults. It warns the operator about the occurrence of faults to make the effective action. The AFTC with a structure of DOB control (AFTC-DOB) [18–20] may have an effect in the case of bias faults. These kinds of faults can be considered as uncertainties or disturbances so that the AFTC-DOB can show acceptable performance depending on the accuracy of the estimator. However, with the loss-effective fault or the combination of loss-effective and bias faults, using merely AFTC-DOB may not reduce the effect of this kind of fault. Hence, AFTC needs a different way to handle the faults due to the occurrence of loss-effective faults. Most of the research studies on FTC [1,9,21,22] focus on the development of the ability to deal with the uncertainties and disturbances or handle the conventional problems of controllers such as the chattering phenomenon or the fast convergence problem. However, they lack the picking phenomenon of the AFTC strategy which has the most impact in degrading the performance of AFTC. Therefore, in this paper, the synchronization technique is proposed to suppress the picking phenomenon when faults occur, especially with the occurrence of loss-effective for the AFTC strategy.

In the field of control, the synchronization control has been widely known with a closed-loop chain mechanism such as the dual-drive gantry mechanism [23], parallel robot manipulator [24], cable-driver parallel manipulator [25], cooperation robot manipulator [26], and so on. These systems have an internal tensor force during the motion due to the closed-loop mechanism. This kind of force can degrade the performance of systems. By using the synchronization technique, the controller makes the position errors at each joint simultaneously approach zero and reduce the internal tensor force. Therefore, the accuracy of systems can be increased. Unlike a closed-loop mechanism, a serial robot manipulator does not contain an internal tensor force. Due to the open-loop mechanism, this internal uncertainty can be compensated by the consideration of a dynamic model. Therefore, the synchronization technique does not show an advantage compared the model-based controller such as DOB in normal operation. However, the ability to make the position error simultaneously equal to and approach zero still remains in an open-loop mechanism with the synchronization technique. In case faults occur, the internal constraints of the synchronization control to keep the error at each joint equal may suppress the effect of the faults. The synchronization controller increases the ability to quickly respond to the controller before it receives the feedback information of faults from FDD or FE. In [27], the authors first proposed the synchronization technique to reduce the effect of faults. In this study, the experimental results showed the outperformance of the proposed controller. However, it can be seen that the proposed controller in [27] did not consider the finite-time convergence issue. In [28], the finite-time synchronization controller was addressed with two proposed controllers. With the first controller in [27,28], it is unclear whether the ability to reduce the effect of faults comes from the synchronization technique or from the integral term of the cross-coupling technique in the design-sliding surface. In [28], the second proposed controller showed the effectiveness of the synchronization technique without the cross-coupling technique. Motivated by that proposed controller, in this paper, a novel AFTC with a finite-time synchronous fast terminal sliding mode control is proposed. In addition, the analysis of synchronization parameters is presented to show the effect of the synchronization technique in fault-tolerant control. In this paper, the picking phenomenon is the specific problem of the AFTC strategy which is analyzed and considered. Then, it can be tackled by using the synchronization technique to guarantee the acceptable performance of a robot manipulator when faults occur. Due to the internal constraints of position error at each joint of the synchronization technique, the proposed controller can reduce the picking effect before the FE receives the information of faults.

In this paper, AFTC using the combination of an extended state observer and novel finite-time synchronous faster terminal sliding mode control is proposed. Firstly, an extended state observer (ESO) [29] is adopted to estimate the lumped uncertainties, disturbances, and faults. The observer has a role as the FDD includes fault detection, fault isolation, and fault estimation within one step. In addition, by using a high-gain technique, the ESO does not require the upper bound information of faults and simple adjustment of the observer parameter. Therefore, the ESO can easily apply in real robot manipulator systems. Next, an AFTC with a novel synchronous fast terminal sliding mode control is proposed without the cross-coupling technique. The synchronization error in [28] was used to guarantee the finite-time convergence of position error at each joint. Finally, the proposed controller is verified on a 3-DOF robot manipulator in simulations and experiments. The results of the proposed control are compared to AFTC with conventional fast terminal sliding mode control [30]. In addition, the analysis effect of synchronization parameters to the fault-tolerant control is presented in a real implementation system. The contribution of this paper can be summarized as follows:

(1) An active fault-tolerant control based on the combination of novel finite-time synchronous fast terminal sliding mode control and an extended state observer is proposed. The proposed control has the ability to quickly respond to faults, before the controller has the feedback information of faults from fault estimation due to the internal constraints of the synchronization technique. Therefore, the proposed control can reduce the picking phenomenon and suppress the effect of faults to show an acceptable performance when it is compared to AFTC with conventional finite-time fast terminal sliding mode control [30]. The proposed controller has the ability to maintain the accepted performance with the occurrence of faults in the system until the maintenance.

(2) By using a novel finite-time synchronous fast terminal sliding surface without the coupling technique, the proposed controller has fast computation, avoids oscillation due to the integral term, and guarantees the finite-time convergence of the position error at each joint compared [27].

(3) The analysis of synchronization parameters is presented to show the effect and behavior of the synchronization technique in fault-tolerant control. Unlike the conventional AFTC controller, the proposed controller can make the error at each joint simultaneously approach zero. This feature can reduce the effect of fault in the robot systems.

(4) Comparing the proposed controller in this paper with the first controller in [28], it can be seen that without the conventional coupling technique, the proposed AFTC can reduce the time computation, avoid the oscillation due to the integral term, and guarantee the finite-time convergence of position errors and fast convergence. In addition, the proposed control in this paper converges faster than the two synchronous sliding mode surfaces in [28]. Therefore, in this proposed controller, the fault-tolerant controller has the ability to reduce the picking phenomenon due to the synchronous technique, as well as the time convergence, faster than in [28]. The proposed controller in this paper has more advantages than two controllers in [28] in both time convergence and time computation.

(5) The implementation of advantage control in real robot systems faces many challenges such as the noise of sensors and output actuators, time of computation, the drawbacks of the sliding mode technique in real systems, and so on. However, in this paper, the experimental results on a real 3-DOF robot manipulator demonstrated the effectiveness and ability to apply to real systems of the proposed controller.

The rest of this paper is organized as follows. In Section 2, the dynamic model of robot manipulators and faults is presented. The design of fault estimation based on an extended state observer is shown in Section 3. In Section 4, a novel finite-time synchronous fast terminal sliding mode control is proposed. The simulation results and some discussions about the effect of kinds of faults are given in Section 5. In Section 6, the experimental results are shown to verify the effectiveness of synchronization technique and the effect of synchronization parameter selection in fault-tolerant control. Finally, the conclusions are given in Section 7.

2. Dynamics Model of Robot Manipulators and Faults

2.1. Dynamics Model of Robot Manipulators

The dynamics of an n-degrees-of-freedom robot manipulator was defined as

$$M(q)\ddot{q} + C(q,\dot{q})\dot{q} + G(q) = \tau \tag{1}$$

where $\ddot{q}, \dot{q}, q \in \Re^n$ are the vectors of joint acceleration, velocity and position, respectively. $M(q) \in \Re^{n \times n}, C(q,\dot{q}) \in \Re^{n \times n}$ and $G(q) \in \Re^n$ represent the inertia matrix, the centripetal and Coriolis matrix, and the vector of gravitation force, respectively. $\tau \in \Re^n$ is the vector of torque at the joints.

In practice, the dynamic model of robots is not known exactly, so the system in (1) can be written as

$$(\mathbf{M}(q) + \Delta M(q))\ddot{q} + (\mathbf{C}(q,\dot{q}) + \Delta C(q,\dot{q}))\dot{q} + (\mathbf{G}(q) + \Delta G(q)) + \delta = \tau \tag{2}$$

where $\Delta M, \Delta C$ and ΔG are unknown dynamic uncertainties and δ is external disturbance. $\mathbf{M}(q), \mathbf{C}(q,\dot{q}), and \mathbf{G}(q)$ are estimation values of $M(q), C(q,\dot{q}), and G(q)$. Thus, (2) can be simply rewritten as

$$\mathbf{M}(q)\ddot{q} + \mathbf{C}(q,\dot{q})\dot{q} + \mathbf{G}(q) + \psi = \tau \tag{3}$$

where $\psi = \Delta M\ddot{q} + \Delta C\dot{q} + \Delta G + \delta$.

2.2. Mathematical Calculation of Joint Actuator Faults

In a robot manipulator, the motions of the robot are generated by a joint actuator which consist of an electronic motor, a motor driver, and a gear. In practice, the occurrence of faults/failure in the electric motor, the gear, the bearing, and the motor driver can degrade the performance of a robot manipulator. For instance, in the electric motor, the increased friction between the stator and the rotor or fault in the driver motor may decrease the motor torque. This can be considered as the loss-effective fault. Generally, the two kinds of faults which commonly occur in a robot manipulator are bias and loss-effective fault, which can be described as

$$\tau^t = (I - \rho(t))\tau + \mathbf{f}(t) \ (t > t_f) \tag{4}$$

where $\tau^t \in \Re^n$ is the vector of torque at the output joint actuator. $\tau \in \Re^n$ is the vector of torque at the output controller. $\rho(\tau) = diag(\rho_i(t)) \in \Re^{n \times n}, 0 \leq \rho_i(t) < 1, (i = 1, 2, \ldots, n)$ denotes the loss-effective rate. $\mathbf{f}(t) \in \Re^n$ is the vector of the bias faults. t_f is the time of the occurrence of faults. $I \in \Re^{n \times n}$ is the identity matrix.

Substituting (4) into (3), the dynamics model of an n-degrees-of-freedom robot manipulator with actuator faults can be written as

$$\mathbf{M}(q)\ddot{q} + \mathbf{C}(q,\dot{q})\dot{q} + \mathbf{G}(q) + \psi = (I - \rho(t))\tau + \mathbf{f}(t) \tag{5}$$

Remark 1. *In this paper, only the loss-effective and bias faults are investigated. The loss-effective fault with $\rho_i(t) = 1$ and lock-in-place fault are not considered because the robot does not have the redundancy actuator, so the robot cannot tolerate those kinds of faults.*

3. Fault Estimation with Extended State Observer

In this section, an extended state observer to estimate the lumped uncertainties, disturbances, and faults is presented.

The dynamic model of the robot manipulator of (5) can be rewritten as

$$\ddot{q} = \mathbf{M}^{-1}(q)(\tau - \mathbf{H}(q,\dot{q})) - \mathbf{M}^{-1}(q)\zeta \tag{6}$$

where $\mathbf{H}(q,\dot{q}) = \mathbf{C}(q,\dot{q})q + \mathbf{G}(q)$, $\zeta = \rho(t)\tau + \psi - \mathbf{f}(t)$ represents uncertainties, disturbances, and faults or failures.

The dynamic model of (6) can be rewritten in the state space as

$$\begin{cases} \dot{x}_1 = x_2 \\ \dot{x}_2 = f(x_1, x_2, \tau) + \phi \end{cases} \tag{7}$$

where $x_1 = q \in \Re^n$, $x_2 = \dot{q} \in \Re^n$, $f(x_1, x_2, \tau) = \mathbf{M}^{-1}(q)(\tau - \mathbf{H}(q,\dot{q}))$, and $\phi = -\mathbf{M}^{-1}(q)\zeta$.

An extended state observer [29] is given as

$$\begin{cases} \dot{\hat{x}}_1 = \hat{x}_2 + \frac{\alpha_1}{\varepsilon}(x_1 - \hat{x}_1) \\ \dot{\hat{x}}_2 = f(\hat{x}_1, \hat{x}_2, \tau) + \frac{\alpha_2}{\varepsilon^2}(x_1 - \hat{x}_1) + \hat{\phi} \\ \dot{\hat{\phi}} = \frac{\alpha_3}{\varepsilon^3}(x_1 - \hat{x}_1) \end{cases} \tag{8}$$

where $\hat{x}_1, \hat{x}_2$ and $\hat{\phi}$ are estimates of x_1, x_2 and ϕ, respectively, α_1, α_2 and α_3 are positive constants, polynomial $s^3 + \alpha_1 s^2 + \alpha_2 s + \alpha_3$ is Hurwitz, and $0 < \varepsilon < 1$.

Lemma 1 [31]. *Considering a function $f \in \Re^n$, f is $\gamma-$Lipschitz with the respective argument*

$$\|f(x) - f(y)\| \leq \gamma\|x - y\| \quad \forall x, y \in \Re^n \tag{9}$$

Theorem 1. *Considering the system (7) with observer (8) and satisfying $0 < \varepsilon < 1$, $\left|\dot{\phi}\right| \leq L$, and exist $P = P^T > 0$, then $\widehat{x}_1(t) \to x_1(t)$, $\widehat{x}_2(t) \to x_2(t)$ and $\widehat{\phi}(q,\dot{q},t) \to \phi(q,\dot{q},t)$ as $t \to \infty$.*

Proof . The error dynamics from (7) and (8) can be written as

$$\begin{cases} \dot{\tilde{e}}_1 = \tilde{e}_2 + \frac{\alpha_1}{\varepsilon}\tilde{e}_1 \\ \dot{\tilde{e}}_2 = \Delta f + \frac{\alpha_2}{\varepsilon^2}\tilde{e}_1 + \tilde{e}_3 \\ \dot{\tilde{e}}_3 = \dot{\phi} - \frac{\alpha_3}{\varepsilon^3}\tilde{e}_1 \end{cases} \tag{10}$$

where $\tilde{e}_1 = x_1 - \hat{x}_1$, $\tilde{e}_2 = x_2 - \hat{x}_2$, $\tilde{e}_3 = \phi - \hat{\phi}$ and $\Delta f = f(x_1, x_2, \tau) - f(\hat{x}_1, \hat{x}_2, \tau)$.

We define the Lyapunov function as

$$V = \varepsilon\tilde{e}^T P\tilde{e} \tag{11}$$

where $\tilde{e} = [\tilde{e}_1, \tilde{e}_2, \tilde{e}_3]^T$

The time derivative of V in Equation (11) is

$$\dot{V} = \varepsilon\dot{\tilde{e}}^T P\tilde{e} + \varepsilon\tilde{e}^T P\dot{\tilde{e}} \tag{12}$$

where

$$
\varepsilon\dot{\tilde{e}} = \begin{bmatrix} \varepsilon\dot{\tilde{e}}_1 \\ \varepsilon\dot{\tilde{e}}_2 \\ \varepsilon\dot{\tilde{e}}_3 \end{bmatrix} = \begin{bmatrix} -\frac{\alpha_1}{\varepsilon^2}\tilde{e}_1 + \frac{1}{\varepsilon}\tilde{e}_2 \\ \varepsilon\Delta f - \frac{\alpha_2}{\varepsilon^2}\tilde{e}_1 + \tilde{e}_3 \\ \varepsilon\dot{\phi} - \frac{\alpha_3}{\varepsilon^2}\tilde{e}_1 \end{bmatrix}
$$

$$
= \begin{bmatrix} -\alpha_1\tilde{e}_1 + \tilde{e}_2 \\ -\alpha_2\tilde{e}_1 + \tilde{e}_3 \\ -\alpha_3\tilde{e}_1 + \varepsilon\dot{\phi} \end{bmatrix} = A\tilde{e} + \varepsilon B_1 \Delta f + \varepsilon B_2 \dot{\phi}
\tag{13}
$$

and

$$
A = \begin{bmatrix} -\alpha_1 & 1 & 0 \\ -\alpha_2 & 0 & 1 \\ -\alpha_3 & 0 & 0 \end{bmatrix}, B_1 = \begin{bmatrix} 0 \\ 1 \\ 0 \end{bmatrix} \text{ and } B_2 = \begin{bmatrix} 0 \\ 0 \\ 1 \end{bmatrix}.
\tag{14}
$$

Substituting (13) into (12)

$$
\begin{aligned}
\dot{V} &= \varepsilon\dot{\tilde{e}}^T P\tilde{e} + \varepsilon\tilde{e}^T P\dot{\tilde{e}} \\
&= \left(A\tilde{e} + \varepsilon B_1\Delta f + \varepsilon B_2\dot{\phi}\right)^T P\tilde{e} + \tilde{e}^T P(A\tilde{e} + \varepsilon B_1\Delta f + \varepsilon B_2\dot{\phi}) \\
&= \tilde{e}^T A^T P\tilde{e} + \varepsilon(B_1\Delta f)^T P\tilde{e} + \varepsilon(B_2\dot{\phi})^T P\tilde{e} + \tilde{e}^T PA\tilde{e} + \varepsilon\tilde{e}^T PB_1\Delta f + \varepsilon\tilde{e}^T PB_2\dot{\phi} \\
&= \tilde{e}^T(A^T P + PA)\tilde{e} + 2\varepsilon\tilde{e}^T PB_1\Delta f + 2\varepsilon\tilde{e}^T PB_2\dot{\phi} \\
&= -\tilde{e}^T Q\tilde{e} + 2\varepsilon\tilde{e}^T PB_1\Delta f + 2\varepsilon\tilde{e}^T PB_2\dot{\phi}
\end{aligned}
\tag{15}
$$

where

$$
A^T P + PA = -Q
\tag{16}
$$

From Lemma 1, we can obtain

$$
\Delta f = f(x_1, x_2, \tau) - f(\hat{x}_1, \hat{x}_2, \tau) \leq \gamma\|\tilde{e}_{12}\|
\tag{17}
$$

where $\tilde{e}_{12} = [\tilde{e}_1, \tilde{e}_2]^T$ and we also obtain

$$
\|\tilde{e}_{12}\| = \sqrt{\tilde{e}_1^2 + \tilde{e}_2^2} \leq \sqrt{\tilde{e}_1^2 + \tilde{e}_2^2 + \tilde{e}_3^2} = \|\tilde{e}\|
\tag{18}
$$

so that

$$
2\varepsilon\tilde{e}^T PB_1\Delta f \leq 2\varepsilon\gamma\|PB_1\|\|\tilde{e}\|^2
\tag{19}
$$

Substituting (18) into (14), we obtain

$$
\begin{aligned}
\dot{V} &= \varepsilon\dot{\tilde{e}}^T P\tilde{e} + \varepsilon\tilde{e}^T P\dot{\tilde{e}} \\
&= -\tilde{e}^T Q\tilde{e} + 2\varepsilon\tilde{e}^T PB_1\Delta f + 2\varepsilon\tilde{e}^T PB_2\dot{\phi} \\
&\leq -\tilde{e}^T Q\tilde{e} + 2\varepsilon\gamma\|PB_1\|\|\tilde{e}\|^2 + 2\varepsilon L\|PB_2\|\|\tilde{e}\| \\
&\leq -(\lambda_{\min}(Q) - 2\varepsilon\gamma\|PB_1\|)\|\tilde{e}\|^2 + 2\varepsilon L\|PB_2\|\|\tilde{e}\|
\end{aligned}
\tag{20}
$$

To guarantee the stability of the system, we impose $\dot{V} < 0$ so the convergence is given as

$$
\|\tilde{e}\| \leq \frac{2\varepsilon L\|PB_2\|}{\lambda_{\min}(Q) - 2\varepsilon\gamma\|PB_1\|}
\tag{21}
$$

$\square$

Remark 2. *The value of L does not need to be exactly known. However, this value always exists to make sure the maximum torque generated at the actuators is able to attenuate ϕ. Therefore, robot manipulators can be controllable.*

4. Active Fault-Tolerant Control with Finite-Time Synchronous Fast Terminal Sliding Mode Control

In this section, the proposed active fault-tolerant control with finite-time synchronous fast terminal sliding mode control (AFTC S-FTSMC) is presented. Some definitions will be useful in the rest of the paper.

Definition 1. *We define* $\lceil x \rfloor^{\Lambda} = \left[|\ x_1\ |^{\lambda_1} sign(x_1), |\ x_2\ |^{\lambda_2} sign(x_2), \ldots, |\ x_n\ |^{\lambda_n} sign(x_n) \right]^{T} \in \Re^n$ *where* $\lambda_i (i = 1, 2, \ldots, n) > 0$ *and* $\Lambda = diag(\lambda_i)$. $x = [x_1, x_2, \ldots, x_n]^{T} \in \Re^n$ *and* $y = [y_1, y_2, \ldots, y_n]^{T} \in \Re^n$.

Definition 2. *We define* $x \cdot y = [x_1 y_1, x_2 y_2, \ldots, x_n y_n]^{T} \in \Re^n$.

Definition 3. *The time derivative of* $\lceil x \rfloor^{\Lambda}$ *is*

$$\frac{d}{dt} \lceil x \rfloor^{\Lambda} = \Lambda |x|^{\Lambda - 1} \cdot \dot{x} = \left[\lambda |x_1|^{\lambda - 1} \dot{x}_1, \lambda |x_2|^{\lambda - 1} \dot{x}_2, \ldots, \lambda |x_n|^{\lambda - 1} \dot{x}_n \right]^{T},$$

and $x^{\Lambda - I} = \left[x_1^{\lambda_1 - 1}, x_2^{\lambda_2 - 1}, \ldots, x_n^{\lambda_n - 1} \right] \in \Re^n$ *where* $I = diag(1) \in \Re^{n \times n}$.

The synchronization error [28] was given as:

$$
\begin{aligned}
\varepsilon_1 &= (1 + \psi_1 \psi_n) e_1 - \psi_1 e_2 - \psi_1 e_n \\
\varepsilon_2 &= (1 + \psi_2 \psi_1) e_2 - \psi_2 e_3 - \psi_2 e_1 \\
&\vdots \\
\varepsilon_n &= (1 + \psi_n \psi_{n-1}) e_n - \psi_n e_2 - \psi_n e_1
\end{aligned}
\tag{22}
$$

where $e_i (i = 1, 2, \ldots, n)$ is the error at each joint, and $\psi_i (i = 1, 2, \ldots, n)$ is the corresponding positive gains. In matrix form

$$\varepsilon = Te \tag{23}$$

where $\varepsilon = [\varepsilon_1, \varepsilon_2, \ldots, \varepsilon_n]^{T} \in \Re^n$, $e = [e_1, e_2, \ldots, e_n]^{T} \in \Re^n$, $T \in \Re^{n \times n}$, and

$$
T = \begin{bmatrix}
(1 + \psi_1 \psi_n) & -\psi_1 & 0 & \cdots & -\psi_1 \\
-\psi_2 & (1 + \psi_2 \psi_1) & -\psi_2 & \cdots & 0 \\
0 & -\psi_3 & (1 + \psi_3 \psi_2) & \cdots & 0 \\
\vdots & & & & \\
-\psi_n & -\psi_n & 0 & \cdots & (1 + \psi_n \psi_{n-1})
\end{bmatrix}
\tag{24}
$$

A novel synchronous fast terminal sliding mode surface is proposed as

$$S = \dot{e} + \alpha \varepsilon + \beta \lceil \varepsilon \rfloor^{\Lambda} \tag{25}$$

where $\alpha = diag(\alpha_i) \in \Re^{n \times n}$ and $\beta = diag(\beta_i) \in \Re^{n \times n}$ are positive matrix gain. $\Lambda = diag(\lambda_i) \in \Re^{n \times n}$ with $0 < \lambda_{1i} < 1$.

The proposed active fault-tolerant control with synchronous fast terminal sliding mode (AFTC S-FTSMC) is given as

$$\tau = \tau_{eq} + \tau_0 + \tau_{ob} \tag{26}$$

where $\tau_{eq} = \mathbf{M}(q)(\ddot{q}_d + \alpha \dot{e} + \beta \Lambda |\varepsilon|^{\Lambda_2 - I} \cdot \dot{\varepsilon} + \mathbf{H}(q, \dot{q}))$, $\tau_0 = \mathbf{M}(q) K_1 sign(S)$, $\tau_{ob} = -\mathbf{M}(q) \hat{\phi}$, and $K_1 = diag(k_{1i}) \in \Re^{n \times n}$.

Theorem 2. *The system described in (5), using the controller specified in (25), guarantees that* $e \rightarrow 0$ *is finite time.*

Proof. The Lyapunov function can be selected as

$$V = \frac{1}{2}S^T S \tag{27}$$

The time derivative of V in (26) is

$$\begin{aligned}
\dot{V} &= S^T \dot{S} \\
&= S^T \left(\ddot{e} + \alpha \dot{e} + \Lambda \beta |\varepsilon|^{\Lambda - I} \cdot \dot{\varepsilon} \right) \\
&= S^T \left(\ddot{q}_d - \ddot{q} + \alpha \dot{e} + \Lambda \beta |\varepsilon|^{\Lambda - I} \cdot \dot{\varepsilon} \right) \\
&= S^T \left(\ddot{q}_d - \mathbf{M}^{-1}(q)(\tau - \mathbf{H}(q, \dot{q})) + \mathbf{M}^{-1}(q)\zeta(q, \dot{q}, \tau, t) + \alpha \dot{e} + \Lambda \beta |\varepsilon|^{\Lambda - I} \cdot \dot{\varepsilon} \right)
\end{aligned} \tag{28}$$

Substituting (25) into (27), we have

$$\begin{aligned}
\dot{V} &= -S^T K_1 sign(S) \\
&\le -\sigma_1 V^{\frac{1}{2}} < 0
\end{aligned} \tag{29}$$

where $\sigma_1 = \lambda_{\min}(K_1)$. When the sliding mode achieves $S = 0$ and converges, then $E = 0$ and $\dot{E} = 0$, and we have:

$$\begin{aligned}
\dot{e}_i &= -\alpha_i \varepsilon_i - \beta_i \lceil \varepsilon_i \rceil^\lambda \\
&= -\alpha_i((1 + \psi_i \psi_{i-1})e_i - \psi_i e_{i-1} - \psi_i e_{i+1}) \\
&\quad - \beta_i |(1 + \psi_i \psi_{i-1})e_i - \psi_i e_{i-1} - \psi_i e_{i+1}|^\lambda sign((1 + \psi_i \psi_{i-1})e_i - \psi_i e_{i-1} - \psi_i e_{i+1})
\end{aligned} \tag{30}$$

The system in (29) has equilibrium points at $e_i = 0$ ($i = 1, 2, \ldots, n$; $n + 1 = 1$). According to the definition of terminal attractors [32], we have

$$\left| \frac{\partial \dot{e}_i}{\partial e_j} \right| = -\alpha_i \left| \frac{\partial \varepsilon_i}{\partial e_j} \right| - \beta_i \lambda |(1 + \psi_i \psi_{i-1})e_i - \psi_i e_{i-1} - \psi_i e_{i+1}|^{\lambda - 1} \left| \frac{\partial \varepsilon_i}{\partial e_j} \right| = \infty \tag{31}$$

where $j = (i - 1, i, i + 1)$ $(i = 1, 2, \ldots, n)$.

From (5), we have $e_i \to 0 (i = 1, 2, \ldots, n)$ at finite time. Therefore, Theorem 1 is proven. $\square$

Remark 3. *The time convergence shows that*

$$t = t_r + t_{e_i} \tag{32}$$

Time convergence $S \to 0$:

$$t_r \le \frac{2V^{\frac{1}{2}}(0)}{\sigma_1} \tag{33}$$

Time convergence $e_i \to 0$

$$t_{e_i} = \frac{1}{\alpha_i(1 - \lambda_i)} \left(\ln \left(\alpha_i(ae_i(0) + be_{i-1}(0) + ce_{i+1}(0))^{1 - \lambda_i} \right) - \ln(\beta_i) \right) \tag{34}$$

where $a = 1 + \psi_i \psi_{i-1}$, $b = -\psi_i$, and $c = -\psi_i$.

Remark 4. *A singularity may occur at $|\varepsilon|^{\Lambda - I}$ in (25) when the synchronization error approaches or crosses zero. By using saturation function $sat(x^{\Lambda - I}, u_s), x = |\varepsilon|$ where $u_s > 0$, the singularity can be avoided and the system retains finite-time stability* [33].

5. Simulation Results

This section, the comparison of the proposed AFTC S-FTSMC and an AFTC with the combination of an conventional fast terminal sliding mode and an extended state observer ([30]+ESO) for the 3-DOF manipulator, is presented with different kinds of faults. In this simulation, the geometry parameters from the catalog of a FARA-AT2 robot manipulator were used to establish the 3-DOF robot manipulator. In Table 1, the dynamics parameters of the robot manipulator in Solidworks are shown. From Solidworks, a 3-DOF robot manipulator was imported to Matlab environment through Simmechanics with full information of the dynamics parameters. The robot in the simulation environment was shown in Figure 1.

Table 1. Dynamics Parameters of 3-DOF Robot Manipulator.

Links	Length (m)	Weight (kg)	Center of Mass (m)	Inertia (kg·m^2)
Link 1	0.15	56.5	[0.05 0 0.09]	Ix = [0.86 0 0.52] Iy = [0 1 0] Iz = [−0.52 0 0.86]
Link 2	0.255	35.6	[−0.01 0 0.1]	Ix = [−0.01 0 1] Iy = [0 −1 0] Iz = [1 0 1]
Link 3	0.41	58.9	[0.05 0 0.08]	Ix = [1 0 0.08] Iy = [−0.01 1 0.08] Iz = [−0.08 −0.08 0.99]

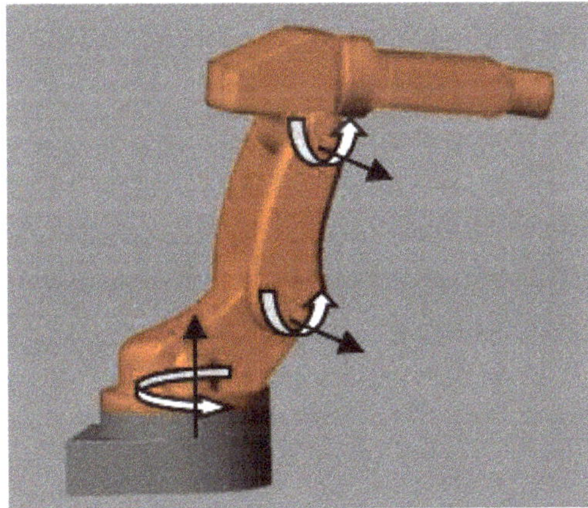

Figure 1. The simulation of 3-DOF Robot Manipulator in MATLAB/Simulink.

For this trajectory-tracking simulation, the desired position trajectories at each joint are given as

$$\begin{cases} q_1 = 0.5\cos(t/2) - 0.5 \\ q_2 = 0.3 * \cos(t) - 0.3 \\ q_3 = 0.2 * \cos(t) - 0.2 \end{cases} \tag{35}$$

The disturbance at each joint is assumed to be

$$\begin{cases} \delta_{1f} = 0.2sgn(\dot{q}_1) + 0.3\dot{q}_1 \\ \delta_{2f} = 0.2sgn(\dot{q}_2) + 0.3\dot{q}_2 \\ \delta_{3f} = 0.2sgn(\dot{q}_3) + 0.3\dot{q}_3 \end{cases} \tag{36}$$

The related parameters for the ESO were chosen to be $\alpha_1 = 8, \alpha_2 = 28, \alpha_3 = 7$, and $\varepsilon = 0.01$. The controller [30]+ESO is given as:

$$\tau_{[25]+ESO} = \tau_0 + \tau_{smc} + \tau_{ob} \tag{37}$$

where $\tau_{eq} = \mathbf{M}(q)(\ddot{q}_d + d\dot{e} + c\Lambda|e|^{\Lambda-I} \cdot \dot{e}) + \mathbf{H}(q,\dot{q})$, $\tau_o = \mathbf{M}(q)K_1 sign(S)$, and $\tau_{ob} = -\mathbf{M}(q)\hat{\phi}$, and $K_1 = diag(k_{1i}) \in \Re^{n \times n}$. The conventional fast terminal sliding surface [30] was selected as:

$$S = \dot{e} + de + c\lceil e \rfloor^{\Lambda} \tag{38}$$

The parameters for the [30]+ESO were suitably chosen as $c = diag(0.5; 0.5; 0.5)$, $d = diag(0.72; 0.72; 0.72)$, $K_1 = diag(80; 80; 110)$ and $\Lambda = diag(0.6; 0.6; 0.6)$.

The parameters for the AFTC S-FTSMC were chosen as $\psi_1 = \psi_2 = \psi_3 = 2$, $\alpha = diag(0.5; 0.5; 0.5)$, $\beta = diag(0.72; 0.72; 0.72)$, $\Lambda = diag(0.6; 0.6; 0.6)$, and $K_1 = diag(80; 80; 110)$.

To avoid a singularity, the terms containing power $\Lambda - I$ in (25) and (36) were replaced with the saturation function.

$$sat(u_f, u_s) = \begin{cases} u_s & if\ u_f \geq u_s \\ u_f & if\ u_f < u_s \end{cases}$$

where $u_s = 0.3$ is a positive constant, and $u_f = \Lambda|x|^{\Lambda-I} \cdot \dot{x}$ with $x = e$ and ε.

To avoid chattering, the signum function in (25) and (36) was replaced with the saturation function.

$$sat(s) = \begin{cases} sgn(s) & if|s| \geq \lambda \\ \frac{s}{\lambda} & if|s| < \lambda \end{cases}$$

where $\lambda = 0.8$.

5.1. Simulation 1

In this subsection, the comparison of two AFTCs (AFTC S-FTSMC and [30]+ESO) is described with the combination of loss-effective and bias fault at joint 2 at the fifth second. The total torque function at each joint was assumed to be:

$$\begin{cases} \tau_1^t = \tau_1 \\ \tau_2^t = (1 - \rho_2(t))\tau_2 + f_2(t)\ t \geq 5 \\ \tau_3^t = \tau_3 \end{cases} \tag{39}$$

where $\rho_2(t) = 0$ and $f_2(t) = 200 \sin(\pi(t-5)/2)$.

In Figure 2, the estimation of a fault is shown. It can be seen that the FE performance does not have high accuracy, but with that magnitude error in the FE, the sliding mode control technique can handle uncertainties or disturbances. It can be seen in Figure 3. Before the fifth second, both controllers have similar accuracy because in normal operation mode, the synchronization control does not show the effectiveness in an open-loop mechanism that serial robot manipulators do. After the fifth second, the proposed control AFTC S-FTSMC can significantly reduce the effect of the fault when compared with AFTC with [30]+ESO, because in this case, the internal constraints of the synchronization technique can suppress the effect of the fault to make the position error at each joint tend to be equal. However, the accuracy of [30]+ESO is still inside the order 10^{-3} radian, so it can be said that [30]+ESO showed acceptable performance in this case. From these results, both controllers have the ability to guarantee an acceptable performance with this kind of fault.

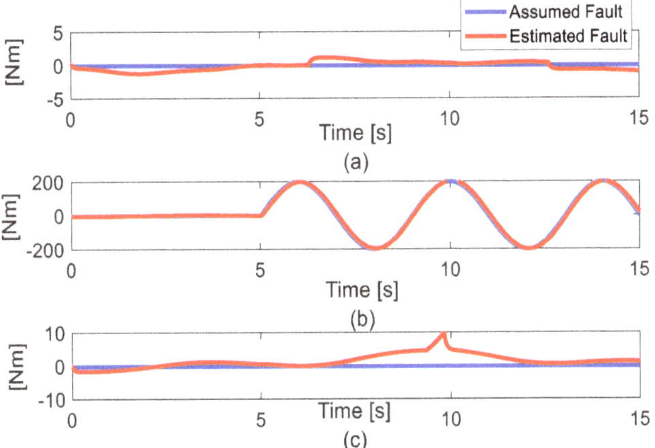

Figure 2. Fault estimation results. (**a**) Joint 1; (**b**) Joint 2; (**c**) Joint 3.

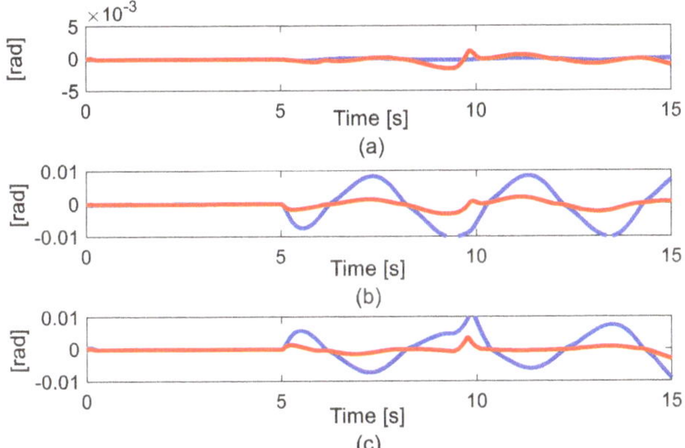

Figure 3. Tracking error results. Blue line is [30]+ESO controller. Red line is the proposed controller. (**a**) Joint 1; (**b**) Joint 2; (**c**) Joint (3).

5.2. Simulation 2

In this subsection, the total two kinds of actuator fault with a similar magnitude of fault to Simulation 1 are considered. The assumed fault function is shown as

$$
\begin{cases}
\tau_1^t = \tau_1 \\
\tau_2^t = (1 - \rho_2(t))\tau_2 + f_2(t) \ t \geq 5 \\
\tau_3^t = \tau_3
\end{cases}
\tag{40}
$$

where $\rho_2(t) = -0.4\sin(\pi(t))$ and $f_2(t) = 90\sin(\pi(t)/2)$.

In Figure 4, the magnitude of fault is similar to the case in Simulation 1, but in this case, the effect of actuator loss-effective fault is considered. In Figure 5, after the fifth second, by using AFTC with [30]+ESO, the picking values occur in joint 2 and joint 3 at 10.42 s and 14.48 s, respectively. The acceptable performance of AFTC with [30]+ESO cannot be guaranteed in this case. In contrast, the proposed AFTC S-FTSMC still maintains the accuracy inside order 10^{-3} rad. This causes the effectiveness of synchronization control which can suppress the effect of the fault. However, it can be seen that the tracking error in joint 1 slightly increases due to the synchronization technique which also makes the error at each joint tend to be equal. This characteristic is the disadvantage of the synchronization technique.

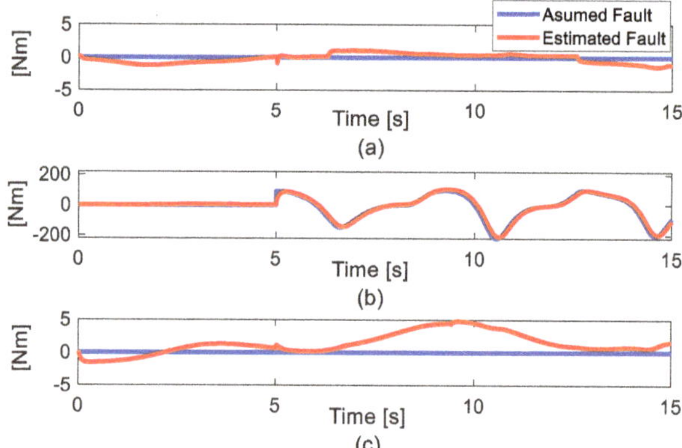

Figure 4. Fault estimation results. (**a**) Joint 1; (**b**) Joint 2; (**c**) Joint (3).

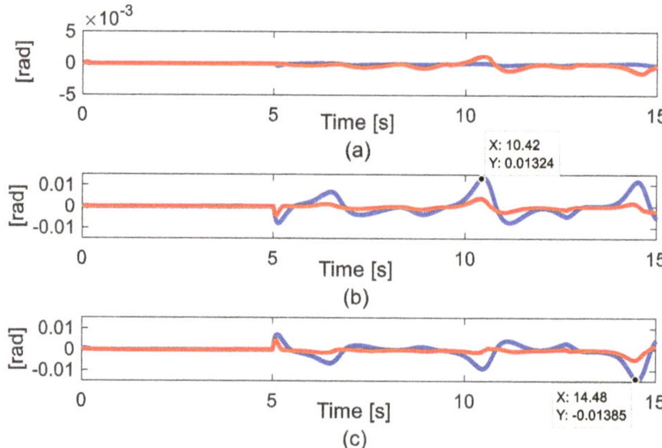

Figure 5. Tracking error results. Blue line is [30]+ESO controller. Red line is the proposed controller. (**a**) Joint 1; (**b**) Joint 2; (**c**) Joint (3).

Remark 5. *From the results in Simulation 1 and 2, it can be seen that the different kinds of faults can highly affect the performance of the active fault-tolerant control strategy despite having similar magnitude at the actuators. Therefore, to prevent the effect of faults, AFTC needs a different solution for the way it deals with uncertainties and disturbances. AFTC with the conventional controller technique may not be enough to guarantee an acceptable performance in the robot manipulators when faults occur. In contrast, with the synchronization technique, the proposed AFTC showed effectiveness in tolerating faults despite this technique having no effect in the normal operation mode.*

6. Experimental Results

6.1. Experiment Setup

The experimental setup is shown in Figure 6 and uses a 3-DOF FARA-AT2 robot manipulator. This robot manipulator has 6 DOF, but for these experiments, joints 4, 5, and 6 were blocked. The 3-DOF FARA-AT2 robot had a CSMP series motor at each joint. The gear box at each joint was 120:1, 120:1, and 100:1 at joints 1, 2, and 3, respectively. The encoder at each joint was a 2048 line count incremental encoder. The controller was run on Labview-FPGA, NI-PXI-8110, and NI-PXI-7842R PXI cards with the frequency control set at 500 Hz. The NI-PXI-8110 was run on a Windows operating system.

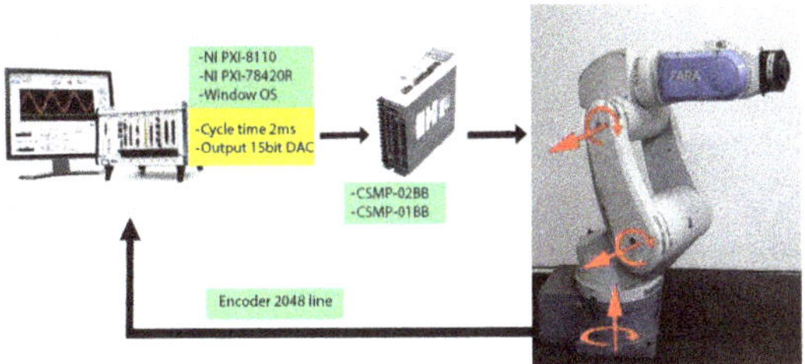

Figure 6. Experiment setup for 3-DOF FARA-AT2 robot manipulator.

The desired trajectory at each joint is given as:

$$q_{id}(t) = \frac{\pi}{6}\sin(\frac{\pi t}{1600})(i = 1,2,3) \tag{41}$$

The related parameters of the fault estimation (8) can be suitably selected as $\alpha_1 = 2, \alpha_2 = 3, \alpha_3 = 0.3$, and $\varepsilon = 0.01$.

The parameters of AFTC with [30]+ESO in (36) were suitably selected as $c = diag(0.2; 0.2; 0.2)$, $d = diag(0.1; 0.1; 0.1)$, $\Lambda = diag(0.62; 0.62; 0.62)$ and $K_1 = diag(30; 30; 50)$.

The parameters of the proposed AFTC S-FTSMC (25) were suitably chosen as $\alpha = diag(0.2; 0.2; 0.2)$, $\beta = diag(0.1; 0.1; 0.1)$, $\Lambda = diag(0.62; 0.62; 0.62)$, $K_1 = diag(28; 25; 45)$ and $\psi_1 = \psi_2 = \psi_3 = 3.5$.

To avoid a singularity, the terms containing power $\Lambda - I$ in (25) and (36) were replaced with the saturation function.

$$sat(u_f, u_s) = \begin{cases} u_s & if\ u_f \geq u_s \\ u_f & if\ u_f < u_s \end{cases}$$

where $u_s = 10$ is a positive constant, and $u_f = \Lambda|x|^{\Lambda - I} \cdot \dot{x}$ with $x = e$ and ε.

To avoid chattering, the signum function in (25) and (36) were replaced with the saturation function.

$$sat(s) = \begin{cases} sgn(s) & if\ |s| \geq \lambda \\ \frac{s}{\lambda} & if\ |s| < \lambda \end{cases}$$

where $\lambda = 1.6$.

6.2. Experiment 1

In this experiment, the total two kinds of actuator faults are considered to show the advantage of the proposed AFTC when compared to AFT with [30]+ESO. The assumed fault function is shown as

$$\begin{cases} \tau_1^t = \tau_1 \\ \tau_2^t = (1 - \rho_2(t))\tau_2 + f_2(t)\ t \geq 10 \\ \tau_3^t = \tau_3 \end{cases} \tag{42}$$

where $\rho_2(t) = 0.4\sin(\pi(t - 10)/2.4)$ and $f_2(t) = 60\sin(\pi(t - 10)/2.4)$.

In Figure 7, the fault estimation results are shown. Unlike in Simulation 2's results, the high magnitude of uncertainties occurred in the real implementation estimation results. However, the magnitude of the assumed fault dominated over the uncertainties. Hence, the effect of the total uncertainties and faults is mainly affected by faults. In Figure 8, the tracking error at each joint is presented. Before the 10th second, as mentioned in Section 1, the synchronization technique has no advantage in an open-loop chain mechanism such as the serial robot manipulator in normal operation mode. Therefore, the error at each joint of two controllers has similar accuracy. It is shown that the AFTC with [30]+ESO still maintains the ability to deal with a high magnitude of uncertainties. After the 10th second, in Figure 8b, the difference accuracy of two controllers is shown. The proposed AFTC S-FTSMC outperformed the AFTC with [30]+ESO. It can be seen that the effectiveness of the synchronization technique in the AFTC strategy is verified. In addition, in Figure 8c, the error of the proposed AFTC S-FTSMC is slightly increased due to the synchronization method. However, this increase still guarantees an acceptable performance of the robot manipulator.

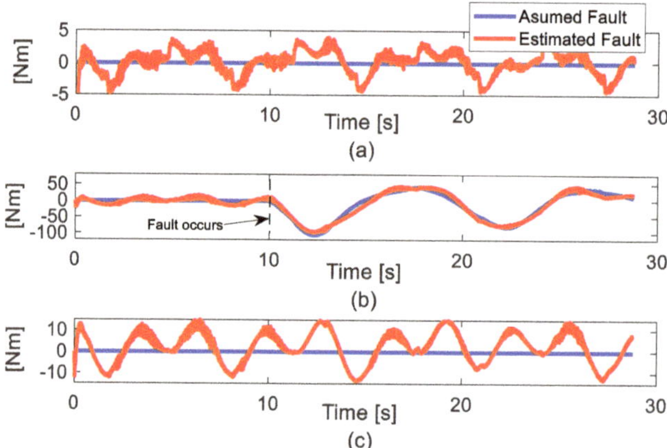

Figure 7. Fault estimation results. (**a**) Joint 1; (**b**) Joint 2; (**c**) Joint (3).

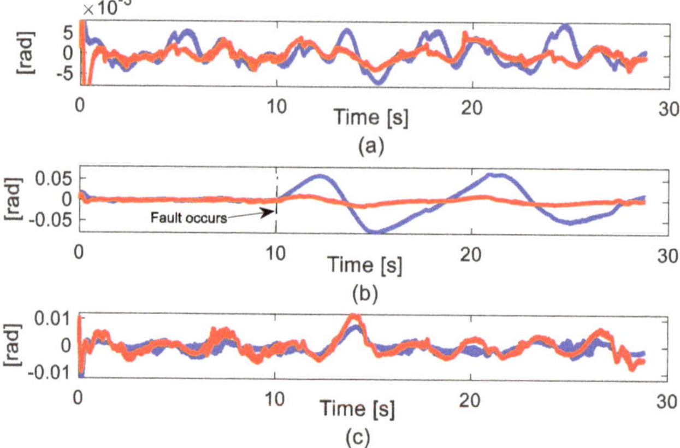

Figure 8. Tracking error results. Blue line is [30]+ESO controller. Red line is the proposed controller. (**a**) Joint 1; (**b**) Joint 2; (**c**) Joint (3).

6.3. Experiment 2

In this experiment, three different group values of synchronization parameters $\psi_1 = \psi_2 = \psi_3 = \psi(Psi)$ where $\psi = 1.5$, 3.5 and 4.5 were considered to show the effect of these parameters in the proposed controller.

Before the 10th second, in Figure 9a–c, the position errors at each joint tend to be equal so that they have similar behavior due to the synchronization technique. With different values of group ψ, the synchronization occurs during the tracking trajectory of the robot manipulator. However, with a higher value of ψ, the system has better synchronization as shown in Figure 9b,c. It can be seen that from $\psi = 3.5$ to $\psi = 4.5$, the position error at each joint has not much change. In this implementation with a 3-DOF FARA-AT2 robot manipulator, the value of ψ is limited to 4.5. If ψ is selected to be higher, oscillation will occur in the robot manipulator system. In Figure 9d, the error of AFTC with [30]+ESO showed a different characteristic at each joint. However, in Figure 9b–d, the accuracy still maintains. After the 10th second, in Figure 9d without the synchronization technique, the position error at each joint shows dependent characteristics. In contrast, in Figure 9a–c, the synchronization guarantees that the position error at each joint tends to be equal. This characteristic of the synchronization technique is very helpful in tolerating faults because the internal constraints of synchronization can suppress the effect of faults in the system.

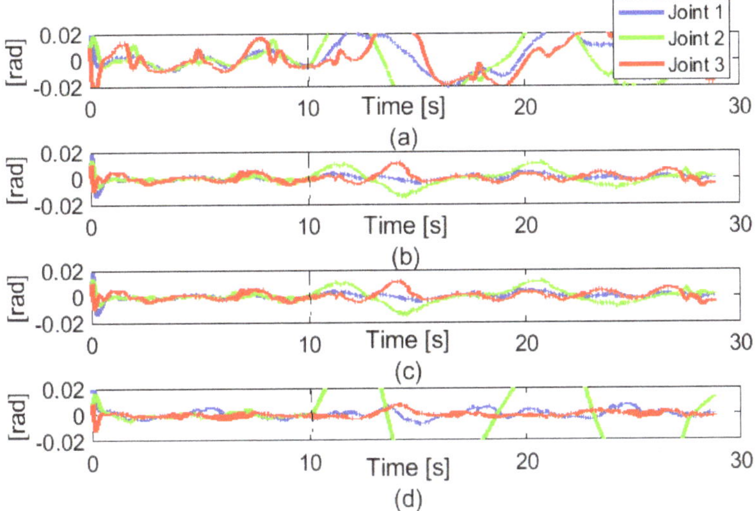

Figure 9. Tracking error results with difference group value of ψ. (**a**) $\psi_1 = \psi_2 = \psi_3 = \psi = 1.5$; (**b**) $\psi_1 = \psi_2 = \psi_3 = \psi = 3.5$; (**c**) $\psi_1 = \psi_2 = \psi_3 = \psi = 4.5$; (**d**) Non-synchronization.

7. Conclusions and Future Works

In this paper, active fault-tolerant control with the combination of a novel finite-time synchronous fast terminal sliding mode control and an extended state observer was proposed. Due to the internal constraints of the synchronization technique, the proposed controller has the ability to suppress the effect of faults and reduce the picking phenomenon in systems when faults occur. Without the conventional coupling technique, the proposed active fault-tolerant controller can reduce the computation and fast convergence and avoid the oscillation due to the integral term and guarantees the finite-time convergence of position errors. From the simulation and experimental results, it is shown that the effect of different kinds of faults cannot prevent uncertainties and disturbances like the control technique can. Therefore, the synchronization technique is a suitable technique to tolerate faults. It effectively treats the loss-effective faults at the actuators of robot manipulators.

In future work, selecting the optimal synchronization parameters using an optimal method or intelligent technique such as neural network and fuzzy could be considered to improve the performance of the synchronous technique in fault-tolerant control. In addition, the data-driver technique based on a deep neural network could be useful to detect faults on the system early, which is very important in tolerating the faults and maintaining acceptable performance when the faults occur. The combination of the model and signal base will be applied to a real robot system to detect faults early, then make a decision for the output of the robot actuator to tolerate the faults.

Author Contributions: All authors contributed equally to this article and accepted the final report. All authors have read and agreed to the published version of the manuscript.

Funding: This research was supported by Basic Science Research Program through the National Research Foundation of Korea (NRF) funded by the Ministry of Education (NRF-2016R1D1A3B03930496).

Conflicts of Interest: All authors announce that they have no conflict of interest in relation to the publication of this article.

References

1. Le, Q.D.; Kang, H.-J. Real Implementation of an Active Fault Tolerant Control Based on Super Twisting Technique for a Robot Manipulator. In Proceedings of the International Conference on Intelligent Computing, Nanchang, China, 3–6 August 2019; Huang, D.S., Huang, Z.K., Hussain, A., Eds.; Springer: Cham, Switzerland, 2019; pp. 294–305.
2. Shen, Q.; Yue, C.; Goh, C.H.; Wang, D. Active Fault-Tolerant Control System Design for Spacecraft Attitude Maneuvers with Actuator Saturation and Faults. *IEEE Trans. Ind. Electron.* **2019**, *66*, 3763–3772. [CrossRef]
3. Zhang, Q.; Wang, C.; Su, X.; Xu, D. Observer-based terminal sliding mode control of non-affine nonlinear systems: Finite-time approach. *J. Franklin Inst.* **2018**, *355*, 7985–8004. [CrossRef]
4. Wang, R.; Wang, J. Passive actuator fault-tolerant control for a class of overactuated nonlinear systems and applications to electric vehicles. *IEEE Trans. Veh. Technol.* **2013**, *62*, 972–985. [CrossRef]
5. Stefanovski, J.D. Passive fault tolerant perfect tracking with additive faults. *Automatica* **2018**, *87*, 432–436. [CrossRef]
6. Benosman, M.; Lum, K.-Y. Passive actuators' fault-tolerant control for affine nonlinear systems. *IEEE Trans. Control Syst. Technol.* **2010**, *18*, 152–163. [CrossRef]
7. Le, Q.D.; Kang, H.-J. Implementation of sensorless contact force estimation in collaborative robot based on adaptive third-order sliding mode observer. *Syst. Sci. Control Eng.* **2022**, *10*, 507–516. [CrossRef]
8. Qi, W.; Aliverti, A. A multimodal wearable system for continuous and real-time breathing pattern monitoring during daily activity. *IEEE J. Biomed. Health Inform.* **2019**, *24*, 2199–2207. [CrossRef]
9. Xu, S.S.-D.; Chen, C.-C.; Wu, Z.-L. Study of nonsingular fast terminal sliding-mode fault-tolerant control. *IEEE Trans. Ind. Electron.* **2015**, *62*, 3906–3913. [CrossRef]
10. Shen, Q.; Wang, D.; Zhu, S.; Poh, E.K. Integral-Type Sliding Mode Fault-Tolerant Control for Attitude Stabilization of Spacecraft. *IEEE Trans. Control Syst. Technol.* **2015**, *23*, 1131–1138. [CrossRef]
11. Shen, Q.; Jiang, B.; Cocquempot, V. Fuzzy logic system-based adaptive fault-tolerant control for near-space vehicle attitude dynamics with actuator faults. *IEEE Trans. Fuzzy Syst.* **2013**, *21*, 289–300. [CrossRef]
12. Cao, Y.; Song, Y.D. Adaptive PID-like fault-tolerant control for robot manipulators with given performance specifications. *Int. J. Control* **2018**, *7179*, 377–386. [CrossRef]
13. Zhang, J.X.; Yang, G.H. Robust adaptive fault-tolerant control for a class of unknown nonlinear systems. *IEEE Trans. Ind. Electron.* **2017**, *64*, 585–594. [CrossRef]
14. Jiang, T.; Khorasani, K.; Tafazoli, S. Parameter estimation-based fault detection, isolation and recovery for nonlinear satellite models. *IEEE Trans. Control Syst. Technol.* **2008**, *16*, 799–808. [CrossRef]
15. Edwards, C.; Spurgeon, S.K.; Patton, R.J. Sliding mode observers for fault detection and isolation. *Automatica* **2000**, *36*, 541–553. [CrossRef]
16. Han, J.; Zhang, H.; Wang, Y.; Zhang, K. Fault Estimation and Fault-Tolerant Control for Switched Fuzzy Stochastic Systems. *IEEE Trans. Fuzzy Syst.* **2018**, *26*, 2993–3003. [CrossRef]
17. Bouibed, K.; Seddiki, L.; Guelton, K.; Akdag, H. Actuator and sensor fault detection and isolation of an actuated seat via nonlinear multi-observers. *Syst. Sci. Control Eng.* **2014**, *2*, 150–160. [CrossRef]
18. Lin, X.; Yao, X. Improved disturbance-observer-based fault-tolerant control for the linear system subject to unknown actuator faults and multiple disturbances. *Int. J. Control* **2021**, *94*, 2730–2740. [CrossRef]
19. Zhang, X.; Zhu, Z.; Yi, Y. Anti-Disturbance Fault-Tolerant Sliding Mode Control for Systems with Unknown Faults and Disturbances. *Electronics* **2021**, *10*, 1487. [CrossRef]
20. Han, J.; Zhang, H.; Wang, Y.; Liu, Y. Disturbance observer based fault estimation and dynamic output feedback fault tolerant control for fuzzy systems with local nonlinear models. *ISA Trans.* **2015**, *59*, 114–124. [CrossRef]
21. Van, M.; Ge, S.S.; Ren, H. Finite Time Fault Tolerant Control for Robot Manipulators Using Time Delay Estimation and Continuous Nonsingular Fast Terminal Sliding Mode Control. *IEEE Trans. Cybern.* **2017**, *47*, 1681–1693. [CrossRef]
22. Yu, X.; Jiang, J. Hybrid Fault-Tolerant Flight Control System Design. *IEEE Trans. Control Syst. Technol.* **2011**, *20*, 871–886. [CrossRef]
23. Li, C.; Yao, B.; Wang, Q. Modeling and Synchronization Control of a Dual Drive Industrial Gantry Stage. *IEEE/ASME Trans. Mechatron.* **2018**, *23*, 2940–2951. [CrossRef]
24. Ren, L.; Mills, J.K.; Sun, D. Experimental comparison of control approaches on trajectory tracking control of a 3-DOF parallel robot. *IEEE Trans. Control Syst. Technol.* **2007**, *15*, 982–988. [CrossRef]
25. Shang, W.; Zhang, B.; Zhang, B.; Zhang, F.; Cong, S. Synchronization Control in the Cable Space for Cable-Driven Parallel Robots. *IEEE Trans. Ind. Electron.* **2019**, *66*, 4544–4554. [CrossRef]
26. Cui, R.; Yan, W. Mutual synchronization of multiple robot manipulators with unknown dynamics. *J. Intell. Robot. Syst. Theory Appl.* **2012**, *68*, 105–119. [CrossRef]

27. Le, Q.D.; Kang, H.-J. Implementation of Fault-Tolerant Control for a Robot Manipulator Based on Synchronous Sliding Mode Control. *Appl. Sci.* **2020**, *10*, 2534. [CrossRef]
28. Le, Q.D.; Kang, H.-J. Finite-Time Fault-Tolerant Control for a Robot Manipulator Based on Synchronous Terminal Sliding Mode Control. *Appl. Sci.* **2020**, *10*, 2998. [CrossRef]
29. Khalil, H.K.; Praly, L. High-gain observers in nonlinear feedback control. *Int. J. Robust Nonlinear Control* **2014**, *24*, 993–1015. [CrossRef]
30. Yu, X.; Zhihong, M. Fast terminal sliding-mode control design for nonlinear dynamical systems. *IEEE Trans. Circuits Syst. I Fundam. Theory Appl.* **2002**, *49*, 261–264. [CrossRef]
31. Zemouche, A.; Boutayeb, M. On LMI conditions to design observers for Lipschitz nonlinear systems. *Automatica* **2013**, *49*, 585–591. [CrossRef]
32. Zak, M. Introduction to terminal dynamics. *Complex Syst.* **1993**, *7*, 59.
33. Feng, Y.; Yu, X.; Han, F. On nonsingular terminal sliding-mode control of nonlinear systems. *Automatica* **2013**, *49*, 1715–1722. [CrossRef]

Article

Optimal Design of a Novel Leg-Based Stair-Climbing Wheelchair Based on the Kinematic Analysis of the Stair Climbing States

Diego Delgado-Mena, Emiliano Pereira *, Cristina Alén-Cordero, Saturnino Maldonado-Bascón and Pedro Gil-Jiménez

Department of Signal Processing and Communications, Universidad de Alcalá, 28001 Madrid, Spain
* Correspondence: emiliano.pereira@uah.es; Tel.: +34-91-886-6711

Abstract: This work presents a method to find the optimal configuration of a leg-based stair-climbing wheelchair. This optimization begins with the definition of a high-level control architecture, in which the kinematics restrictions related to the specific obstacles are considered. Then, the reference trajectories for all the actuators are generated as a function of the physical parameters of the mechanism, the dynamic restrictions of the actuators (velocity and acceleration) and the sensor errors. This work illustrates, based on a set of configurations, how the total time to climb up and climb down a defined stair depends on all these parameters, also reporting the best set of parameters that reduces the time and makes the mechanism more stable for a given scenario. The optimization in this work is performed with a brute-force search within a grid of parameters with a resolution of 1 mm. Thus, as the local minima is located, the complexity of the problem is revealed.

Keywords: motion planning; mobile robots; actuator dynamics and control

Citation: Delgado-Mena, D.; Pereira, E.; Alén-Cordero, C.; Maldonado-Bascón, S.; Gil-Jiménez, P. Optimal Design of a Novel Leg-Based Stair-Climbing Wheelchair Based on the Kinematic Analysis of the Stair Climbing States. *Actuators* **2022**, *11*, 289. https://doi.org/10.3390/act11100289

Academic Editors: Marco Carricato and Edoardo Idà

Received: 22 August 2022
Accepted: 27 September 2022
Published: 9 October 2022

Publisher's Note: MDPI stays neutral with regard to jurisdictional claims in published maps and institutional affiliations.

1. Introduction

Stair-climbing mechanisms have been researched and developed during the last decades [1]. These mechanisms have been used to assist disabled people [2], to transport other devices such as robots or wheelchairs [3], or to assist devices [4]. All these mechanisms can be classified into (see Table I in [1]): (i) track-based stair-climbing mechanisms, (ii) wheel cluster-based stair-climbing mechanisms, (iii) leg-based stair-climbing mechanisms and (iv) hybrid stair-climbing mechanisms.

Track-based stair-climbing mechanisms have been successfully commercialized. These mechanisms are based on the interlocking effect between the track's outer teeth and the steps' sharp corners. TopChair-S [5] proposes a solution based on a caterpillar mechanism, which has a cost of around EUR 15,500. Another commercial solution is the PW-4x4Q Stair-Climbing Wheelchair [2], which is based on large wheels whose relative height can be modified. Its cost is around EUR 12,500. The performance of these solutions depends on the grip of the material on the obstacle, which can deteriorate over time, making the cost of the solution even more expensive due to the maintenance required.

The wheel cluster-based stair-climbing mechanisms are relatively compact and can easily switch to the wheeled mobile mode when running on level ground. Examples of these mechanisms can be found in [6,7], where a cluster of three wheels is proposed. In [6], a mechanism with only one motor and a transmission system per locomotion unit is proposed. The wheelchair passively changes its locomotion, from rolling on wheels ("advancing mode") to walking on legs ("automatic climbing mode"), according to local friction and dynamic conditions. In [7], a track-based stair-climbing mechanism is combined with the cluster of three wheels in order to improve the wheelchair's stability. This mechanism has been recently built and used, as reported in [8].

A good example of a hybrid stair-climbing mechanism is the one proposed in [9], which was optimally designed in [10]. This mechanism can be adapted to different steps and obstacles, generating smooth and comfortable trajectories for the user. The control of the mechanism and the improvement in the trajectory generation have been studied in later works [11–13].

Leg-based stair-climbing mechanisms can be classified into biped and parallel mechanisms. For example, in [14], a biped stair-climbing mechanism is developed based on a Stewart platform. This mechanism can walk up and down a stair with a riser height of 150 mm, continuously carrying a 60 kg load. A stair-climbing vehicle named "Zero Carrier" with eight legs was proposed in [3]. In [15], the concept of an eight-legged wheelchair, aiming at improving the limitations of the Zero Carrier design, was proposed. The eight legs are grouped into two independent frames of four legs each. Both frames can change the relative horizontal position between them. Thus, the height of the legs can be substantially reduced with respect to the design proposed in [3]. However, the mechanism needed to move the frame horizontally may be inconvenient when heavy loads must be carried.

According to [1], although these leg-based stair-climbing vehicles are complex, have high costs and unconventional appearances, they are able to achieve the core function of stair ascent and descent but also provide some innovations in climbing wheelchair design. This motivates the work in [16], where a novel leg-based stair-climbing mechanism was presented. This mechanism, which is based on a patent [17], introduces some modifications, such as a novel configuration of the linear actuators. Thus, the first prototype developed and built in [16] increases the flexibility of the mechanism, allowing the wheelchair to climb up and down without changing the orientation of the chair and ensuring the horizontal position of the user at any moment. This first prototype presents some advantages with respect to other leg-based stair-climbing mechanisms. Thus, the horizontal position of the user can be guaranteed with a relative low stroke in the linear actuators, which is one of the problems of the solution proposed in [3]. Besides this, a relative displacement among the four frame legs is not necessary, which is the main problem of [15].

One of the main challenges of these climbing mechanisms is the control of the actuators in order to generate safe trajectories. In addition, the control of the actuators and the strategy, which is used to climb the obstacles, are required for the process of mechanism geometry optimization. Some previous works have analyzed variables, such as the area, the velocity, the ergonomic and/or the adaptability to different obstacles. For example, the hybrid stair-climbing mechanism presented in [9] was optimized and controlled in [9,10]. In [11,12], trajectory generation is proposed in order to improve the stair-climbing time and the user's comfort, taking into account the most important constraints inherent to the system behavior, such as the geometry of the architectural barrier, the re-configurable nature of the discontinuous states, state-transition diagrams, comfort restrictions and physical limitations regarding the actuators, the speed and the acceleration.

Leg-based stair-climbing mechanisms, such as the one defined in [16,17], present the problem of controlling the linear actuators. The amount of linear actuators that must be synchronized, is one of the main limitations when these mechanisms must be controlled. Thus, a control architecture (high-level planning) is needed, which simplifies the practical implementation of the actuators' control and increases the safety of the mechanism. A control architecture for the design proposed in [16] was recently presented in [18]. However, in this early version, we did not consider any dynamic restrictions, i.e., maximum speed and maximum acceleration and deceleration. In addition, for the vertical actuators, the weight the actuator must push when elevating/inclining the structure was not considered. All these parameters have now been taken into account in this improved version.

Thus, the first contribution is the improvement in the control architecture presented in [18] by programming a low control level designed to generate the velocity profiles. The resulting trajectory generation improves the ergonomics of the mechanism. The second contribution is the proposal of an optimal design methodology, which adjusts the physical parameters of the wheelchair to minimize the total time required to climb up and down a given stair. Interesting and not obvious conclusions are presented in the result section. These conclusions should be considered when building a physical prototype.

The work presented starts with a brief definition of the prototype. In this section, the direct and inverse kinematics equations are defined. Then, the control architecture is presented. The work continues with the optimal design methodology. Then, an application example is presented. The work ends with some conclusions and future works.

2. Leg-Based Stair-Climbing Mechanism

The leg-based stair-climbing mechanism was presented in [16]. This section summarizes the mechanical design and the kinematics equations used in the control architecture. Let us simplify the explanation by considering the problem in 2D, where the obstacle and the wheelchair can be represented, as shown in Figure 1. Whenever the stair to climb is straight, this simplification does not affect the general 3D solution. The linear actuators L_1, L_2, L_3 and L_4 can change the height or inclination of the whole wheelchair or the vertical position of its corresponding ending wheel. The linear actuator L_9 changes the shape of the frame from a rectangle to a rhomboid. This mechanism allows us to reduce the lengths of L_1, L_2, L_3 and L_4 needed to climb up or down a step, as will be shown later. Figure 1a includes the length of the linear actuators (L_1, L_2, L_3, L_4 and L_9), the wheel radius (r_1, r_2, r_3 and r_4) and the structure's frame dimensions (a, b, c and d). In addition, Figure 1b,c shows the coodinates of the wheels and angles α and β. Note that $\alpha + \beta = 90°$, with α being negative and positive in Figure 1b,c, respectively.

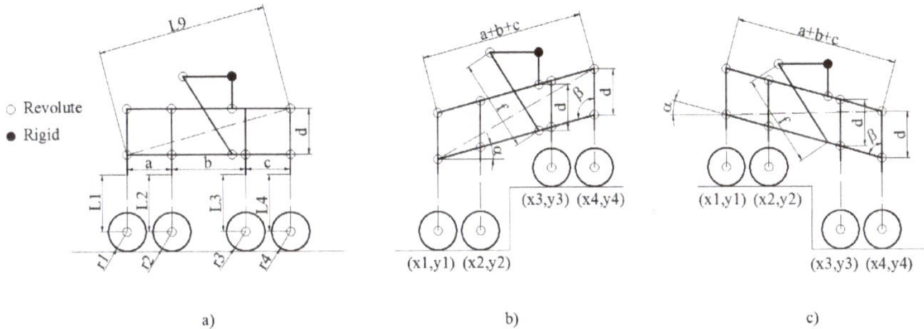

Figure 1. Working principle. It can be seen that the user is always kept horizontal when the wheelchair is horizontal, is climbing up or climbing down an obstacle.

In this paper, we propose to control the wheelchair's horizontal motion with electrical motors in wheels 1 and 2, leaving wheels 3 and 4 without traction. To ensure traction, one traction wheel, wheel 1 or wheel 2, must always be placed on the ground. Moreover, to ensure stability, one of the front wheels, wheel 3 or wheel 4, must also be placed on the ground so that there is always one wheel of each pair on the ground. In addition, all the actuators must be orthogonal to the ground at any moment, which is achieved by controlling the height of the wheels with L_1, L_2, L_3 and L_4, and the angle β with L_9. The rectangle of the structure (Figure 1a) changes into a rhomboid (Figure 1b,c) with an angle equal to β in order to ensure the orthogonality of all the legs. Thus, the horizontal bar of the 3-bars mechanism, where the chair is placed, is kept horizontal. The angle β can be calculated as follows:

$$\beta = \cos^{-1}\left(\frac{d^2 + L_H^2 - L_9^2}{2 \cdot L_H \cdot d}\right),\tag{1}$$

where L_H is equal to $a + b + c$. The horizontal and vertical coordinates of each wheel can be expressed with respect to one of them. For example, if wheel 1 is considered as the reference, these coordinates can be obtained as follows:

$$x_2 = x_1 + a \cdot \sin \beta,$$
$$x_3 = x_1 + (a + b) \cdot \sin \beta, \tag{2}$$
$$x_4 = x_1 + L_H \cdot \sin \beta,$$

$$y_2 = y_1 + (L_2 - L_1) + a \cdot \cos \beta,$$
$$y_3 = y_1 + (L_3 - L_1) + (a + b) \cdot \cos \beta, \tag{3}$$
$$y_4 = y_1 + (L_4 - L_1) + L_H \cdot \cos \beta.$$

If the stair geometry and the location of the wheelchair with respect to the stair are known, the lengths of L_1, L_2, L_3, L_4 and L_9 can be obtained from Equations (1)–(3), as follows (i.e., inverse kinematic model):

$$L_2 = L_1 + y_2 - a \cdot \cos \beta$$
$$L_3 = L_1 + y_3 - (a + b) \cdot \cos \beta$$
$$L_4 = L_1 + y_4 - L_H \cdot \cos \beta \tag{4}$$
$$L_9 = \sqrt{d^2 + L_H^2 - 2 \cdot L_H \cdot d \cdot \cos \beta}$$

The wheelchair parameters must be fixed considering the stairs' dimensions. Thus, the wheels' radius, the lengths a, b and the angle β are related to the tread size (TS) and riser height (RH). The first restriction must guarantee that the pairs formed by wheels 1–2 and 3–4 can be placed on one step. Thus, the four following restrictions must be achieved:

$$a \cdot \sin \beta + r_1 < TS - \delta_H \text{ (climb down),}$$
$$a \cdot \sin \beta + r_2 < TS - \delta_H \text{ (climb up),}$$
$$c \cdot \sin \beta + r_3 < TS - \delta_H \text{ (climb down),} \tag{5}$$
$$c \cdot \sin \beta + r_4 < TS - \delta_H \text{ (climb up),}$$

where δ_H is a security parameter. Parameter δ_H is considered to ensure that a wheel can be placed in the following step (climb up and climb down) and to prevent a collision with the next step (climb up). The second restriction must ensure that the distance between wheels is greater than 0. Thus, the three following restrictions must be achieved:

$$a \cdot \sin \beta > r_1 + r_2,$$
$$b \cdot \sin \beta > r_2 + r_3, \tag{6}$$
$$c \cdot \sin \beta > r_3 + r_4,.$$

In the application example, a minimum distance between wheels is considered. In the case of $b \cdot \sin \beta$, the minimum distance depends on the weight/force of the user and the gravity center of the system. The objective is to avoid any unbalance while the wheelchair climbs up or climbs down. The unbalanced problem also depends on the slope of the stair (RH/TS). This work considers that this slope is less than or equal to the maximum value of β (Equation (1)). Thus, the values of L_1, L_2, L_3 and L_4 must be achieved with the following restrictions:

$$y_2 - y_1 = L_2 - L_1 + a \cdot \cos \beta > RH + \delta_V \text{ (climb up)}$$
$$y_4 - y_3 = L_4 - L_3 + c \cdot \cos \beta > RH + \delta_V \text{ (climb up),} \tag{7}$$

where δ_V, like δ_H, is a security parameter. Parameter δ_V is added to the vertical displacement of the wheel when it climbs up a step. Parameters RH and TS are defined taking into account building codes for each country (e.g., in Spain, the minimum values for RH and TS are 175 and 280 mm, respectively [19]). Figure 2 shows two examples of the wheelchair when it climbs up (left side) and climbs down (right side). The restrictions and the variables defined in Equations (5) and (7) can be seen in this figure. Note that with these restrictions, the wheelchair can climb up and climb down a stair with any number of steps.

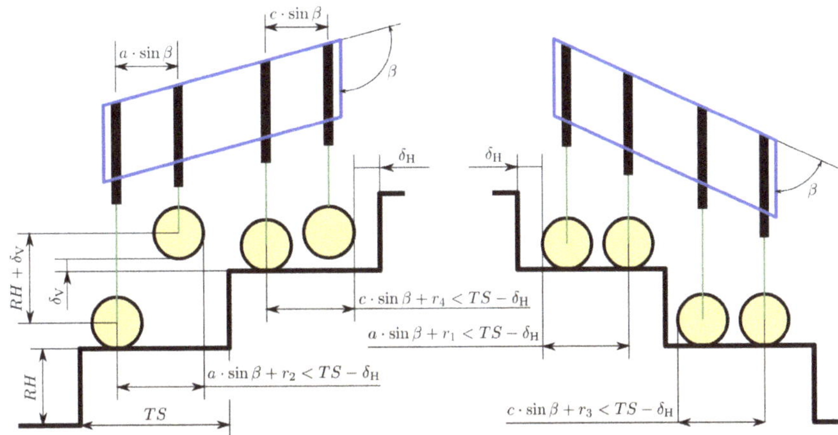

Figure 2. Wheelchair restrictions when the mechanism climbs up and climbs down.

3. Control Architecture

In this section, the strategy to control the actuators is presented. The objective is to explain how the coordinates of the wheels change in order to climb up or climb down a stair. Figures 3 and 4 show the strategies followed when climbing up and down a stair, respectively. The trajectory of each wheel is calculated by defining intermediate points, which are denoted as wheel states. Four and three states are defined for climbing up and climbing down, respectively.

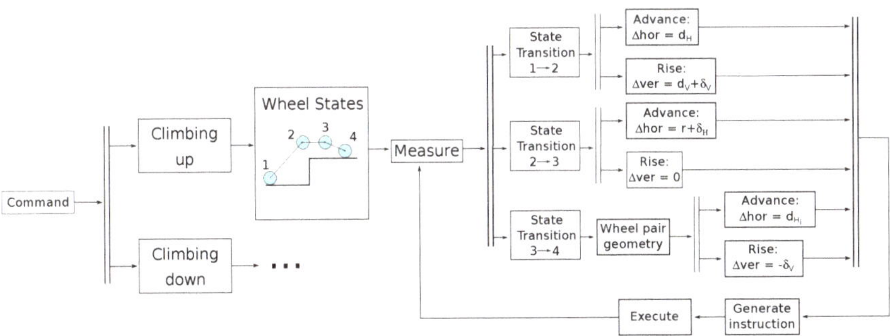

Figure 3. Strategy followed in order to climb up [18].

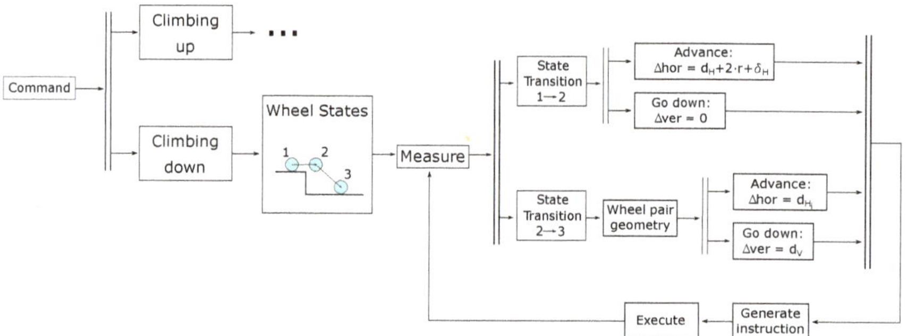

Figure 4. Strategy followed in order to climb down [18].

Then, the wheel that changes its height in each iteration is decided by the command *Measure*. *Measure* gets the index of the wheel that is closest to its nearest step (denoted i), providing the horizontal d_H and the vertical d_V distances (see Figure 5).

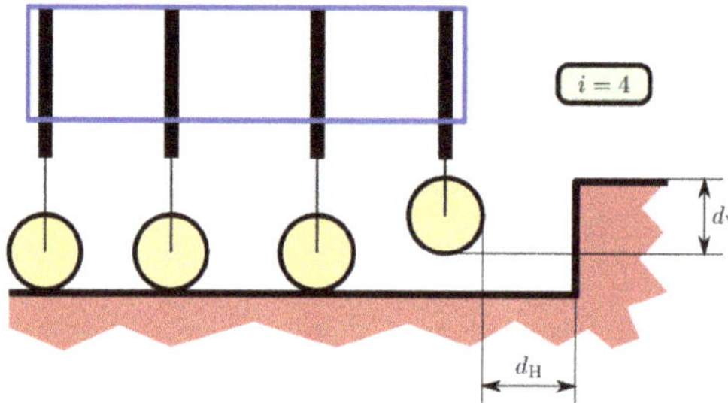

Figure 5. Practical example of command *Measure*, where the closest wheel is $i = 4$ [18].

The commands *Advance*, *Rise* and *GoDown* are used to move the wheels through the state transitions. Note that *Advance/Rise* and *Advance/GoDown* are executed in parallel to save time. *Advance* is achieved with the motors of wheels 1 and 2. *Rise* and *GoDown* can be achieved with the linear actuators defined in Equation (4).

Figure 6 shows the software architecture of the control software. The first level is *Wheelsx4*, which corresponds to the wheel level. Each wheel is controlled considering the equations described in Section 2. The second level is *Pair*, where the front and back pair are separated. The first pair controls the actuators of wheels 4 and 3 (L_4 and L_3), and the second pair the wheels 2 and 1 (L_2 and L_1). The third level is *Base* (i.e., wheelchair level), which coordinates the horizontal movement with the vertical movement. This level considers the kinematics equations and restrictions described in Section 2 and the control strategy. In addition, the procedure to change the vertical position of any wheel is also decided at this level. Note that the height of each wheel depends on the reference wheel (i.e., L_1), the length of its actuator and the angle β (i.e., L_9). The commands *Rise* and *GoDown* consider the three levels mentioned above to change the height of each wheel.

The previous control architecture, which was published in [18], did not consider any acceleration/deceleration restrictions. However, the modification of the control architecture proposed in this work does consider these restrictions in all the actuators. In addition, different values of acceleration and velocity can be set. The low-level control considers the following restrictions:

- $actuator_{up}$: Speed for an actuator when it is elevating the wheel.
- $actuator_{dw}$: Speed for an actuator when it is taking the wheel down.
- $elevate_{up}$: Speed when the actuators are elevating the structure.
- $elevate_{dw}$: Speed when the actuators are taking the structure down.
- $incline_{up}$: Speed when the actuators are inclining the structure up.

- *incline$_{dw}$*: Speed when the actuators are inclining the structure down.
- *speed*: Maximum horizontal speed.
- *acceleration*: Maximum horizontal acceleration.
- *deceleration*: Maximum horizontal deceleration.

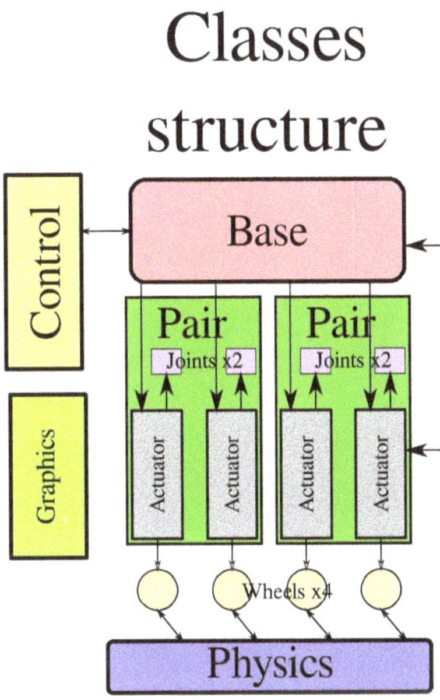

Figure 6. Software structure of the control architecture [18].

The computation of the horizontal velocity profiles (i.e., the low control level) is described in the following Technical Report (https://github.com/pedrogil1919/Structure/blob/master/Structure/docs/dynamics/calculus.odt, accessed on 20 September 2022). The link (https://youtube.com/playlist?list=PL-cQTqyWA2d1upFVvzsNcJ0bn3QE4KyfV, accessed on 20 September 2022) includes some videos to show the trajectory generation when the wheelchair climbs a stair with several steps. Figures 7 and 8 show two snapshots of these videos, corresponding to times equal to 44 and 73.5 s, respectively. These figures show: (i) current structure position (top-left), (ii) structure inclination (mm), measured as the difference in height between the front and the rear extremes of the structure, and horizontal velocity of the wheelchair (mm/s) (bottom-left) and (iii) actuator position (L_1, L_2, L_3 and L_4) (mm) (right). It can be seen that the trajectories of Figures 7 and 8 are smoother than Figures 13 and 14 of [18]. Therefore, the ergonomics of this new version of the control architecture have been improved.

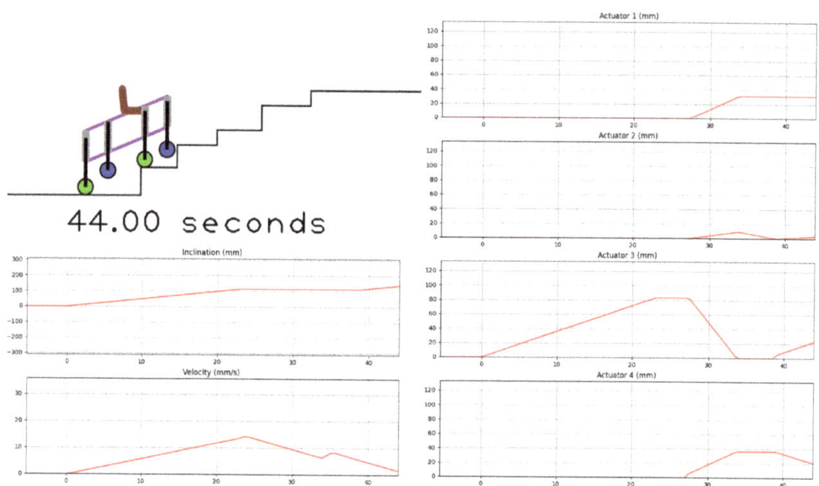

Figure 7. Data previously generated before 44.00 s of the trajectory generator for a stair with diferent step sizes (positive *HR*).

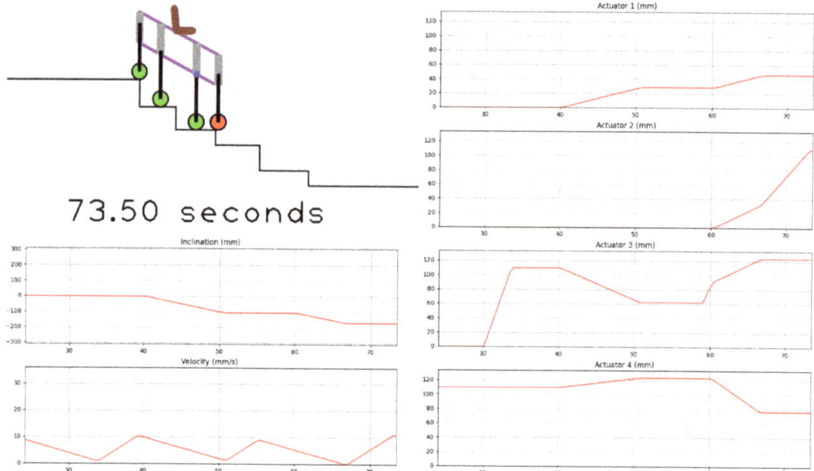

Figure 8. Data previously generated before 73.5 s of the trajectory generator for a stair with diferent step sizes (negative *HR*).

In the following subsections, the three levels are explained in detail.

3.1. Individual Wheel Level

The climb up (see Figure 9) and climb down (see Figure 10) trajectories of an individual wheel are described herein. Both trajectories are divided in the states defined above, which are explained in detail in this subsection. The nomenclature is: (i) Δ_{x_i} and Δ_{y_i} are the horizontal and vertical displacements of the wheel i in each instruction, (ii) r is the radius of the wheel and (ii) δ_H and δ_V are additional displacements included to prevent wheel collisions with the stair, mainly due to sensor precision and geometric tolerances.

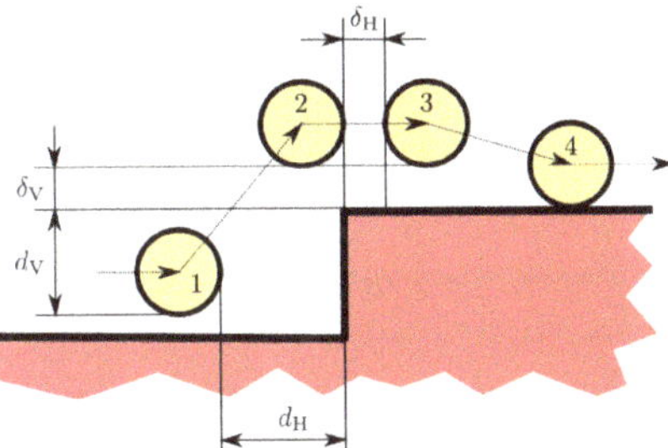

Figure 9. Individual Wheel level - climb up [18].

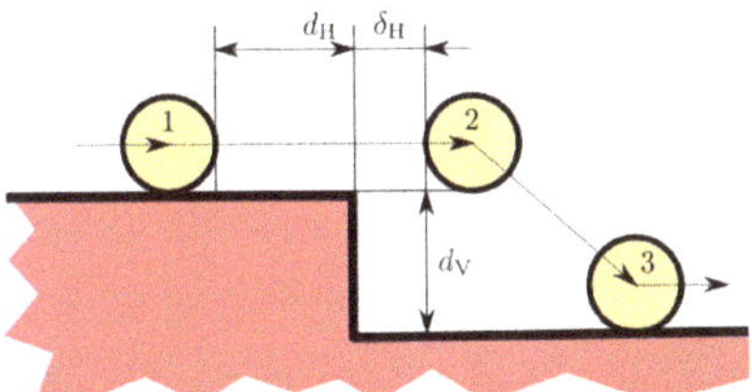

Figure 10. Individual Wheel level—climb down [18].

3.1.1. Climb Up—Figure 9

- State 1. The command *Measure* obtains the distance d_H and d_V of the closest wheel (i).
- State 2. $\Delta_x = d_{H_i}$ and $\Delta_{y_i} = d_V + \delta_V$.
- State 3. $\Delta_x = r + \delta_H$ and $\Delta_{y_i} = 0$.
- State 4. $\Delta_{y_i} = -\delta_V$. The horizontal position Δ_x can be increased if possible since this strategy reduces the trajectory time. The value of Δ_x depends on the wheel pair and wheelchair level.

3.1.2. Climb Down—Figure 10

- State 1. The command *Measure* obtains the distance d_H and d_V of the closest wheel (i).
- State 2. $\Delta_x = d_{H_i} + \delta_H + 2r$ and $\Delta_{y_i} = 0$.
- State 3. $\Delta_{y_i} = -d_V$. The horizontal position Δ_x can be increased if possible since this strategy reduces the trajectory time.

Note that the rest of the wheels move accordingly without the risk of collision with any obstacle, since they are further from any obstacle than wheel i, as described above.

3.2. Wheel Pair Level

This subsection explains the wheel pair geometry considerations shown in Figure 3 (state transition from 3 to 4) and Figure 4 (state transition from 2 to 3). Note that the wheelchair in Figure 1 can be considered as two independent wheel pairs. Thus, the first wheel (4 or 2) climbs up (or climbs down) the step first. This subsection denotes the first and second wheels of the pairs as f (front) and r (rear), respectively. The objective is to decide the maximum value of d_{H_f} in the last *Advance* instruction, which depends on d_{H_r}.

Figures 11 and 12 show the distances d_{H_f} and d_{H_r} in state 3 (climb up) and state 2 (climb down), respectively. In both cases, it must be guaranteed that $d_{H_f} < d_{H_r}$. Therefore, the maximum velocity for the last command *Advance* is limited by this restriction.

Wheel pair-level also checks that, at any time, at least one wheel of each pair is on the ground to ensure wheelchair stability. Note that when wheels 1 and 4 are in the air, and the wheelchair is supported only on wheels 2 and 3, the wheelchair is in the state of least stability. This problem will be addressed later.

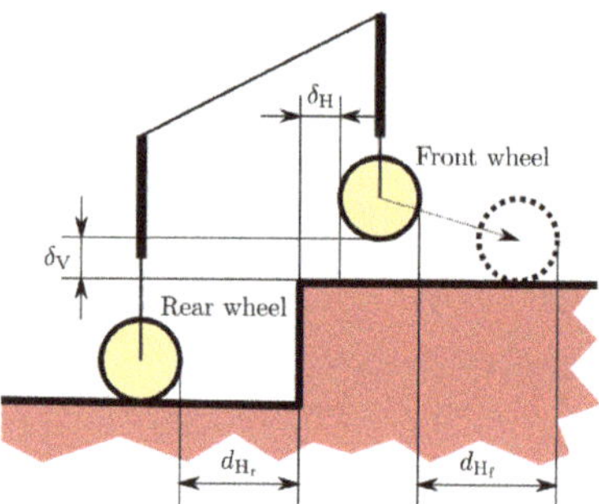

Figure 11. Wheel pair level—climb up.

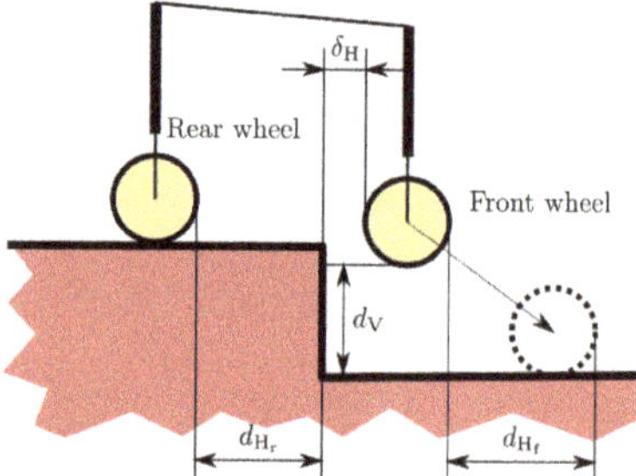

Figure 12. Wheel pair level—climb down.

3.3. Wheelchair Level

This level coordinates both wheel pairs, computing the horizontal velocity of the structure in order to ensure that there is no collision between wheels and obstacles. In addition, this level coordinates the length of the actuators L_1, L_2, L_3, L_4 and L_9 in the *Rise* and *GoDown* commands. This coordination depends on the wheel that is currently changing its height. That is, if when trying to shift an actuator, there is not enough space for the actuator to complete the motion, the structure must be elevated/inclined to gain more space. Thus, this movement is implemented as follows:

- Wheels 4 and 1: The space is gained by changing β, i.e., inclining the structure. Then, if there is still not enough room for the actuator to achieve the height required, the wheelchair is elevated until the actuator can achieve it.
- Wheels 2 and 3: As opposed to wheels 4 and 1, first elevate the wheelchair. If the total height can not be achieved, the structure is inclined (change β) until the actuator can achieve it.

Note that on some occasions, the whole height can not be achieved, due to, for instance, to a too high stair step. In this case, the instruction can not be completed, and so, the stair can not be climbed, requiring a redesign of the structure dimensions.

4. Optimization

The optimization is carried out as follows:

- Stair definition. The number of steps to climb up and climb down and the variables RH and TS are defined.
- Actuator dynamic restrictions. The following variables are defined: $actuator_{up}$, $actuator_{dw}$, $elevate_{up}$, $elevate_{dw}$, $incline_{up}$, $incline_{dw}$, $speed$, $acceleration$ and $deceleration$.
- Wheelchair constant parameters. The following variables are defined: wheel ratios (r_1, r_2, r_3 and r_4), sensor errors (δ_H, δ_V), maximum value of inclination (β), wheelchair length (L_H) and minimum values for a, b and c.

- Calculate the maximum values for parameters a and c from Equation (5) and the variables defined above.
- Define the resolution for the intervals of a and c.
- Calculate the total time used to climb up and down the stair defined above for each possible pair values of a and c.
- Plot the total time as a function of a and c.
- Decide the best configurations of a and c.

The optimization, control architecture and leg-based stair-climbing mechanism have been programmed in Python. This program, which models a non-linear problem, can calculate the total time used to climb up and down a defined stair in a few seconds. All the code can be downloaded from the following public repository, (https://github.com/pedrogil1919/Structure, accessed on 20 September 2022).

Application Example

The objective of this application example is to show that the configuration of the parameters a, b and c is not obvious. The application example considers the following configuration:

- Stair definition: Number of steps to climb up and climb down equal to 5 steps, $RH = 175$ mm and $TS = 280$ mm.
- Wheels radius: $r_1 = r_2 = r_3 = r_4 = 60$ mm
- Actuator dynamic restrictions:

    - $actuator_{up} = 20$ mm/s
    - $actuator_{dw} = 30$ mm/s
    - $elevate_{up} = 5$ mm/s
    - $elevate_{dw} = 10$ mm/s
    - $incline_{up} = 4$ mm/s
    - $incline_{dw} = 8$ mm/s
    - $speed = 30$ mm/s
    - $acceleration = 0.8$ mm/s^2
    - $deceleration = 1.8$ mm/s^2

- Wheelchair constant parameters (see Table 1):
- The resolution grid for parameters a and c for the brute-force search chosen is equal to 1 mm.

The first set of figures (Figure 13) plots the total time required to climb up and down the stair as a function of parameters a (horizontal axis) and c (vertical axis), considering a wheelchair with $L_H = 700$ mm, minimum value for $a = c = 140$ mm, minimum value for $b = 340$ mm, actuators length $L_1 - L_4 = 250$ mm and two intervals of β. Note that the maximum value for a is reached when b and c are minimum and vice versa. In the following figures, the position for (a, c) painted in white is a configuration where the system can not give a valid result. This can be a forbidden dimension, according to the restrictions defined above (top-right triangle), or a wheelchair configuration where the control algorithm can not find a valid trajectory to climb up or down the stair (white dots inside the bottom-left triangle).

Table 1. Optimization parameters.

Figure	L_H	$a_{min} = c_{min}$	b_{min}	$a_{max} = c_{max}$	β	$L_1 - L_4$
Figure 13a	700 mm	140 mm	340 mm	220 mm	$(\pi/2, 0)$	250 mm
Figure 13b	700 mm	140 mm	340 mm	220 mm	$[\pi/4, \pi/2)$	250 mm
Figure 13c	700 mm	125 mm	340 mm	235 mm	$(\pi/2, 0)$	250 mm
Figure 14a	700 mm	140 mm	340 mm	220 mm	$(\pi/2, 0)$	185 mm
Figure 14b	700 mm	140 mm	340 mm	220 mm	$(\pi/2, 0)$	355 mm
Figure 14c	700 mm	140 mm	340 mm	220 mm	$(\pi/2, 0)$	136 mm
Figure 15a	750 mm	140 mm	390 mm	220 mm	$(\pi/2, 0)$	250 mm
Figure 15b	900 mm	140 mm	540 mm	220 mm	$(\pi/2, 0)$	250 mm
Figure 15c	1000 mm	140 mm	640 mm	220 mm	$(\pi/2, 0)$	250 mm

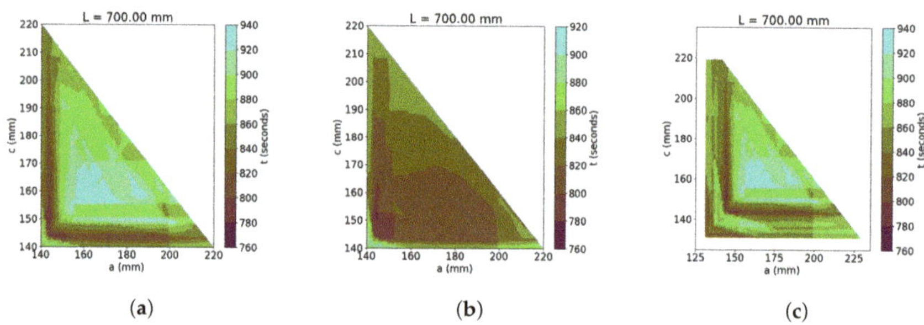

(a)   (b)   (c)

Figure 13. Wheelchair with L_H = 700 mm, minimum value of b = 270 mm and $L_1 - L_4$ = 250 mm. Influence of a_{min}, c_{min} and β in total time. (**a**) $a_{min} = c_{min}$ = 140 mm and $|\beta| \in (\pi/2, 0)$, (**b**) $a_{min} = c_{min}$ = 140 mm and $|\beta| \in [pi/4, pi/2)$, (**c**) $a_{min} = c_{min}$ = 125 mm and $|\beta| \in (\pi/2, 0)$.

Figure 13a shows the results when the minimum value of $a = c$ = 140 mm and there is no limitation in the angle β. The minimum time is around 760 s and the best configurations are achieved with a around 143–152 mm and c around 143–160 mm. In addition, note that points close to the main diagonal of the colormap graph correspond with the smallest values for b (distance between wheels 2 and 3), but we know that the larger this value, the more stable the wheelchair is when wheels 1 and 4 are in the air, and the wheelchair is supported on wheels 2 and 3. Therefore, the objective is to find points as close to the bottom-left corner of the graph as possible. Thus, we conclude that it is better to consider a and c around 143 mm.

Figure 13b limits the minimum absolute value of the angle β to $\pi/4$ and keeps the same configuration as in Figure 13a. Figure 13b shows that, for the considered control architecture, it is better to limit this angle. The reason is that there are more values of a and c where a lower total time is achieved than in Figure 13a. Figure 13b shows that the best values of a and c are around 143–152 mm, which is also convenient for increasing the stability of the wheelchair when wheels 2 and 3 support the structure.

Figure 13c shows the influence of the minimum distance between wheels 1 and 2 and between wheels 3 and 4. The minimum value of a and c is now considered equal to 125 mm. The main effect of this reduction is that there are more configurations where the control architecture can climb up and climb down the stairs. However, the optimal configurations are identical to Figure 13a.

The second set of figures (Figure 14) is plotted by considering three different values of L_H when there is no limitation on angle β. The rest of the configuration is L_H = 700 mm, minimum values of b = 340 mm and $a = c$ = 140 mm, which are the same as in Figure 13a with different values of $L_1 - L_4$. Figure 14a considers the actuator lengths equal to $RH + \delta_V$. Thus, the wheelchair can climb up or climb down one defined step without changing angle β. Figure 14b considers the actuator lengths equal to $2RH + 2\delta_V$. Thus, the wheelchair can climb up or climb down two steps without changing angle β. If Figures 13a and 14a,b are compared, it can be noted that the best configurations are with a and c of around 143 mm. In addition, the total time in Figure 14a,b is larger than in Figure 13a. Then, if the cost of the actuators is not considered, the best configuration when $L_H = 700$ mm is with $L_1 - L_4 = 250$ mm. Finally, we have included one more figure to show the influence of $L_1 - L_4$ in the total time. In Figure 14c, we test the system for an actuator length smaller than RH, more specifically, equal to 136 mm. Although total times are greater than in the previous examples, and the number of valid configurations is less, this example shows that the wheelchair can even climb stairs with a raiser height taller than its actuators.

Thus, the main conclusions of the last six figures are that angle β should be limited, so the length of actuator L_9, and that the actuator lengths should be between the riser heights of one and two steps. Thus, the control architecture has found the best trade-off between the lengths of $L_1 - L_4$ and the length of L_9.

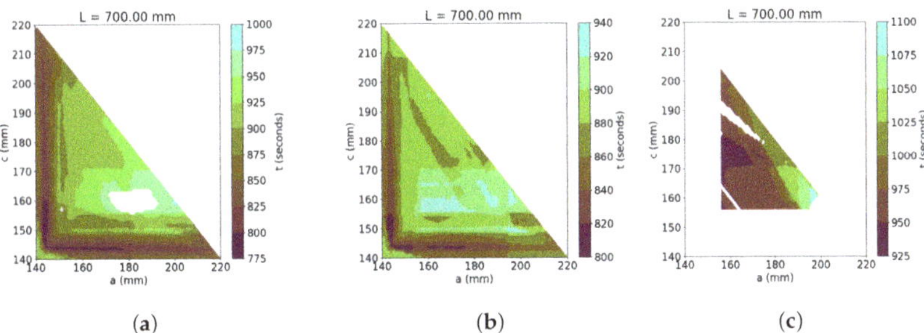

(a)   (b)   (c)

Figure 14. Wheelchair with $L_H = 700$ mm, minimum values of $b = 270$ mm and $a = c = 140$ mm. Influence of actuator lengths on total time. (**a**) Length of actuators $L_1 - L_4 = 185$ mm, (**b**) Length of actuators $L_1 - L_4 = 355$ mm, (**c**) Length of actuators $L_1 - L_4 = 136$ mm.

Finally, the last comparison is shown in Figure 15. Figure 15a–c shows the influence of L_H when it is considered on a wheelchair with $L_1 - L_4 = 250$ mm, minimum values of $a = c = 140$ mm, maximum values of $a = c = 220$ mm and no limitation in β. These figures can be compared with Figure 13a. The objective is to show the influence of L_H on the total time. The increment of L_H, keeping the minimum and maximum values of a and c, can obtain configurations with bigger values of b, which are more stable. The first conclusion is that if L_H is increased, the total time is also increased. The second conclusion is that there are more local areas where a minimum total time can be achieved. This can be better observed in Figure 15b when $L_H = 900$ mm. If $L_H = 750$ mm, there are two areas. The first one is with a or c equal to the minimum (140 mm). Thus, the best configuration must be $a = c = 140$ mm ($b = 470$ mm). The other area is with a around 190 mm and with c around 150 mm. However, this configuration is less stable. If the criterium of stability is considered, the best configurations for $L_H = 900$ mm (Figure 15b) are with a around 140 mm and c around 175 mm ($b = 585$ mm). The configuration with a around 175 mm and c around 155 mm can also be considered ($b = 570$ mm). Finally, Figure 15c has two local areas. The first one is with a around 205 mm and with c around 140 mm ($b = 655$ mm). The second area is with a around 180 mm and with c around 150 mm ($b = 670$ mm). Therefore, if L_H is increased, the stability of the mechanism is better, but the total time is also increased.

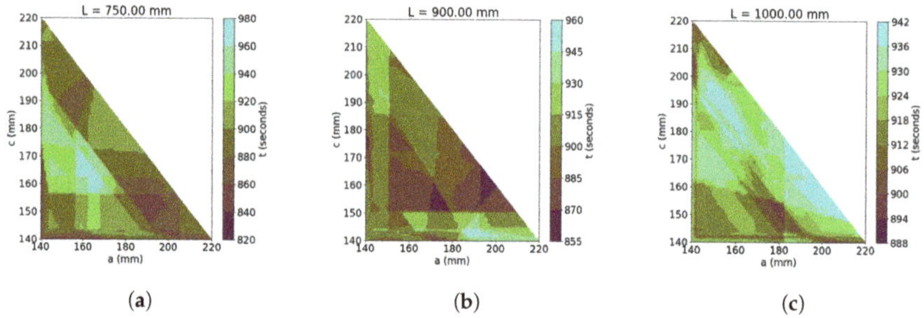

(a)   (b)   (c)

Figure 15. Wheelchair length of actuators $L_1 - L_4 = 250$ mm, minimum values of $a = c = 140$ mm and $|\beta| \in [0, 90]$. Influence of L_H on total time. (**a**) $L_H = 750$ mm and minimum value of $b = 320$ mm, (**b**) $L_H = 900$ mm and minimum value of $b = 470$ mm, (**c**) $L_H = 1000$ mm and minimum value of $b = 570$ mm.

The nine optimization examples carried out in this section show that the optimal wheelchair should have: (i) actuator lengths of around 250 mm, which is a value between $RH + \delta_V$ and $2RH + \delta_V$, (ii) there are local minima close to the minimum values of a and c, which motivate the optimization of the mechanism, (iii) variable L_H does not significantly affect the total time and increases the stability of the wheelchair and (iv) limitation of the angle β simplifies the optimization. Thus, the wheelchair proposed in Figure 13b is the best, with $a = c = 145$ mm.

5. Conclusions

In this work, we have shown that the optimal configuration of the leg-based stair-climbing wheelchair is not obvious. Although we only considered the sum of the total time to climb up and the total time to climb down the same stair, parameters a and c depend on the minimum distances between wheels, the angle β, the sensor errors (δ_V and δ_H) and the total length of the mechanism (L_H). The main conclusions are:

- Angle β and actuator lengths $L_1 - L_4$ should be limited. Thus the control architecture can better find an optimal trajectory, reducing the total time. In addition, the reduction in $L_1 - L_4$ makes the mechanism more competitive from an economical point of view.
- The sensor errors affect the range of parameters of a and c that can climb up and climb down the stairs, but the total time is not significantly affected. Therefore, the control architecture can include these uncertainties.
- The length of the mechanism (L_H) increases its stability and the total time is not significantly increased.

The control architecture and optimization must be considered in the future in a multi-objective optimization with all the parameters defined in this work. Thus, the optimization should consider: the total time, the stability of the mechanism and the energy needed by the actuators. In addition, the experimental validation of the optimized prototype will be considered in future work.

Author Contributions: Conceptualization, E.P. and P.G.-J.; methodology, E.P. and P.G.-J.; software, D.D.-M. and P.G.-J.; validation, D.D.-M., E.P. and P.G.-J.; formal analysis, E.P. and P.G.-J.; investigation, E.P., S.M.-B. and P.G.-J.; resources, E.P., C.A.-C., S.M.-B. and P.G.-J.; data curation, E.P. and P.G.-J.; writing—original draft preparation, D.D.-M., E.P. and P.G.-J.; supervision, P.G.-J.; project administration, P.G.-J.; funding acquisition, P.G.-J. All authors have read and agreed to the published version of the manuscript.

Funding: This work has been supported by the following research projects: CM/JIN/2019-022 of CAM-UAH and PID2019-104323RB-C31 of Spain's Ministry of Science and Innovation.

Institutional Review Board Statement: Not applicable.

Informed Consent Statement: Not applicable.

Data Availability Statement: Not applicable.

Conflicts of Interest: The authors declare no conflict of interest.

References

1. Tao, W.; Xu, J.; Liu, T. Electric-powered wheelchair with stair-climbing ability. *Int. J. Adv. Robot. Syst.* **2008**, *14*, 1–13. [CrossRef]
2. Wheelchair88 Limited. PW-4x4Q Stair Climbing Wheelchair, All Terrain 4 Wheel Drive Power Chair. 2017. Available online: https://www.wheelchair88.com/product/pw-4x4q/ (accessed on 2 August 2022).
3. Yuan, J.; Hirose, S. Research on leg-wheel hybrid stair-climbing robot, Zero Carrier. In Proceedings of the IEEE International Conference on Robotics and Biomimetics, Shenyang, China, 22–26 August 2004; pp. 654–659.
4. Ikeda, H.; Toyama, T.; Maki, D.; Sato, K.; Nakano, E. Cooperative step-climbing strategy using an autonomous wheelchair and a robot. *Robot. Auton. Syst.* **2021**, *135*, 103670. [CrossRef]
5. Heinrich, A. Topchair-S Wheelchair Has No Problem with Stairs, New Atlas. 2016. Available online: https://newatlas.com/topchairs-stair-climbing-wheelchair/41421/ (accessed on 2 August 2022).
6. Quaglia, G.; Nisi, M. Design of a self-leveling cam mechanism for a stair climbing wheelchair. *Mech. Mach. Theory* **2017**, *112*, 84–104. [CrossRef]
7. Quaglia, G.; Franco, W.; Oderio, R. Wheelchar.q, a motorized wheelchair with stair climbing ability. *Mech. Mach. Theory* **2011**, *46*, 1601–1609.
8. Giuseppe, Q.; Walter, F.; Matteo, N. Stair-Climbing Wheelchair.q05: From the Concept to the Prototype. In *Advances in Service and Industrial Robotics: Proceedings of the 27th International Conference on Robotics in Alpe-Adria Danube Region (RAAD 2018), Patras, Greece, 6–8 June 2018*; Aspragathos, N., Koustoumpardis, P., Moulianitis, V., Eds.; Springer: Cham, Switzerland, 2018; Volume 67.
9. Morales, R.; Feliu, V.; González, A.; Pintado, P. Kinematic model of a new staircase climbing wheelchair and its experimental validation. *Int. J. Robot. Res.* **2006**, *25*, 825–841. [CrossRef]
10. Gonzalez, A.; Ottaviano, E.; Ceccarelli, M. On the kinematic functionality of a four-bar based mechanism for guiding wheels in climbing steps and obstacles. *Mech. Mach. Theory* **2009**, *44*, 1507–1523. [CrossRef]
11. Morales, R.; Chocoteco, J.; Feliu, V.; Sira-Ramirez, H. Obstacle surpassing and posture control of a stairclimbing robotic mechanism. *Control. Eng. Pract.* **2013**, *21*, 604–621. [CrossRef]
12. Chocoteco, J.; Morales, R.; Feliu, V.; Sánchez, L. Trajectory Planning for a Stair-Climbing Mobility System Using Laser Distance Sensors. *IEEE Syst. J.* **2016**, *10*, 944–956. [CrossRef]
13. Chocoteco, J.A.; Morales, R.; Feliu-Batlle, V. Enhancing the Trajectory Generation of a Stair-Climbing Mobility System. *Sensors* **2017**, *11*, 2608. [CrossRef]
14. Sugahara, Y.; Hashimoto, K.; Kawase, M.; Ohta, A.; Sunazuka, H.; Tanaka, C.; Lim, H.; Takanishi, A. Walking pattern generation of a biped walking vehicle using a dynamic human model. In Proceedings of the IEEE/RSJ International Conference on Intelligent Robots and Systems, Beijing, China, 9–15 October 2006; pp. 2497–2502.
15. Wang, H.; He, L.; Li, Q.; Zhang, W.; Zhang, D.; Xu, P. Research on Kind of Leg-Wheel Stair-Climbing Wheelchair. In Proceedings of the IEEE International Conference on Mechatronics and Automation, Tianjin, China, 3–6 August 2014; pp. 2101–2105.
16. Pereira, E.; Gómez-Moreno, H.; Alén-Cordero, C.; Gil-Jiménez, P.; Maldonado-Bascón, M. A Novel Approach for a Leg-based Stair-climbing Wheelchair based on Electrical Linear Actuators. In Proceedings of the 16th International Conference on Informatics in Control, Automation and Robotics, Prague, Czech Republic, 29–31 July 2019.
17. Kluth, H. Stair Climbing Wheelchair. U.S. Patent US4569409A, 7 July 1982.
18. Delgado-Mena D.; Pereira, E.; Alén-Cordero, C.; Maldonado-Bascón, S.; Gil-Jiménez, P. Control architecture for a novel Leg-Based Stair-Climbing Wheelchair. In Proceedings of the European Conference on Mobile Robots (ECMR), Bonn, Germany, 31 August–3 September 2021; pp. 1–6.
19. Fomento. Documento Básico SUA, Seguridad de Utilización y Accesibilidad, Ministerio de Fomento Secretaría de Estado de Infraestructuras, Transporte y Vivienda Dirección General de Arquitectura, Vivienda y Suelo, 2010.

Article

Cable-Driven Parallel Robot Actuators: State of the Art and Novel Servo-Winch Concept

Edoardo Idà * and Valentina Mattioni

Department of Industrial Engineering, University of Bologna, Via Terracini 24, 40131 Bologna, Italy
* Correspondence: edoardo.ida2@unibo.it

Abstract: Cable-Driven Parallel Robots (*CDPRs*) use cables arranged in a parallel fashion to manipulate an end-effector (*EE*). They are functionally similar to several cranes that automatically collaborate in handling a shared payload. Thus, *CDPRs* share several types of equipment with cranes, such as winches, hoists, and pulleys. On the other hand, since *CDPRs* rely on model-based automatic controllers for their operations, standard crane equipment may severely limit their performance. In particular, to achieve reasonably accurate feedback control of the *EE* pose during the process, the length of the cable inside the workspace of the robot should be known. Cable length is usually inferred by measuring winch angular displacement, but this operation is simple and accurate only if the winch transmission ratio is constant. This problem called for the design of novel actuation schemes for *CDPRs*; in this paper, we analyze the existing architectures of so-called servo-winches (i.e., servo-actuators which employ a rotational motor and have a constant transmission ratio), and we propose a novel servo-winch concept and compare the state-of-the-art architectures with our design in terms of pros and cons, design requirements, and applications.

Keywords: cable-driven parallel robots; wire-driven parallel robots; tendon-driven parallel robots; actuators; winch; design

Citation: Idà, E.; Mattioni, V. Cable-Driven Parallel Robot Actuators: State of the Art and Novel Servo-Winch Concept. *Actuators* **2022**, *11*, 290. https://doi.org/10.3390/act11100290

Academic Editor: Zhuming Bi

Received: 19 September 2022
Accepted: 8 October 2022
Published: 11 October 2022

Publisher's Note: MDPI stays neutral with regard to jurisdictional claims in published maps and institutional affiliations.

1. Introduction

Large-scale handling of bulky loads is a widespread necessity throughout the world. Manufacturing plant logistics, infrastructure construction and maintenance are just two of the most prominent examples where anyone can observe several overhead cranes, truck-mounted and fixed-installation cranes, working independently, and entirely manually operated. Conversely, Cable-Driven Parallel Robots (*CDPRs* in short) work like fully-automated collaborative cranes. They are parallel robotic manipulators where rigid links are replaced by extendable cables. The latter are wound and unwound by linear or rotational actuation units (called *winches* in the following) and routed using guidance devices toward a shared end-effector (*EE* in short), on which they are attached in a parallel fashion [1].

CDPRs potentially have a large and reconfigurable workspace. First, because very long cables can be coiled on rotary winches. Furthermore, cables operate in a structurally efficient manner, being subject only to tensile loads. In addition, if properly designed, actuation units and guidance systems can be rearranged discretely [2] and continuously [3], allowing for rapid changes in workspace size and shape. However, since multiple cables act in parallel on the same load, part of the work they exert is spent keeping each other in tension [4]. Nevertheless, they may be more efficient than industrial robots, as the latter have to carry their weight around [5]. Additionally, if the task and worksite characteristics are specified, cables can also be balanced with counterweights [6].

Despite their advantages, the use of *CDPRs* in the industry is still limited due to their design, control, and safety challenges. Controllability and safety, on the one hand, can be enhanced by employing more cables than the *EE* degrees of freedom (*DoFs* in short). However, this can cause cables to collide with each other and their surrounding environment, limiting the robot's workspace or forcing the use of rigorous design techniques to avoid it [7]. In this case, workspace accessibility may be improved by suspending the *EE*, using a redundant number of cables [8,9]. Regardless, the likelihood of cable-to-cable interference could increase [10]. Therefore, in order to simplify the design and rationalize the cost of the robot, simpler suspended *CDPR* have been proposed, with fewer cables than *EE*'s *DoFs*, which, on the other hand, require dedicated control schemes for their effective use since the *EE* is unconstrained [11,12].

The earliest example of a *CDPR* is the famous *Skycam*® [13], which is still used as a camera motion device for overhead sport event shooting. However, research interest in the possible applications of this technology only began a few years later, when Higuchi et al. [14] highlighted the numerous advantages of cranes' automatic cooperation. Only in the nineties the *RoboCrane*® was introduced [15]: this equipment was the first to allow both position and orientation of its *EE* to be automatically controlled with six cables. Since then, numerous applications have been proposed and successfully implemented by researchers: large-scale additive manufacturing [16], laser-based manufacturing [1], contour crafting [17], marine handling systems [18], warehouse retrieval systems [6], large-scale handling systems [19,20], facade cleaning [21] and installation [7] systems, motion simulators [22], large-aperture telescopes [23], measurement devices [24–27], rehabilitation devices [28], and haptic interfaces [29].

Different *cable-driven* applications usually have highly different requirements: even though the principal mean of transmission is a cable, its actuation unit and guidance system are engineered according to other principles. Ref. [30] reports a comprehensive study of cranes, winches, and hoists to be used in civil engineering applications. On the opposite spectrum than civil applications, there are cable-driven hands and fingers, where cables are used to actuate joints remotely, so that most of the actuation weight is as distant as possible to where the force application is needed [31,32]. Miniaturization, force capability, and motion accuracy is instead required in so-called *tendon-driven* continuum robots, where a remote cable actuation is needed to control the deformation of slender links [33,34], and in *tensegrity-based* robots [35]. Lastly, the growing interest in mobile robot applications has motivated researchers to develop lightweight and small winches with high-force capabilities [36,37].

For industrial applications, guidance systems are usually a combination of fixed and swiveling pulleys [1], whose geometry and installation configuration are dictated by geometric and loading conditions of the operation (many research prototypes have even simpler guidance systems, such as eyelets where cables may slide through [38]). Conversely, the design of the actuation unit is driven by application requirements in terms of rated power, cable tension, and speed, but also by the requirements of the control system. The most common one is the ability to feedback control the *EE* pose. To succeed in such a task, one may rely on exteroceptive measurement devices directly providing *EE* pose information [39,40], state estimators [41,42], or forward kinematics based on cable length estimation. The latter approach is widely used thanks to well-established techniques in the solution to the forward kinematic problem and thanks to the fact that no sensors other than the ones embedded in the actuators for their low-level feedback control need to be added to the robot (additional sensors can be added to speed-up computation, improve accuracy [43,44], or if embedded sensors are not sufficient [45]). However, accurate pose information is achievable through forward kinematics if and only if (i) a cable model suitable to the application requirement is used [46–48], and (ii) there is a clear correlation between actuator displacement $\Delta\theta$ and cable displacement Δl, namely the actuation unit *transmission ratio* $K = \Delta l / \Delta\theta$. If the latter condition is not satisfied, it is unlikely that the use of a suitable cable model would work without additional sensors. This fact motivated researchers to characterize existing types of cable actuation systems or develop new ones suitable for robotic purposes [49].

Concerning actuation, the most straightforward way to wind a cable is using a smooth drum connected to a motor [50,51]. Unfortunately, it is not trivial to determine how the cable is wound over the drum, as the axial a and radial r winding distances are not a function of the motor angle (Figure 1a). As an alternative, cables can be overlapped (i) on very short drums [52] (Figure 1b), (ii) on smooth or grooved drums by a self-reversing screw (Figure 1c), or even (iii) on variable-radius drums [53,54]; all these choices allow for a correlation between cable and motor displacement. Unfortunately, the transmission ratio K is a function of the absolute motor angle, which may not be known at start-up time, and, furthermore, a varying r implies varying tension-speed limits for a given motor-rated power. The possibly simpler and commercially available solution for a constant and known K is to use a hoist (Figure 1d), and a linear actuator for its control [55]. However, if long cables need to be used, the installation space, transmission ratio, and cable wear increase alongside the number of pulleys in the hoist [49].

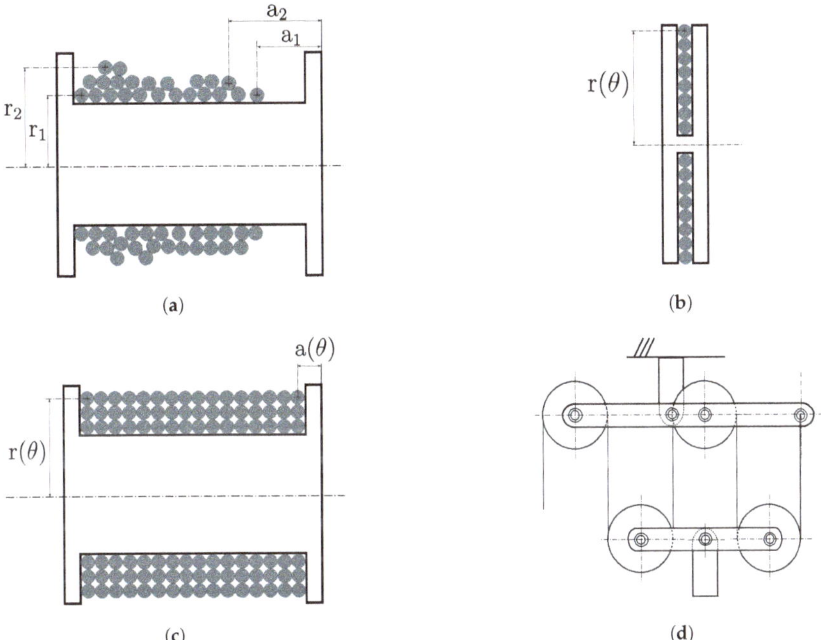

Figure 1. Examples of cable actuation units. (**a**) Smooth drum. (**b**) Short drum with overlapping cable. (**c**) Drum where the cable is overlapped with a self-reversing screw. (**d**) Hoist.

To the best of the authors' knowledge, the first example of a *servo-winch*, namely a rotary winch characterized by a constant and known transmission ratio, was proposed in [56]. Since then, many other solutions have been proposed in the literature, mainly to compensate for the lack of existing commercial devices. Generally speaking, to build a *CDPR* prototype for research or industrial purposes, a dedicated winch must be designed in-house. A comparison of existing architectures of servo-winches is not available in the literature, and thus, one can choose between different designs, based solely on experience, if any. The reason for choosing one servo-winch is, to say the least, unclear and not shared among the cable-robotic community. Thus, this paper aims to (i) provide a comprehensive description of the state-of-the-art solutions for servo-winch design, (ii) propose a novel servo-winch design, and, lastly, (iii) provide guidelines for the selection of optimal servo-winch architecture based on the benefits and drawbacks of the existing solutions. The remainder of the paper is structured as follows: Section 2 analyzes the state of the art in the servo-winch design, while Section 3 describes a novel design, called *Spline Winch*; design comparison and architecture selection guidelines are discussed in Section 4. Finally, conclusions are drawn.

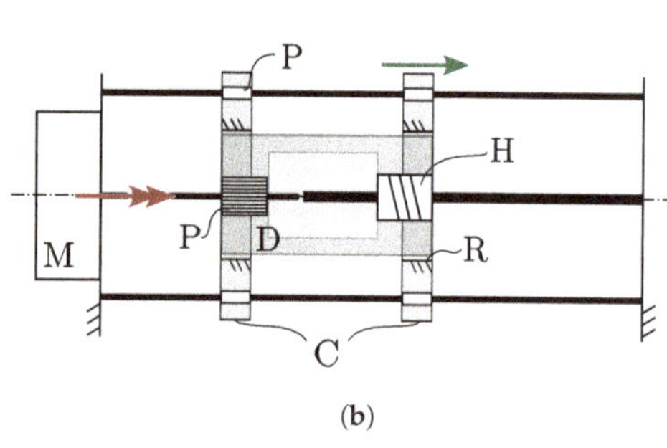

(a)

Figure 5. Winch with ball spline. (**a**) Picture of the winch developed at Irma l@B (University of Bologna). (**b**) The motor (M) is fixed to the frame and coupled to the drum (D) using a ball spline (P) to transmit the torque. The drum is mounted onto two plates (which form the carriage C), and can translate thanks to a screw/nut pair (H).

3.1. Kinematics

The motor (M) is fixed to the winch frame, while the drum (D) can (see Figure 5):

- Rotate since a spline shaft is rigidly attached to the motor axis, and a spline nut is attached to the drum; the spline shaft/nut pair is effectively a prismatic joint (P), designed so as to transmit torque while allowing axial translation;
- Translate since a screw shaft is rigidly attached to the winch frame, and a screw nut is attached to the drum; this is the classical helical pair (H) used in all winch designs.

The drum is supported via two plates (or carriages, C): a revolute joint (R) and two prismatic joints (P) are embedded into each plate, so that the drum can freely rotate w.r.t. the plates, and the drum-plates assembly can translate w.r.t. the winch frame.

Since the drum rototranslates, the overall transmission ratio K of the spline winch is the same as the rototranslating-drum one, namely:

$$K = \sqrt{K_S^2 + r_D^2} \ [\text{m/rad}] \tag{4}$$

3.2. Mechanical Design

The proposed Spline Winch was designed and built at IRMA L@B (Figure 6). Its mechanical design is detailed hereafter and shown in Figure 6b. Its frame consists of two aluminum plates (1), connected by extruded aluminum profiles (see Figure 6a). Two floating plates (2) are connected with four ball bushings (3), that allows the translation w.r.t. to two rods (4); the latter are connected to the frame through rigid couplings (5). The motor is coupled to the ball spline shaft (6) through a bellow coupling (torsionally stiff but flexurally compliant, see Figure 6a since motor and coupling are not represented in the cross section). Instead of a regular spline shaft, a ball spline shaft is chosen due to its zero-backlash and low friction properties; this component is widespread and cost-effective due to its frequent use in machining equipment tool-change systems. A ball screw shaft (8) is also attached to the winch frame, on the opposite side w.r.t. the ball spline shaft.

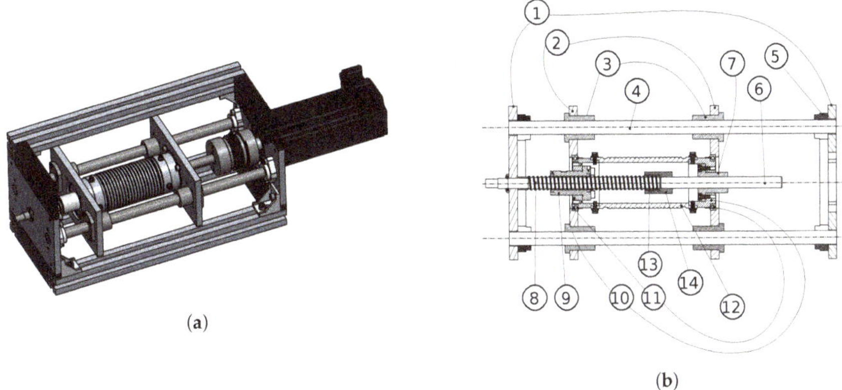

Figure 6. CAD model of the Spline Winch. (**a**) Axonometric view (part of the frame is removed for display reasons). (**b**) Cross-section of the winch (motor and motor coupling are removed for display reasons).

The ball spline (7) and ball screw (9) nuts are rigidly attached to two drum covers (10), which are free to rotate w.r.t. the floating plates (2) thanks to radial bearings (11), and are rigidly attached to a tube drum (12). The choice of decomposing the drum in two covers allows to (i) critically reduce weight, (ii) save machining waste (for a single component drum of the same weight, most of the raw material would only be waste), modify the winch transmission ratio by only machining a new tube—the last feature is particularly interesting for research prototypes, whose performance requirements may vary over time.

At last, the ball spline and ball screw shaft are aligned with a coupling (14), which embeds a bushing (13) to allow the rotational motion of the spline w.r.t. the screw.

The winch is structurally optimal, since the shafts (4) resist the external load imposed by the cable, and the overall moving mass and inertia are inherently reduced w.r.t. to rototranslating-drum and translating-motor winches. With $r_d = 29$ mm and $h = 5$ mm, the winch has a transmission ratio $K = 29.01$ mm/rad. Thanks to the $P_{m,r} = 750$ W of rated power, and $T_{m,r} = 2.39$ Nm of rated torque, the winch can nominally balance a tension of $\tau = 82.38$ N, while displacing the cable at 9.103 m/s. The applications intended for this winch are highly dynamic ones.

4. Design Comparison and Application Guidelines

After briefly revising existing servo-winch architectures, and introducing the Spline Winch, a critical comparison of each winch's pros and cons is in order. By the end of this section, we aim to provide some winch architecture selection guidelines, depending on application requirements. In the following, friction in the components is neglected. This choice was deemed necessary not because friction is, in fact, negligible but because we want to highlight several important factors which are fundamental regardless of friction. The reader is referred to Chapter 8.6.2 [61] for details about single component selection for optimizing winch frictional behavior. Additionally, the effect of the gravitational force on the winch dynamics is not explicitly accounted for, because it varies depending on the winch installation configuration. This effect will only be evaluated qualitatively in the following.

We start by observing that all winch architectures share some components, such as a drum, some rods w.r.t. whom the drum can slide, a translating component (whether the drum or the spooling helper), and a screw/nut system for transforming rotational motion in a linear one. We can then divide the architectures into two groups, namely one group characterized by the rototranslation of the drum (rototranslating-drum, translating-motor, and Spline Winch design), and one group represented by the decoupling of rotation and translation for achieving constant cable direction exiting the drum, namely the spooling-helper design.

For each winch, the overall dynamics is:

$$T_M = J^\star \ddot{\theta} + K\tau \tag{5}$$

where T_M is the motor torque, K is the winch transmission ratio, and $J^\star$ the overall transmission inertia reduced to the motor axis. From a dynamic point of view, the first group of winches shares similar load distributions when transforming the motor torque into cable tension. The rototranslating drum is selected to provide a schematic representation (see Figure 7a). The drum rotational dynamics is given by (see Figure 7b):

$$J\ddot{\theta} + T_s + r_D \tau_{CS} = r_s F \tag{6}$$

where J is the inertia of all the rotating components, T_s is the screw reaction torque, τ_{CS} is the tension component projected onto the drum cross-section, $r_s F = T_M$ is the product between the shaft radius and the shaft reaction force, which equals the torque exerted by the motor. If we consider the relationship between the axial force Q and the torque T_S exchanged in the screw/nut pair to be:

$$T_S = K_S Q, \qquad K_S = \frac{h}{2\pi}\text{m/rad} \tag{7}$$

The translational dynamics of the drum are (Figure 7c):

$$Q = \tau_{AX} + M K_S \ddot{\theta} \tag{8}$$

where M is the mass of all translating components, and τ_{AX} is the component of τ directed as the winch axis.

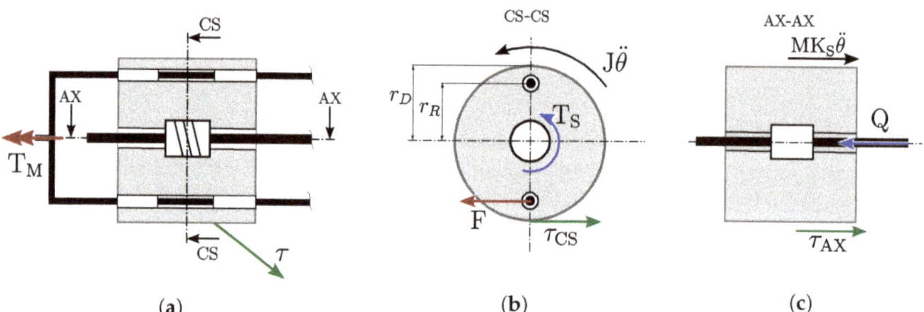

Figure 7. Dynamic loads in the rototranslating-drum design. (**a**) Overview. (**b**) Rotational dynamics. (**c**) Translational dynamics.

If we now remember helix geometrical properties, we have:

$$\tau = \sqrt{\tau_{AX}^2 + \tau_{CS}^2}, \qquad \frac{\tau_{AX}}{h} = \frac{\tau_{CS}}{2\pi r_D} \tag{9}$$

and substitute Equations (7) and (8) in (6), after some algebraic manipulation we finally get:

$$T_m = r_S F = \left(J + K_S^2 M\right)\ddot{\theta} + \left(\sqrt{K_S^2 + r_d^2}\right)\tau \tag{10}$$

which compared with Equation (5) gives:

$$J^\star = J + K_S^2 M, \qquad K = \sqrt{K_S^2 + r_d^2} \tag{11}$$

According to the provided analysis, it is possible to compare some of the winches' performances:

- The rototranslating-drum design, though conceptually simple, suffers from three main drawbacks:
  - To transmit torque to the drum, the shafts that pass through the winches are subject to a radial force which may be critically higher than the cable tension by design since $r_d > r_S$; this means that these shafts need to be bulky enough, which in turns means that the drum radius (and thus the transmission ratio) cannot vary freely.
  - If an *open-end* design of the shafts is employed, such as the one proposed in Figure 2a, the torsional load of the winch may deform the rods without actually transmitting force to the drum.
  - The manufacturing tolerance of the shaft, the drum, and the bushing inside the drum, need to be very high in order to avoid the winch stalling [38]; this, in turn, highlights that the mechanical design should be everything but simple.

  Its primary advantage is the possibility to freely install the winch in any configuration since its dynamics is only affected by the drum weight, which can be very low.

- The translating motor winch has three significant advantages, namely:
  - It can be easily miniaturized since it has no components passing through the drum other than the screw;
  - It is mechanically straightforward (most of the components for its manufacturing are commercially available), and thus also cheap;
  - It is structurally efficient since the rods withstanding the external load (but possibly also the motor weight) can be placed in a convenient position, and be as sturdy as needed since $r_R > r_d$.

  On the other hand, its main characteristic is also its main drawback: the motor (and gearbox, if used) mass needs to be translated with the drum, which means that:
  - According to Equation (11), the overall transmission inertia may be critically high since M includes both the drum and the winch mass, thus severely limiting winch dynamics;
  - If the winch is installed with its axis vertical, the weight of both the motor and the drum has to be compensated by the motor torque, which is not very efficient.

- As previously mentioned, the Spline Winch attempts to summarize the rototranslating-drum and translating-motor winches' advantages, while not suffering from the drawbacks:
  - As the rototranslating-drum design, it can be freely installed because it does not have to carry the motor weight around, even though it needs to compensate for the two additional translating plates (and bearings) as a trade-off;
  - As the translating-motor design, it can be miniaturized (small screw and spline shaft are commercially available). The additional mechanical complexities are the spline shaft and the motor-shaft-spline-shaft coupling, which is commercially available and structurally efficient.

  It does not suffer from any rototranslating-drum and translating-motor design drawbacks, but it strictly requires two more components: the spline shaft and the motor-shaft-spline-shaft coupling. This means that it may not be as cheap and small as the translating-motor design.

The dynamic of the spooling-helper winch is slightly different, due to the decoupled nature of rotating and translating components (see Figure 8a). The drum rotational dynamics are given by (see Figure 8b):

$$J\ddot{\theta} + T_{SB} + r_D \tau_{CS} = T_M \tag{12}$$

where $T_{SB} = T_S$ is the torque transmitted through the synchronous belt to the screw, which are equal if we neglect friction and elasticity. The translational dynamics of the spooling helper are instead (Figure 8c):

$$Q = \tau - \tau_{AX} + MK_S\ddot{\theta} \tag{13}$$

If we substitute Equations (7), (9), and (13) in (12), after some algebraic manipulation we finally get:

$$T_m = \left(J + K_S^2 M\right)\ddot{\theta} + \left(\sqrt{K_S^2 + r_d^2} - K_S\right)\tau \tag{14}$$

which compared with Equation (5) gives:

$$J^\star = J + K_S^2 M, \qquad K = \sqrt{K_S^2 + r_d^2} - K_S \tag{15}$$

The inertia of the rotating components includes the motor, the drum, the synchronous pulleys, and the screw, while the only translating part is the spooling helper. This winch is the only one optimizing the cable-to-footprint ratio quantity since the drum does not translate and can occupy all the winch length. One minor disadvantage is the necessity to use one more pulley than other designs since the spooling helper necessitates one to deflect the cable from the drum to a direction parallel to the helper translation. One possibly major disadvantage, if cable tension is measured on the spooling helper (as it is usually done in these winches), is that the dynamic bandwidth of the winch is severely limited due to loadcell translational motion with the helper. A summary of the discussed pros and cons can be found in Table 1.

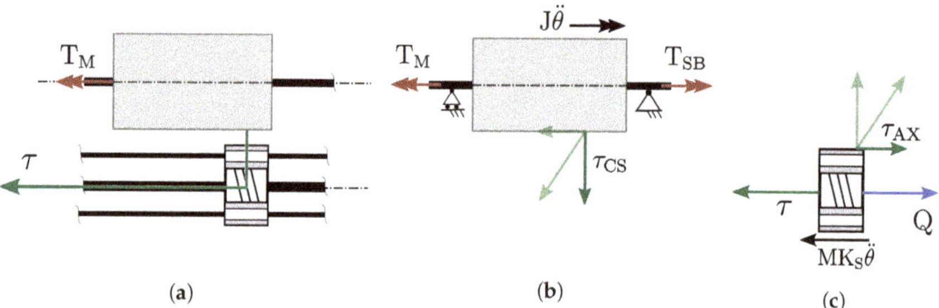

Figure 8. Dynamic loads in the spooling helper design. (**a**) Overview. (**b**) Rotational dynamics. (**c**) Translational dynamics.

Table 1. Summary of the various design pros and cons. Scale: (++) very positive, (+) positive, (o) neutral, (-) negative, (–) very negative.

	Rototranslating Drum	Translating Motor	Spooling Helper	Spline Winch
Mechanical simplicity	–	++	-	+
Free configuration installation	+	–	++	o
No limits on transmission ratio	–	+	++	+
Dynamic capabilities	o	–	++	++
Built-in sensor capabilities	o	o	–	o
Cost	-	–	+	+

5. Conclusions

This paper presented the state of the art in servo-winch design for cable-driven robots. A novel design concept was introduced and critically compared to the existing and proposed architectures from an application point of view. It was shown that the rototranslating-drum concept presents no significant advantages, even though it was the first one historically developed. The translating-motor concept is an optimal choice for low-cost, not-highly dynamical applications, where installation orientation requirements are not strict. At the same time, the novel Spline Winch is the go-to choice for vertical-winch axis installations and high-dynamic applications. The spooling helper solution optimizes the quantity of stored cable w.r.t. winch footprint. However, highly dynamical operations should be avoided if a load cell is embedded in the helper for measuring cable tension.

Author Contributions: Conceptualization, E.I.; methodology, E.I.; software, V.M.; validation, V.M.; formal analysis, E.I.; investigation, E.I.; resources, E.I.; data curation, V.M.; writing—original draft preparation, E.I. and V.M.; writing—review and editing, E.I. and V.M.; visualization, V.M.; supervision, E.I.; project administration, E.I.; funding acquisition, E.I. All authors have read and agreed to the published version of the manuscript.

Funding: This research received no external funding.

Institutional Review Board Statement: Not applicable.

Informed Consent Statement: Not applicable.

Data Availability Statement: Not applicable.

Acknowledgments: The authors would like to thank Ing. Federico Zaccaria for the help with CAD modeling of the novel winch concept proposed in this paper.

Conflicts of Interest: The authors declare no conflict of interest.

References

1. Mattioni, V.; Idà, E.; Carricato, M. Design of a Planar Cable-Driven Parallel Robot for Non-Contact Tasks. *Appl. Sci.* **2021**, *11*, 9491. [CrossRef]
2. Gagliardini, L.; Caro, S.; Gouttefarde, M.; Girin, A. Discrete reconfiguration planning for Cable-Driven Parallel Robots. *Mech. Mach. Theory* **2016**, *100*, 313–337. [CrossRef]
3. Heyden, T.; Woernle, C. Dynamics and flatness-based control of a kinematically undetermined cable suspension manipulator. *Multibody Syst. Dyn.* **2006**, *16*, 155. [CrossRef]
4. Mattioni, V.; Idà, E.; Carricato, M. Force-distribution sensitivity to cable-tension errors in overconstrained cable-driven parallel robots. *Mech. Mach. Theory* **2022**, *175*, 104940. [CrossRef]
5. Kraus, W.; Spiller, A.; Pott, A. Energy efficiency of cable-driven parallel robots. In Proceedings of the 2016 IEEE International Conference on Robotics and Automation (ICRA), (ICRA) OR International Conference on Robotics and Automation, Stockholm, Sweden, 16–21 May 2016; pp. 894–901.
6. Bruckmann, T.; Sturm, C.; Fehlberg, L.; Reichert, C. An energy-efficient wire-based storage and retrieval system. In Proceedings of the 2013 IEEE/ASME International Conference on Advanced Intelligent Mechatronics, Wollongong, Australia, 9–12 July 2013; pp. 631–636.
7. Iturralde, K.; Feucht, M.; Illner, D.; Hu, R.; Pan, W.; Linner, T.; Bock, T.; Eskudero, I.; Rodriguez, M.; Gorrotxategi, J.; et al. Cable-driven parallel robot for curtain wall module installation. *Autom. Constr.* **2022**, *138*, 104235. [CrossRef]
8. Bettega, J.; Richiedei, D.; Trevisani, A. Using Pose-Dependent Model Predictive Control for Path Tracking with Bounded Tensions in a 3-DOF Spatial Cable Suspended Parallel Robot. *Machines* **2022**, *10*, 453. [CrossRef]
9. Lin, D.; Mottola, G.; Carricato, M.; Jiang, X.; Li, Q. Dynamically-Feasible Trajectories for a Cable-Suspended Robot Performing Throwing Operations. In *ROMANSY 23—Robot Design, Dynamics and Control*; Venture, G., Solis, J., Takeda, Y., Konno, A., Eds.; Springer International Publishing: Cham, Switzerland, 2021; pp. 547–555.
10. Martin, A.; Caro, S.; Cardou, P. Geometric determination of the cable-cylinder interference regions in the workspace of a cable-driven parallel robot. In *Cable-Driven Parallel Robots*; Springer: Berlin, Germany, 2018; pp. 117–127.
11. Idà, E.; Briot, S.; Carricato, M. Robust Trajectory Planning of Under-Actuated Cable-Driven Parallel Robot with 3 Cables. In *Advances in Robot Kinematics 2020*; Lenarčič, J., Siciliano, B., Eds.; Springer International Publishing: Cham, Switzerland, 2021; pp. 65–72.
12. Idà, E.; Carricato, M. A New Performance Index for Underactuated Cable-Driven Parallel Robots. In *Cable-Driven Parallel Robots*; Gouttefarde, M., Bruckmann, T., Pott, A., Eds.; Springer International Publishing: Cham, Switzerland, 2021; pp. 24–36.
13. Cone, L.L. Skycam-an aerial robotic camera system. *Byte* **1985**, *10*, 122.
14. Higuchi, T.; Ming, A.; Jiang-Yu, J. Application of Multi-Dimensional Wire Cranes in Construction. In Proceedings of the 5th International Symposium on Automation and Robotics in Construction (ISARC), International Association for Automation and Robotics in Construction (IAARC), Tokyo, Japan, June 1988; pp. 661–668.;
15. Albus, J.; Bostelman, R.; Dagalakis, N. The NIST robocrane. *J. Natl. Inst. Stand. Technol.* **1992**, *10*, 709–724. [CrossRef]
16. Tho, T.P.; Thinh, N.T. Using a Cable-Driven Parallel Robot with Applications in 3D Concrete Printing. *Appl. Sci.* **2021**, *11*, 563. [CrossRef]

17. Williams, R.L., II; Xin, M.; Bosscher, P. Contour-crafting-cartesian-cable robot system: Dynamics and controller design. In Proceedings of the ASME 2008 International Design Engineering Technical Conferences, New York, NY, USA, 3–6 August 2008; Volume 2, pp. 39–45.

18. Lv, W.; Tao, L.; Ji, Z. Design and control of cable-drive parallel robot with 6-dof active wave compensation. In Proceedings of the 2017 3rd International Conference on Control, Automation and Robotics (ICCAR), Nagoya, Japan, 22–24 April 2017; pp. 129–133.

19. Pott, A.; Meyer, C.; Verl, A. Large-scale assembly of solar power plants with parallel cable robots. In Proceedings of the ISR 2010 (41st International Symposium on Robotics) and ROBOTIK 2010 (6th German Conference on Robotics), Munich, Germany, 7–9 June 2010; pp. 1–6.

20. Culla, D.; Gorrotxategi, J.; Rodríguez, M.; Izard, J.B.; Hervé, P.E.; Cañada, J. Full Production Plant Automation in Industry Using Cable Robotics with High Load Capacities and Position Accuracy. In *ROBOT 2017: Third Iberian Robotics Conference*; Ollero, A., Sanfeliu, A., Montano, L., Lau, N., Cardeira, C., Eds.; Springer International Publishing: Cham, Switzerland, 2018; pp. 3–14.

21. Voss, K.H.J.; van der Wijk, V.; Herder, J.L. Investigation of a Cable-Driven Parallel Mechanism for Interaction with a Variety of Surfaces, Applied to the Cleaning of Free-Form Buildings. In *Latest Advances in Robot Kinematics*; Lenarcic, J., Husty, M., Eds.; Springer: Dordrecht, The Netherlands, 2012; pp. 261–268.

22. Miermeister, P.; Lächele, M.; Boss, R.; Masone, C.; Schenk, C.; Tesch, J.; Kerger, M.; Teufel, H.; Pott, A.; Bülthoff, H.H. The CableRobot simulator large scale motion platform based on cable robot technology. In Proceedings of the 2016 IEEE/RSJ International Conference on Intelligent Robots and Systems (IROS), Daejeon, Korea, 9–14 October 2016; pp. 3024–3029.

23. Nan, R.D.; Di, L.; Jin, C.; Wang, Q.; Zhu, L.; Zhu, W.; Zhang, H.Y.; Yue, Y.; Qian, L. The five-hundred-meter aperture spherical radio telescope (FAST) project. *Int. J. Mod. Phys. D* **2011**, *20*, 989–1024. [CrossRef]

24. Idà, E.; Marian, D.; Carricato, M. A Deployable Cable-Driven Parallel Robot With Large Rotational Capabilities for Laser-Scanning Applications. *IEEE Robot. Autom. Lett.* **2020**, *5*, 4140–4147. [CrossRef]

25. Idà, E.; Briot, S.; Carricato, M. Identification of the inertial parameters of underactuated Cable-Driven Parallel Robots. *Mech. Mach. Theory* **2022**, *167*, 104504. [CrossRef]

26. Previati, G.; Gobbi, M.; Mastinu, G. Measurement of the mass properties of rigid bodies by means of multi-filar pendulums—Influence of test rig flexibility. *Mech. Syst. Signal Process.* **2019**, *121*, 31–43. [CrossRef]

27. Kamali, K.; Joubair, A.; Bonev, I.A.; Bigras, P. Elasto-geometrical calibration of an industrial robot under multidirectional external loads using a laser tracker. In Proceedings of the 2016 IEEE International Conference on Robotics and Automation (ICRA), Stockholm, Sweden, 16–21 May 2016; pp. 4320–4327.

28. Rosati, G.; Gallina, P.; Masiero, S. Design, implementation and clinical tests of a wire-based robot for neurorehabilitation. *IEEE Trans. Neural Syst. Rehabil. Eng.* **2007**, *15*, 560–569. [CrossRef] [PubMed]

29. Murayama, J.; Bougrila, L.; Akahane, Y.K.; Hasegawa, S.; Hirsbrunner, B.; Sato, M. SPIDAR G&G: A two-handed haptic interface for bimanual VR interaction. In Proceedings of the EuroHaptics 2004, Munchen, Germany, 5–7 June 2004; pp. 138–146.

30. Samset, I. *Winch and Cable Systems*; Springer Science & Business Media: Berlin, Germany, 2013; Volume 18.

31. King, J.P.; Bauer, D.; Schlagenhauf, C.; Chang, K.H.; Moro, D.; Pollard, N.; Coros, S. Design. Fabrication, and Evaluation of Tendon-Driven Multi-Fingered Foam Hands. In Proceedings of the 2018 IEEE-RAS 18th International Conference on Humanoid Robots (Humanoids), Beijing, China, 6–9 November 2018; pp. 1–9.

32. Liang, Z.; Wang, B.; Song, Y.; Zhang, T.; Xiang, C.; Guan, Y. Design of a Novel Cable-Driven 3-DOF Series-Parallel Wrist Module for Humanoid Arms. In Proceedings of the 2021 IEEE International Conference on Mechatronics and Automation (ICMA), Takamatsu, Japan, 8–11 August 2021; pp. 709–714.

33. Nguyen, T.D.; Burgner-Kahrs, J. A tendon-driven continuum robot with extensible sections. In Proceedings of the 2015 IEEE/RSJ International Conference on Intelligent Robots and Systems (IROS), Hamburg, Germany, 28 September–3 October 2015; pp. 2130–2135.

34. Li, M.; Kang, R.; Geng, S.; Guglielmino, E. Design and control of a tendon-driven continuum robot. *Trans. Inst. Meas. Control* **2018**, *40*, 3263–3272. [CrossRef]

35. Liu, Y.; Bi, Q.; Yue, X.; Wu, J.; Yang, B.; Li, Y. A review on tensegrity structures-based robots. *Mech. Mach. Theory* **2022**, *168*, 104571. [CrossRef]

36. Li, Z.; Erskine, J.; Caro, S.; Chriette, A. Design and Control of a Variable Aerial Cable Towed System. *IEEE Robot. Autom. Lett.* **2020**, *5*, 636–643. [CrossRef]

37. Heap, W.E.; Keeley, C.T.; Yao, E.B.; Naclerio, N.D.; Hawkes, E.W. Miniature, Lightweight, High-Force, Capstan Winch for Mobile Robots. *IEEE Robot. Autom. Lett.* **2022**, *7*, 9873–9880. [CrossRef]

38. Izard, J.B.; Gouttefarde, M.; Michelin, M.; Tempier, O.; Baradat, C. A Reconfigurable Robot for Cable-Driven Parallel Robotic Research and Industrial Scenario Proofing. In *Cable-Driven Parallel Robots*; Bruckmann, T., Pott, A., Eds.; Springer: Berlin/Heidelberg, Germany, 2013; pp. 135–148.

39. Dallej, T.; Gouttefarde, M.; Andreff, N.; Hervé, P.E.; Martinet, P. Modeling and vision-based control of large-dimension cable-driven parallel robots using a multiple-camera setup. *Mechatronics* **2019**, *61*, 20–36. [CrossRef]

40. Zake, Z.; Chaumette, F.; Pedemonte, N.; Caro, S. Robust 2½D Visual Servoing of A Cable-Driven Parallel Robot Thanks to Trajectory Tracking. *IEEE Robot. Autom. Lett.* **2020**, *5*, 660–667. [CrossRef]

41. Qi, R.; Rushton, M.; Khajepour, A.; Melek, W.W. Decoupled modeling and model predictive control of a hybrid cable-driven robot (HCDR). *Robot. Auton. Syst.* **2019**, *118*, 1–12. [CrossRef]

42. Le Nguyen, V.; Caverly, R.J. Cable-Driven Parallel Robot Pose Estimation Using Extended Kalman Filtering With Inertial Payload Measurements. *IEEE Robot. Autom. Lett.* **2021**, *6*, 3615–3622. [CrossRef]
43. Martin, C.; Fabritius, M.; Stoll, J.T.; Pott, A. A Laser-Based Direct Cable Length Measurement Sensor for CDPRs. *Robotics* **2021**, *10*, 60. [CrossRef]
44. Merlet, J.P.; Papegay, Y.; Gasc, A.V. The Prince's tears, a large cable-driven parallel robot for an artistic exhibition. In Proceedings of the 2020 IEEE International Conference on Robotics and Automation (ICRA), (ICRA) OR International Conference on Robotics and Automation, Paris, France, 31 May–31 August 2020; pp. 10378–10383.
45. Idá, E.; Merlet, J.P.; Carricato, M. Automatic Self-Calibration of Suspended Under-Actuated Cable-Driven Parallel Robot using Incremental Measurements. In *Cable-Driven Parallel Robots*; Pott, A., Bruckmann, T., Eds.; Springer International Publishing: Cham, Switzerland, 2019; pp. 333–344.
46. Behzadipour, S.; Khajepour, A. Stiffness of Cable-based Parallel Manipulators With Application to Stability Analysis. *J. Mech. Des* **2005**, *128*, 303–310. [CrossRef]
47. Merlet, J.P.; Tissot, R. A Panorama of Methods for Dealing with Sagging Cables in Cable-Driven Parallel Robots. In *Advances in Robot Kinematics 2022*; Altuzarra, O., Kecskeméthy, A., Eds.; Springer International Publishing: Cham, Switzerland, 2022; pp. 122–130.
48. Idà, E.; Briot, S.; Carricato, M. Natural Oscillations of Underactuated Cable-Driven Parallel Robots. *IEEE Access* **2021**, *9*, 71660–71672. [CrossRef]
49. Merlet, J.P. Comparison of Actuation Schemes for Wire-Driven Parallel Robots. In *New Trends in Mechanism and Machine Science*; Viadero, F., Ceccarelli, M., Eds.; Springer: Dordrecht, The Netherlands, 2013; pp. 245–254.
50. Kawamura, S.; Kino, H.; Won, C. High-speed manipulation by using parallel wire-driven robots. *Robotica* **2000**, *18*, 13–21. [CrossRef]
51. Fang, S.; Franitza, D.; Torlo, M.; Bekes, F.; Hiller, M. Motion control of a tendon-based parallel manipulator using optimal tension distribution. *IEEE/ASME Trans. Mechatron.* **2004**, *9*, 561–568. [CrossRef]
52. Varziri, M.S.; Notash, L. Kinematic calibration of a wire-actuated parallel robot. *Mech. Mach. Theory* **2007**, *42*, 960–976.
53. Scalera, L.; Gallina, P.; Seriani, S.; Gasparetto, A. Cable-Based Robotic Crane (CBRC): Design and Implementation of Overhead Traveling Cranes Based on Variable Radius Drums. *IEEE Trans. Robot.* **2018**, *34*, 474–485. [CrossRef]
54. Seriani, S.; Gallina, P. Variable Radius Drum Mechanisms. *J. Mech. Robot.* **2015**, *8*, 021016. [CrossRef]
55. Merlet, J.P. Kinematics of the wire-driven parallel robot MARIONET using linear actuators. In Proceedings of the 2008 IEEE International Conference on Robotics and Automation, Pasadena, CA, USA, 19–23 May 2008; pp. 3857–3862.
56. Pham, C.B.; Yang, G.; Yeo, S.H. Dynamic analysis of cable-driven parallel mechanisms. In Proceedings of the 2005 IEEE/ASME International Conference on Advanced Intelligent Mechatronics, Monterey, CA, USA, 24–28 July 2005; pp. 612–617.
57. Feyrer, K. Wire Ropes, Elements and Definitions. In *Wire Ropes*; Springer: Berlin, Germany, 2015; pp. 1–57.
58. Gonzalez-Rodriguez, A.; Castillo-Garcia, F.; Ottaviano, E.; Rea, P.; Gonzalez-Rodriguez, A. On the effects of the design of cable-Driven robots on kinematics and dynamics models accuracy. *Mechatronics* **2017**, *43*, 18–27. [CrossRef]
59. Miermeister, P.; Pott, A.; Verl, A. Dynamic modeling and hardware-in-the-loop simulation for the cable-driven parallel robot IPAnema. In Proceedings of the ISR 2010 (41st International Symposium on Robotics) and ROBOTIK 2010 (6th German Conference on Robotics), Munich, Germany, 7–9 June 2010; pp. 1–8.
60. Rognant, M.; Courteille, E. Improvement of Cable Tension Observability Through a New Cable Driving Unit Design. In *Cable-Driven Parallel Robots*; Gosselin, C., Cardou, P., Bruckmann, T., Pott, A., Eds.; Springer International Publishing: Cham, Switzerland, 2018; pp. 280–291.
61. Pott, A. *Cable-Driven Parallel Robots: Theory and Application*; Springer: Berlin, Germany, 2018; Volume 120.

Communication

Adaptive Terminal Sliding Mode Control of Picking Manipulator Based on Uncertainty Estimation

Caizhang Wu and Shijie Zhang *

College of Electrical Engineering, Henan University of Technology, Zhengzhou 450001, China
* Correspondence: zhangshijie@haut.edu.cn; Tel.: +86-18-6237-18217

Abstract: In this paper, a robust nonsingular fast terminal sliding mode control scheme for the picking manipulator under the condition of load change and nonlinear friction disturbance is presented. Firstly, the dynamic equation of the picking manipulator under the condition of load change and nonlinear friction disturbance is established. Then, in order to avoid the singularity problem existing in the terminal sliding mode and improve the convergence time, a new nonsingular fast terminal sliding mode control strategy is adopted to design the control law of the picking manipulator, which can guarantee the finite time convergence. The adaptive law is used to estimate the uncertainties of the system, and the finite time convergence of the system state is proved by the Lyapunov criterion. In addition, the genetic algorithm is used to identify the friction parameters to realize the nonlinear friction compensation control of the system. Finally, the simulation results of the picking manipulator under different load conditions show that the controller designed in this paper realizes the fast and accurate positioning of the picking manipulator under load change and nonlinear friction, and the control strategy is reasonable and effective.

Keywords: picking manipulator; sliding mode control; adaptive control; friction compensation; genetic algorithm

Citation: Wu, C.; Zhang, S. Adaptive Terminal Sliding Mode Control of Picking Manipulator Based on Uncertainty Estimation. *Actuators* **2022**, *11*, 347. https://doi.org/10.3390/act11120347

Academic Editors: Marco Carricato and Edoardo Idà

Received: 21 October 2022
Accepted: 18 November 2022
Published: 25 November 2022

Publisher's Note: MDPI stays neutral with regard to jurisdictional claims in published maps and institutional affiliations.

1. Introduction

An agricultural picking manipulator is a kind of mechatronics system with variable parameters and strong nonlinearity [1,2]. Under the action of nonlinear uncertain factors such as the weight change of the picking object and friction disturbance, the general proportional integral derivative (PID) controller does not enable the picking manipulator to obtain reliable control performance, so the mechanical holding brake has to be used for positioning in engineering [3,4]. However, this mechanical positioning method will cause greater impact wear, reduce the service life of the picking manipulator, and seriously reduce its overall performance. Therefore, how to design a robust controller for the picking manipulator is an urgent problem to be solved.

The sliding mode control is one of the control methods to deal with nonlinear systems with parameter perturbation and external disturbance [5–7]. This method has the advantages of strong robustness and easy design, so it has been widely used in a variety of fields [8–11]. With the sliding mode control of the linear switching function, the error between the system state and the expected state converges exponentially, and the system state can only approach the expected trajectory, but cannot reach the expected trajectory. Therefore, Refs. [12–14] proposed the terminal sliding mode (TSM) control. By introducing a nonlinear term into the construction of the sliding mode switching function, the tracking error on the sliding mode surface can converge to zero in a finite time, which solves the problem that the traditional sliding mode control can only converge asymptotically under the action of the linear sliding mode surface. However, the control effect of terminal sliding mode control in a singular control region will tend to infinity, which is not conducive to practical application.

The nonsingular terminal sliding mode (NTSM) control strategy is designed in [15–18], which avoids the singular problem of control when constructing the sliding mode switching function, retains the finite time convergence characteristics, and can obtain higher convergence accuracy. Because the NTSM control method has good control performance, it has been widely used. The authors in [19] combine the advantages of the linear sliding mode and NTSM. The hybrid NTSM control is designed to make the convergence of the system faster. Compared with the proportional-integral (PI) controller, it enhances the robustness of the permanent magnet synchronous motor.

Compared with the linear sliding mode, the NTSM has a higher convergence speed when the system state is close to the equilibrium point, but when the system state is far from the equilibrium point, its convergence time is longer and the dynamic characteristics become worse. In order to avoid the control singularity problem and accelerate the convergence speed when the system is far from the sliding mode surface, in this paper, a new nonsingular fast terminal sliding mode (NFTSM) control strategy is used to design the control law. In the actual process of realizing the sliding mode control, how to determine the switching gain of the sliding mode control is a difficult problem. Usually, the upper bound of the uncertainty in the system is unknown. In order to ensure the good robustness of the control system, the switching gain needs to be large enough [20]. However, excessive switching gain will cause control chattering. In order to reduce system chattering, this paper uses an adaptive control strategy to estimate the uncertain upper bound of the system, so there is no need to know the prior knowledge of the uncertain upper bound, which is conducive to practical application.

Recently, a robust sliding mode-based learning control strategy for a class of nonlinear discrete-time descriptor systems with time-varying delay and external disturbance was developed in [21]. The authors in [22] discussed the control problem of nonlinear disturbed polynomial systems using the formalism of output feedback linearization and a subsequent sliding mode control design. The authors in [23] proposed a novel developed a photovoltaic model based on an improved arithmetic optimization algorithm to extract the solar cell parameters. A neuroadaptive learning algorithm for constrained nonlinear systems with disturbance rejection is proposed in [24,25].

In this paper, an adaptive non-singular fast terminal sliding mode (ANFTSM) control method for the picking manipulator is used to design the corresponding position controller, which avoids the singularity problem of the traditional terminal sliding mode control and the slow convergence of the traditional NTSM control. At the same time, the adaptive estimation of the system uncertainty is used without the prior knowledge of the upper bound of the uncertainty, which effectively reduces the chattering caused by the excessive switching gain. Meanwhile, the genetic algorithm is used to identify the parameters of the nonlinear friction torque in the motion process of the picking manipulator, and the corresponding compensation control is designed. The main contributions of this paper are highlighted as follows. (1) The disturbance existing in the manipulator system is estimated adaptively, and the estimated value is used as a feedback signal to provide compensation for the controller, so as to improve the tracking performance of the system and enhance the robustness of the system. (2) A nonsingular fast terminal sliding surface is constructed to solve the singularity problem of the controller and obtain a finite time stability result, ensuring the convergence speed and transient response performance of the system control. (3) A nonsingular fast terminal sliding mode controller based on friction compensation is constructed. By introducing a saturation function instead of a symbolic function, the chattering problem is further weakened and the desired tracking trajectory of each joint is tracked quickly and accurately.

This paper is divided into five sections. The next section introduces the dynamic model of the picking manipulator and the description of the problem to be solved in this paper. The ANFTSM controller design method and stability analysis for the picking manipulator are proposed in Section 3. According to the controller designed in Section 3, the corresponding simulation results are given in Section 4, followed by a brief conclusion in Section 5.

2. Preliminaries

2.1. Dynamic Model of Picking Manipulator

The structure of the picking manipulator is shown in Figure 1. The DC motor drives the manipulator arm to rotate around the trunnion through the reducer, and the equilibrator is used to balance the load torque to reduce the working load of the driving motor.

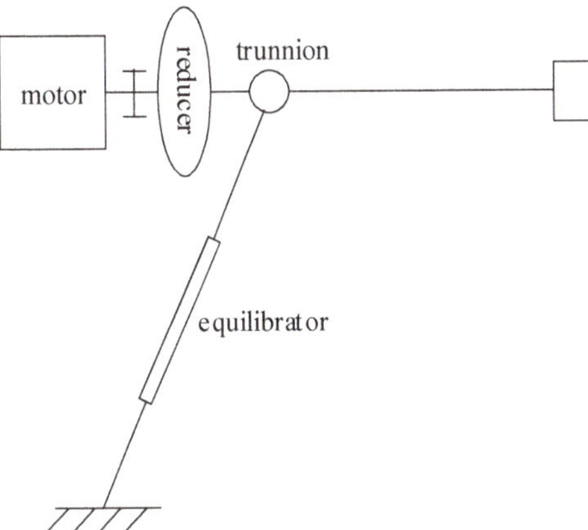

Figure 1. The structural diagram of the picking manipulator.

The picking manipulator is regarded as a single-degree-of-freedom manipulator, and its dynamic equation is [26]

$$J\ddot{\theta} = T \tag{1}$$

where J is the equivalent rotational inertia of the system. θ and $\ddot{\theta}$ are the rotation angle and angular acceleration of the picking manipulator, respectively. T is the equivalent rotation torque acting on the manipulator.

$$T = i_1 \eta_1 k_T I - T_R + T_G - T_f + D \tag{2}$$

where i_1 is the total transmission ratio of the system, η_1 is the transmission efficiency of the reducer, k_T is the motor torque constant, I is the control current of the motor, T_R is the torque of the equilibrator to the manipulator, and T_G is the gravity torque of the manipulator, which depends on the rotation angle of the manipulator. T_f is the friction torque to be identified, and D is an uncertainty term caused by parameter changes, unmodeled dynamics, and external disturbances.

The equilibrator is composed of an oil cylinder and accumulator, in which the oil cylinder pressure is

$$p = p_0 S \left(\frac{V_0}{V_0 - \Delta V} \right)^n \tag{3}$$

where p_0 is the initial pressure of the accumulator, S is the piston area of the cylinder, V_0 is the initial volume of gas, ΔV is the volume of gas change, and n is the polytropic index of the gas. Then, the torque of the equilibrator to the manipulator can be written as

$$T_R = l p_0 S \left(\frac{V_0}{V_0 - \Delta L S} \right)^n \tag{4}$$

where l is the distance from the center of rotation to the balancer, and ΔL is the distance of the piston movement.

2.2. Problem Description

Design an adaptive non-singular fast terminal sliding mode (ANFTSM) controller for the above picking manipulator to ensure that the joint position of the picking manipulator tracks the desired trajectory, and the tracking error can converge to a neighborhood near zero in a finite time.

3. Controller Design of the Picking Manipulator

3.1. ANFTSM Controller

By defining $\theta = x_1, \ddot{\theta} = x_2$, the dynamic equation of the picking manipulator can be simplified into the following second-order nonlinear uncertain system.

$$\begin{cases} \dot{x}_1 = x_2 \\ \dot{x}_2 = \frac{k_T i_1 \eta_1}{J} u + \frac{1}{J}\left(T_G - T_R - T_f\right) + d \\ y = x_1 \end{cases} \tag{5}$$

where $d = D/J$ is an uncertainty term, and $|d| \leq U$, U is a constant. u is the control input, and y is the system output. The following assumptions are assumed to befulfilled henceforth.

Assumption 1: *The states of the picking manipulator system are uniformly bounded.*

Assumption 2: *The uncertainty of the system is continuously differentiable and bounded at all times.*

The angular displacement error of the picking manipulator is defined as

$$\begin{cases} e = x_1 - x_d \\ \dot{e} = \dot{x}_1 - \dot{x}_d \\ \ddot{e} = \ddot{x}_1 - \ddot{x}_d \end{cases}$$

where x_d is a given expected value.

For equation (5), the traditional TSM switching function is generally designed as

$$s = \dot{e} + \alpha e^{q/p} \tag{6}$$

where s is sliding mode switching function, $\alpha > 0$ is a constant, p and q are positive odd number, and $p > q$.

Then, the traditional TSM controller is designed as

$$u = \frac{J}{k_T i_1 \eta_1}(\ddot{x}_d - \frac{1}{J}(T_G - T_R - T_f) - \alpha\frac{q}{p}e^{q/p-1}\dot{e} - (U + \eta)\mathrm{sgn}(s)) \tag{7}$$

where $\eta > 0$ is the design constant, $\mathrm{sgn}(\cdot)$ is a symbol function.

It can be seen from (7) that the control quantity contains a $\alpha\frac{q}{p}e^{q/p-1}\dot{e}$ term, which contains a negative exponential term, that is, the index $q/p - 1 < 0$. Therefore, when $e = 0$, $\dot{e} \neq 0$, the control quantity tends to infinity, which results in singular problems.

To avoid the singularity problem of traditional TSM control, the NTSM switching function can be designed as

$$s = \dot{e} + \frac{1}{\alpha}\dot{e}^{p/q} \tag{8}$$

where $1 < p/q < 2$.

It can be proved that the control action obtained by the sliding mode surface designed by Equation (8) does not contain a negative exponential term, and the control quantity does not have an infinite term, thus solving the singularity problem of traditional TSM control.

However, if $s = 0$, the error convergence rate

$$\dot{e} = (-\alpha e)^{q/p}$$

and $0.5 < q/p < 1$. Therefore, when the system state is far away from the equilibrium point, the error convergence rate becomes slower.

In order to further improve the convergence speed of the NTSM switching function, the following NFTSM switching function is used

$$s = e + \dot{e}^{a_1} + \frac{1}{\alpha}\dot{e}^{a_2} \tag{9}$$

where a_1 and a_2 are design constants, and $1 < a_2 < 2, a_2 < a_1$.

Let $s = 0$, we have

$$\dot{e} = (-\alpha e - \alpha e^{a_1})^{1/a_2}$$

That is, when the system state is far away from the equilibrium point, the error convergence rate is mainly affected by the high-order term of e, and the convergence rate of NFTSM is faster than NTSM. When the system state is close to the equilibrium point, the convergence rate of NFTSM is similar to NTSM. Therefore, during the whole sliding stage, compared with NTSM, NFTSM control can achieve global fast convergence, and the exponential term in Equation (9) is greater than one, which eliminates the singularity problem of the control quantity.

Furthermore, in order to avoid the complex solution of the switching function of Equation (9) when $e < 0$ and $\dot{e} < 0$, the sliding mode switching function can be improved as

$$s = e + k_1|e|^{a_1}\mathrm{sgn}(e) + k_2|\dot{e}|^{a_2}\mathrm{sgn}(\dot{e}) \tag{10}$$

where $k_1 > 0$ and $k_2 > 0$ are design constants.

Taking the derivative of s and substituting it into Equation (5) yields

$$\begin{aligned}
\dot{s} &= \dot{e} + k_1 a_1 |e|^{a_1-1}\dot{e} + k_2 a_2 |\dot{e}|^{a_2-1}\ddot{e} \\
&= \dot{e} + k_1 a_1 |e|^{a_1-1}\dot{e} + k_2 a_2 |\dot{e}|^{a_2-1}(\ddot{x}_2 - \ddot{x}_d) \\
&= \dot{e} + k_1 a_1 |e|^{a_1-1}\dot{e} + k_2 a_2 |\dot{e}|^{a_2-1}(\tfrac{k_{Ti_1}\eta_1}{J}u + \tfrac{1}{J}(T_G - T_R - T_f) + d - \ddot{x}_d)
\end{aligned} \tag{11}$$

In order to further accelerate the convergence speed of the system and weaken the chattering, the exponential approximation law is used to design the controller. The expression is as follows

$$\dot{s} = k_2 a_2 |\dot{e}|^{a_2-1}(-ks - (U+\eta)\mathrm{sgn}(s)) \tag{12}$$

where $k > 0$ is the exponential approach coefficient.

Combining Equations (11) and (12), the control law of the picking manipulator can be designed as

$$u = \frac{J}{k_{Ti_1}\eta_1}(-\frac{1}{k_2 a_2}|\dot{e}|^{2-a_2}\mathrm{sgn}(\dot{e}) - \frac{k_1 a_1}{k_2 a_2}|e|^{a_1-1}|\dot{e}|^{2-a_2}\mathrm{sgn}(\dot{e}) - ks - (U+\eta)\mathrm{sgn}(s) - \frac{1}{J}(T_G - T_R - T_f) + \ddot{x}_d) \tag{13}$$

It can be seen that there is a positive term in Equation (13), and the control input will not produce an infinite quantity, which avoids the singular problem of traditional TSM. However, Equation (13) contains an unknown upper bound U of uncertainty, which makes it impossible to apply the control law directly. Therefore, an adaptive law is proposed in this paper to estimate the upper bound U of uncertainty, which is expressed as

$$\dot{\hat{U}} = \gamma k_2 a_2 |\dot{e}|^{a_2-1}|s| \tag{14}$$

where $\gamma > 0$ is adaptive gain.

To sum up, the adaptive nonsingular fast terminal sliding mode control law (ANFTSM) of the picking manipulator in this paper can be designed as

$$u = \frac{J}{k_T i_1 \eta_1}\left(-\frac{1}{k_2 a_2}|\dot{e}|^{2-a_2}\mathrm{sgn}(\dot{e}) - \frac{k_1 a_1}{k_2 a_2}|e|^{a_1-1}|\dot{e}|^{2-a_2}\mathrm{sgn}(\dot{e}) - ks - (\hat{U}+\eta)\mathrm{sgn}(s) - \frac{1}{J}(T_G - T_R - T_f) + \ddot{x}_d\right) \tag{15}$$

3.2. Stability Analysis

The following lemma is used to prove the stability of the controller designed in this paper.

Lemma 1 [27]. *For nonlinear system*

$$\dot{x} = f(x,t), x \in \mathbb{R}^n \tag{16}$$

Suppose that there exists a continuously differentiable positive definite function $V(x)$ such that the following equation holds

$$\dot{V}(x) \leq -\mu V(x) - \lambda V^m(x) \tag{17}$$

where μ, λ, and m are positive numbers, and $0 < m < 1$. The initial state of the Equation (16) is denoted as $x_0 = x(t_0)$, and t_0 is the initial time of the Equation (16).

Then the system state converges to the equilibrium point in finite time. The convergence time is T, and

$$T \leq \frac{1}{\mu(1-m)} \ln \frac{\mu V^{1-m}(x_0) + \lambda}{\lambda}$$

Theorem 1. *For the picking manipulator Equation (5), if the sliding mode switching function (10) and the approximation law (12) are selected, under the action of the adaptive law (14) and the control law (15) designed in this paper, the sliding mode switching function converges to zero in finite time, and the system position error e and speed error $\dot{e}$ converge to zero in finite time.*

Proof. The estimation error of the adaptive law is defined as $\tilde{U} = \hat{U} - U$, then $\dot{\tilde{U}} = \dot{\hat{U}}$.

Firstly, the estimation error of uncertainty is bounded is proved as follows.

Select the Lyapunov function

$$V_1 = \frac{1}{2}s^2 + \frac{1}{2\gamma}\tilde{U}^2 \tag{18}$$

Taking the derivative of V_1 and substitute into Equation (11) yields

$$\begin{aligned}
\dot{V}_1 &= s\dot{s} + \frac{1}{\gamma}\tilde{U}\dot{\tilde{U}} \\
&= s(\dot{e} + k_1 a_1|e|^{a_1-1}\dot{e} + k_2 a_2|e|^{a_2-1}(\frac{k_T i_1 \eta_1}{J}u + \frac{1}{J}(T_G - T_R - T_f) + d - \ddot{x}_d)) + \frac{1}{\gamma}\tilde{U}\dot{\tilde{U}}
\end{aligned} \tag{19}$$

Then, substitute Equation (19) into Equation (14) and Equation (15), and we can obtain

$$\begin{aligned}
\dot{V}_1 &= sk_2 a_2|\dot{e}|^{a_2-1}(-ks + d - (\hat{U}+\eta)\mathrm{sgn}(s)) + \frac{1}{\gamma}\tilde{U}\gamma k_2 a_2|\dot{e}|^{a_2-1}|s| \\
&= k_2 a_2|\dot{e}|^{a_2-1}(-ks^2 + ds - (\hat{U}+\eta)|s| + (\hat{U}-U)|s|) \\
&\leq -k_2 a_2|\dot{e}|^{a_2-1}(ks^2 + \eta|s|) \\
&\leq 0
\end{aligned} \tag{20}$$

It can be seen from the above Equation (20) that V_1 is bounded, so s and $\tilde{U}$ are bounded, respectively. Then, let $\left|\tilde{U}\right| \leq \varepsilon$, ε is the upper bound of the estimation error $\tilde{U}$.

Furthermore, the following will prove that the sliding mode switching function can converge in finite time.

Reselect the Lyapunov function

$$V_2 = \frac{1}{2}s^2 \tag{21}$$

Taking the derivative of V_2 and substitute into Equation (11) yields

$$\dot{V}_2 = s\dot{s}$$
$$= \dot{e} + k_1 a_1 |e|^{a_1-1}\dot{e} + k_2 a_2 |e|^{a_2-1}\left(\frac{k_T i_1 \eta_1}{J}u + \frac{1}{J}(T_G - T_R - T_f) + d - \ddot{x}_d\right) \tag{22}$$

Substitute Equation (22) into Equations (14) and (15), we have

$$\begin{aligned}
\dot{V}_2 &= sk_2 a_2 |\dot{e}|^{a_2-1}(-ks + d - (\hat{U} + \eta)\text{sgn}(s)) \\
&\leq k_2 a_2 |\dot{e}|^{a_2-1}(-ks^2 + U|s| - (\hat{U} + \eta)|s|) \\
&= k_2 a_2 |\dot{e}|^{a_2-1}(-ks^2 - \tilde{U}|s| - \eta|s|) \\
&\leq k_2 a_2 |\dot{e}|^{a_2-1}(-ks^2 + (\varepsilon - \eta)|s|) \\
&= -k_2 a_2 |\dot{e}|^{a_2-1}(ks^2 + (\eta - \varepsilon)|s|) \\
&= -\lambda_1 V_2 - \lambda_2 V_2^{1/2}
\end{aligned} \tag{23}$$

where

$$\lambda_1 = 2kk_2 a_2 |\dot{e}|^{a_2-1}$$
$$\lambda_2 = \sqrt{2}(\eta - \varepsilon)k_2 a_2 |\dot{e}|^{a_2-1}$$

When $\dot{e} \neq 0$, there is $|\dot{e}|^{a_2-1} > 0$, by selecting parameter $\eta > \varepsilon$, we can obtain

$$\lambda_1 > 0,\ \lambda_2 > 0,\ \dot{V}_2 \leq -\lambda_1 V_2 - \lambda_2 V_2^{1/2}$$

By Lemma 1, it can be seen that the system will converge to $s = 0$ in finite time, and the convergence time satisfies

$$t_r \leq \frac{2}{\lambda_1} \ln \frac{\lambda_1 V_2^{1/2}(x_0) + \lambda_2}{\lambda_2}$$

where $V_2(x_0)$ is the initial value of $V_2(x)$.

When $\dot{e} = 0$, during the system state sliding to the system equilibrium point stage, there is $s \neq 0$. Substitute Equation (15) into Equation (5), we can obtain

$$\ddot{e} = -ks + d - (\hat{U} + \eta)\text{sgn}(s) \neq 0$$

which derives a contradiction. This means that the system state will not remain at this point, and the system state will converge to $s = 0$ in finite time.

When the system state reaches the sliding surface $s = 0$, it can be known from Equation (10) that the system state is determined by the following nonlinear equation

$$e + k_1 |e|^{a_1}\text{sgn}(e) + k_2 |\dot{e}|^{a_2}\text{sgn}(\dot{e}) = 0 \tag{24}$$

According to [28], e and $\dot{e}$ converge to zero in finite time. The convergence time t_s is

$$t_s = \frac{a_2 |e(t_r)|^{1-1/a_2}}{k_1(a_2 - 1)} \cdot F\left(\frac{1}{a_2}, \frac{a_2 - 1}{(a_1 - 1)a_2}; 1 + \frac{a_2 - 1}{(a_1 - 1)a_2}; -k_1 |e(t_r)|^{a_1-1}\right)$$

where $F(\cdot)$ is a Gaussian hypergeometric function. $\square$

3.3. Friction Parameter Identification

Nonlinear friction appears in large quantities in electromechanical servo systems. Its nonlinear time-varying characteristic is one of the important factors affecting the control effect of the system. In order to improve the control performance, effective compensation strategies should be adopted to reduce the influence of friction on the system.

In this paper, the Stribeck friction model is used to identify the friction force during the movement of the picking manipulator. The Stribeck friction model expression can be written as

$$
T_f = \begin{cases} F_c^+ + (F_s^+ - F_c^+)e^{-\left(\frac{\dot{\theta}}{v_s^+}\right)^2} \mathrm{sgn}(\dot{\theta}) + \sigma^+ \dot{\theta}, & \dot{\theta} > 0 \\ F_c^- + (F_s^- - F_c^-)e^{-\left(\frac{\dot{\theta}}{v_s^-}\right)^2} \mathrm{sgn}(\dot{\theta}) + \sigma^- \dot{\theta}, & \dot{\theta} < 0 \end{cases}
\tag{25}
$$

where F_c and F_s are Coulomb friction torque and maximum static friction torque, respectively. v_s is the Stribeck speed, σ is the viscous friction coefficient.

In this paper, the genetic algorithm [24] is used to identify these parameters such as $F_c^+, F_c^-, F_s^+, F_s^-, F_c^+, F_c^-, v_s$ and σ in (25). The specific identification process is as follows.

Firstly, disconnect the connection between the picking manipulator and the equilibrator, and the system dynamics equation is transformed into

$$
J\ddot{\theta} = i_1 \eta_1 k_T I - T_f
\tag{26}
$$

where I is the control current of the system. Let the motor move with a set of constant speed $(\dot{\theta})_{i=1}^N$ and record the average current value $(I)_{i=1}^N$ at different speeds.

It can be seen from Equation (26) that if $\ddot{\theta} = 0$, then we have $T_f = i_1 \eta_1 k_T I$, such that the relationship between current and speed is determined by the sequence $(I)_{i=1}^N$ and $(\dot{\theta})_{i=1}^N$.

Let the identification vector be

$$
x_s = \begin{bmatrix} F_c^+ & F_s^+ & v_s^+ & \sigma^+ & F_c^- & F_s^- & v_s^- & \sigma^- \end{bmatrix}
\tag{27}
$$

and denote the identification error as

$$
e_I\left(x_s, \dot{\theta}\right) = i_1 \eta_1 k_T I - T_f\left(x_s, \dot{\theta}\right)
\tag{28}
$$

Take the objective function as

$$
J_e = \frac{1}{2} \sum_{i=1}^N e_I^2\left(x_s, \dot{\theta}\right)
\tag{29}
$$

Then, the problem of solving the friction parameter in (27) is transformed into solving the minimum value problem of the objective function J_e. The parameter identification process using the genetic algorithm is as follows.

Step 1. Initializing the population $P(0)$ randomly, and $X_i (i = 1, 2, \cdots, M)$ is the individual in the population, t is the evolution algebra, let the maximum evolution algebra is $T = 10,000$, the population size is $M = 300$.

Step 2. Calculating the Individual Fitness Function

$$
f(X_i) = \frac{1}{J_e(X_i)}
\tag{30}
$$

Step 3. Determine whether t is equal to T, if $t = T$, output the results of the identification parameters, otherwise turn to the next step.

Step 4. Saving the random sampling of the best individual for selection operation, and forming the next generation population $P(t)$.

Step 5. Perform the crossover operation with the uniform crossover operator $p_c = 0.9$.

Step 6. Set the adaptive mutation operator $p_c = 0.1 - 0.099t/T$ for mutation operation.

Step 7. Set $t = t + 1$, repeat step (2)–(6), and output the final optimal solution, which is the identification result.

4. Simulation

In this section, a simulation example of a single joint picking manipulator is provided to illustrate the theoretical result. In order to simplify the experimental device, the gas spring is used instead of the balancer. The parameter of Equation (5) is shown in Table 1.

Table 1. Parameters of the picking manipulator.

Parameters	Value
p_0	$0.35/\mathrm{MPa}$
S	$6.8 \times 10^{-5}/\mathrm{m}^2$
V_0	$7.654 \times 10^{-5}/\mathrm{m}^3$
n	1.2
i_1	630
α_1, α_2	0

The structure diagram of the control system is shown in Figure 2.

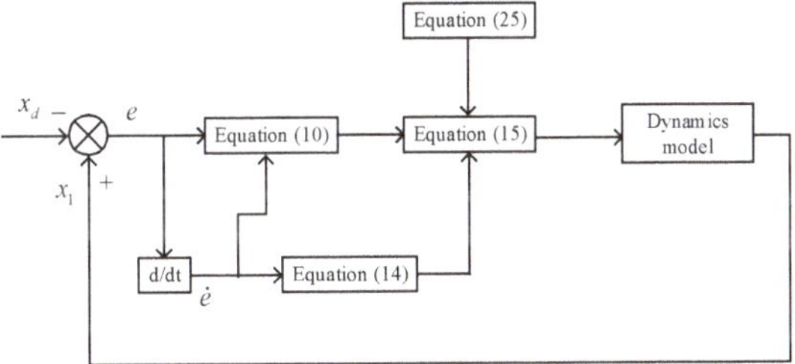

Figure 2. The structure diagram of the control system.

Firstly, in order to realize the compensation control of friction, the genetic algorithm in Section 3 is used to identify the friction parameters of the picking manipulator prototype. The comparison of experimental results and identification results is shown in Figure 3, and the identification results of model parameters are shown in Table 2. It can be seen from Figure 3 that the identification curve is consistent with the experimental curve, indicating that the identified parameters are accurate and reasonable. The identification results can be substituted into the control law for compensation design.

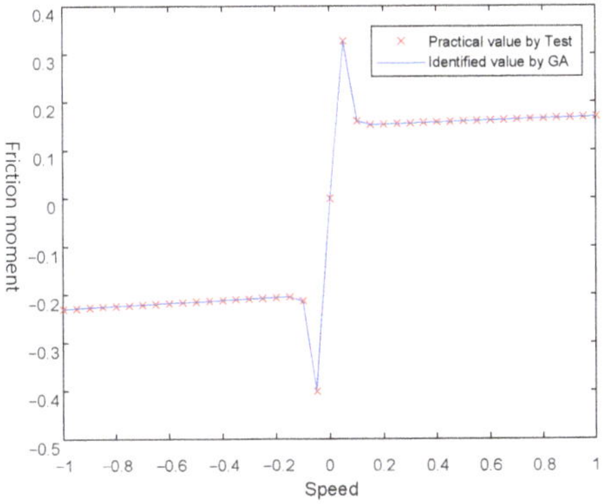

Figure 3. Friction parameter identification curve.

Table 2. Friction parameter identification results.

Parameters	Identification Results
F_c^+	7.2401/Nm
F_s^+	6.0127/Nm
F_c^-	0.3102/Nm
F_s^-	6.2149/Nm
v_s^+	0.3660 rad/s
v_s^-	1.2681 rad/s
σ^+	0.9460 Nms/rad
σ^-	1.7982 Nms/rad

In order to suppress the chattering problem in sliding mode control effectively, the saturation function $sat(s)$ is used instead of $\text{sgn}(s)$.

$$sat(s) = \begin{cases} 1, & s > \Delta \\ s/\Delta, & |s| \leq \Delta \\ -1, & s < -\Delta \end{cases}$$

where $\Delta = 0.05$ is the boundary layer thickness constant.

Other control parameters in this paper are selected as $k_T = 35$ mN·m/A, $\eta_1 = 0.9$, $a_1 = 2$, $a_2 = 1.5$, $k_1 = 200$, $k_2 = 10$, and $\eta = 0.1$.

In this section, two different simulations are carried out, and the load of the picking manipulator is set to 0.5 kg and 1.5 kg, respectively. The control objective is to control the picking manipulator to track the desired trajectory $x_d = 0.1 \sin(t)$, and the initial state of the system is $x(0) = 0$, $\dot{x}(0) = 0$.

In the first case, when the load of the manipulator is 0.5 kg, the simulation results are shown in Figures 4–6.

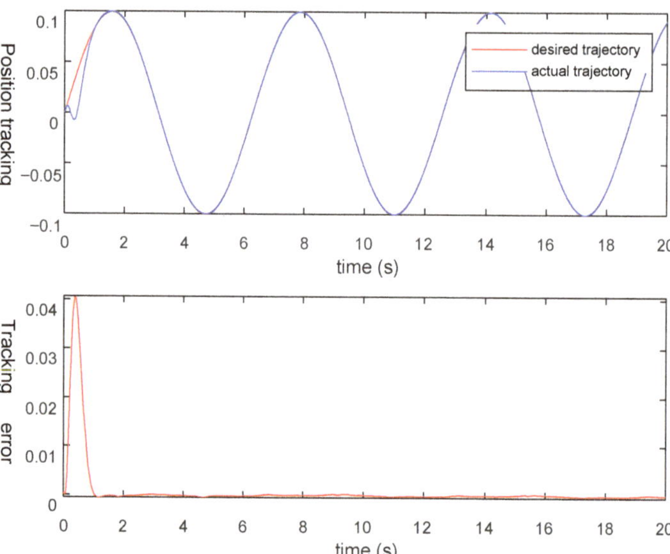

Figure 4. Position tracking curve and tracking error curve.

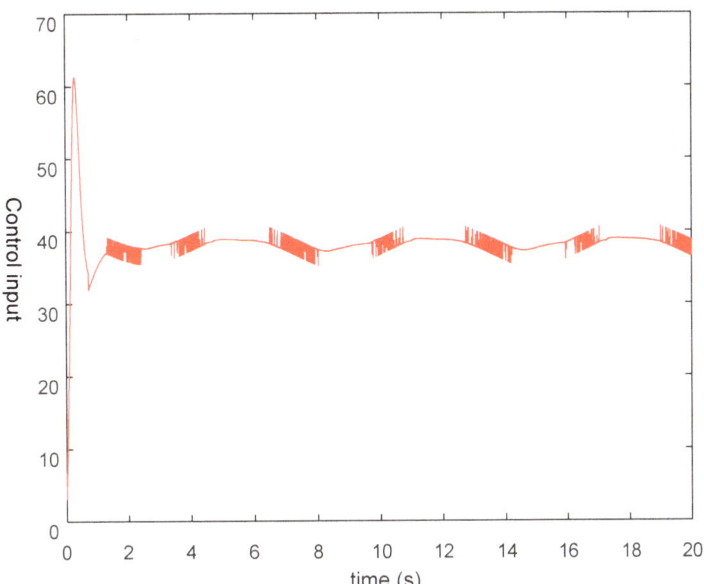

Figure 5. The control input of the system.

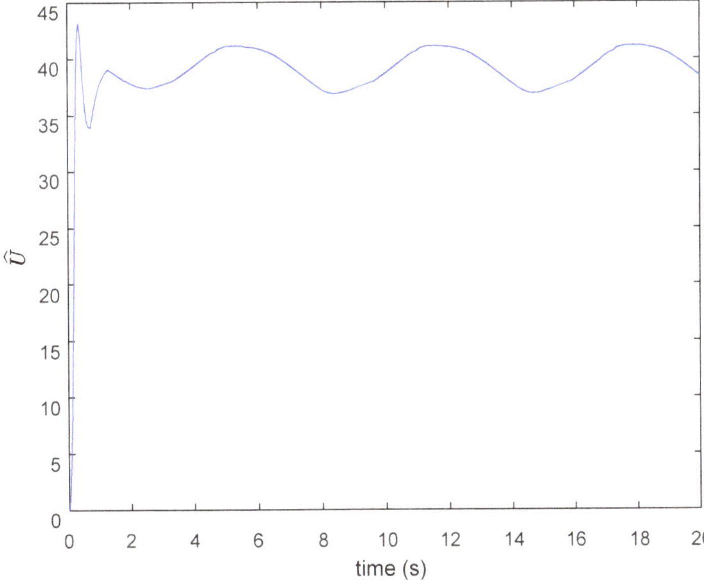

Figure 6. Estimation of the upper bound of uncertainty.

In the second case, when the load of the manipulator is 1.5 kg, the simulation results are shown in Figures 7–9.

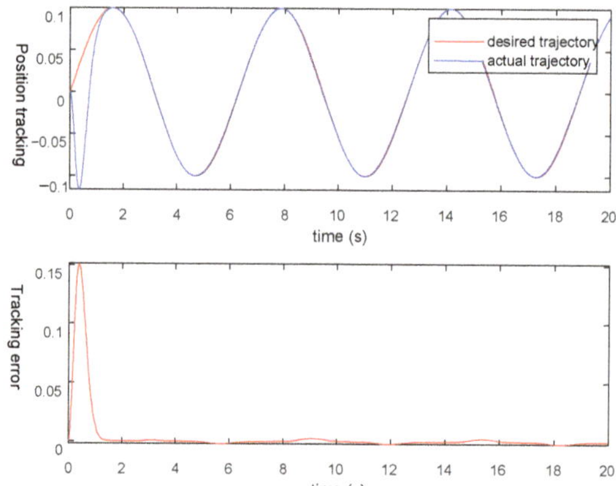

Figure 7. Position tracking curve and tracking error curve.

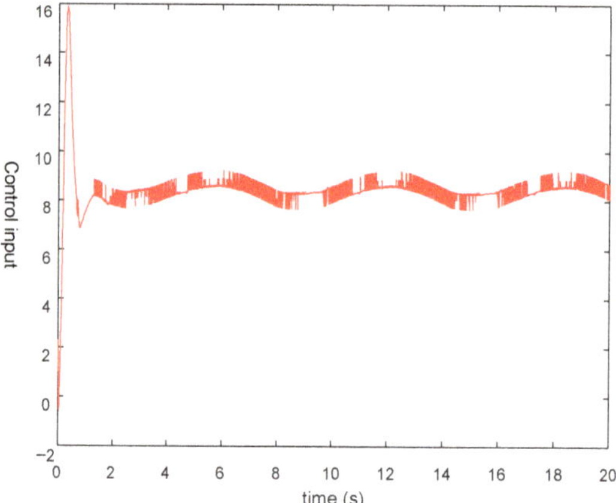

Figure 8. The control input of the system.

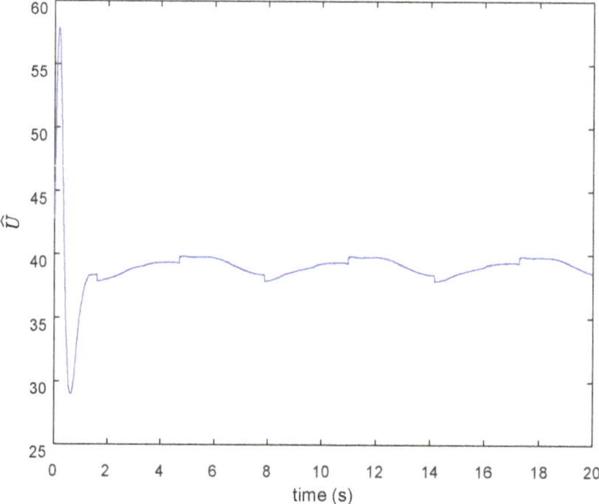

Figure 9. Estimation of the upper bound of uncertainty.

It can be seen from Figures 4 and 7 that the joint trajectory can track the desired trajectory at 1.101 s and 1.502 s, respectively. The tracking error can also converge to the equilibrium position rapidly under the controller designed in this paper. It shows that the ANFTSM control algorithm designed in this paper has strong robustness to load parameter changes and nonlinear friction. It also can be seen from Figures 5 and 8 that the control input does not produce singular problems under different loads. Figures 6 and 9 show the convergence of the adaptive estimation of the uncertainty, which avoids the problem of control chattering caused by taking a too large uncertainty upper bound in the absence of uncertainty prior knowledge, and is more conducive to practical engineering applications.

In order to assess the control strategy synthesized in this paper, a comparative analysis is established to examine the performance of the ANFTSM control algorithm across the traditional sliding mode control scheme. The load of the manipulator is set as 1 kg. The simulation results are depicted in Figures 10 and 11, which represent the system's trajectory tracking curves and the resultant control inputs, respectively.

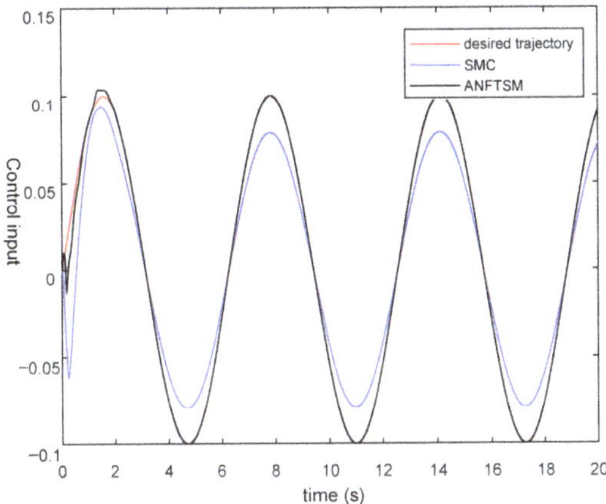

Figure 10. Position tracking curve.

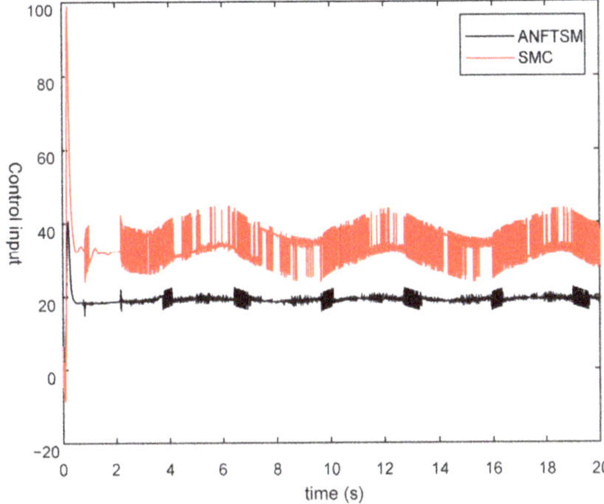

Figure 11. Control input.

It can be seen from Figures 10 and 11 that compared with the traditional sliding mode control method, the proposed method in this paper has great advantages in tracking accuracy and convergence time.

5. Conclusions

In this paper, the dynamic model of the picking manipulator is analyzed, and a new NFTSM control algorithm is used to design the corresponding position controller, which avoids the singularity problem of the traditional terminal sliding mode control and the slow convergence of the traditional NTSM control. At the same time, the adaptive estimation of the system uncertainty is used without the prior knowledge of the upper bound of the uncertainty, which effectively reduces the chattering caused by the excessive switching gain. Aiming at the nonlinear friction disturbance in the motion process of the picking manipulator system, the genetic algorithm is used to identify the friction parameters, and the obtained parameters are used for the compensation control of the picking manipulator. The simulation results show that the controller can achieve fast and accurate positioning of the picking manipulator under different load conditions. The designed controller has good robustness and is easy to apply in engineering practice.

Future development work will focus on applying new algorithms to approximate system uncertainties, such as meta heuristic algorithms. In addition, how to use the optimization algorithm to obtain the optimal adjustment parameters of the controller can be considered.

Author Contributions: Conceptualization, C.W. and S.Z.; methodology, C.W.; software, S.Z.; validation, C.W. and S.Z.; formal analysis, C.W. and S.Z.; investigation, S.Z.; resources, S.Z.; data curation, C.W. and S.Z.; writing—original draft preparation, S.Z.; writing—review and editing, C.W. and S.Z.; visualization, C.W. and S.Z.; supervision, C.W. and S.Z.; project administration, C.W. and S.Z.; funding acquisition, C.W. All authors have read and agreed to the published version of the manuscript.

Funding: This research was funded by National Natural Science Foundation of China, grant number 61078070; the open project of Key Laboratory of Grain Information Processing and Control, grant number KFJJ-2021-110.

Institutional Review Board Statement: Not applicable.

Informed Consent Statement: Not applicable.

Data Availability Statement: Not applicable.

Conflicts of Interest: The authors declare no conflict of interest. The funders had no role in the design of the study; in the collection, analyses, or interpretation of data; in the writing of the manuscript, or in the decision to publish the results.

References

1. Wang, J.; Cheng, Z.; Zhang, F. Design of the Divided No-till Wheat Planter Orchard Picking Manipulator Research. *J. Agric. Mech. Res.* **2020**, *14*, 158–163.
2. Yu, X.; Fan, Z.; Wang, X.; Wan, H.; Wang, P.; Zeng, X.; Jia, F. A lab-customized autonomous humanoid apple harvesting robot. *Comput. Electr. Eng.* **2021**, *96*, 107459. [CrossRef]
3. Fang, Z.; Liang, X.F. Intelligent obstacle avoidance path planning method for picking manipulator combined with artificial potential field method. *Ind. Robot* **2022**, *49*, 835–850. [CrossRef]
4. Schuetz, C.; Baur, J.; Pfaff, J.; Buschmann, T.; Ulbrich, H. Evaluation of a direct optimization method for trajectory planning of a 9-DOF redundant fruit-picking manipulator. In Proceedings of the 2015 IEEE International Conference on Robotics and Automation (ICRA), Seattle, WA, USA, 26–30 May 2015; pp. 2660–2666.
5. Liu, Y.C.; Yan, W.; Zhang, T.; Yu, C.X.; Tu, H.Y. Trajectory Tracking for a Dual-Arm Free-Floating Space Robot with a Class of General Nonsingular Predefined-Time Terminal Sliding Mode. *IEEE Trans. Syst. Man Cybern. Syst.* **2022**, *52*, 3273–3286. [CrossRef]
6. Khurram, A.; Adeel, M.; Jamshed, I. Terminal Sliding Mode Control of an Anthropomorphic Manipulator with Friction Based Observer. In Proceedings of the 2021 International Conference on Robotics and Automation in Industry, Xi'an, China, 30 May –5 June 2021; pp. 143–150.
7. Xu, H.; Li, M.; Lu, C.L. Nonlinear sliding mode control of manipulator based on iterative learning algorithm. *J. Electr. Syst.* **2021**, *17*, 421–437.
8. Su, L.Q.; Guo, X.; Ji, Y. Tracking control of cable-driven manipulator with adaptive fractional-order nonsingular fast terminal sliding mode control. *JVC/J. Vib. Control* **2021**, *27*, 2482–2493. [CrossRef]
9. Lu, Z.P.; Li, Y.; Fan, X.; Li, Y.C. Decentralized Fault Tolerant Control for Modular Robot Manipulators via Integral Terminal Sliding Mode and Disturbance Observer. *Int. J. Control Autom. Syst.* **2022**, *20*, 3274–3284. [CrossRef]

10. Zaare, S.; Soltanpour, M.R. Adaptive fuzzy global coupled nonsingular fast terminal sliding mode control of n-rigid-link elastic-joint robot manipulators in presence of uncertainties. *Mech. Syst. Signal Process.* **2022**, *163*, 108165. [CrossRef]
11. Song, T.Z.; Fang, L.; Wang, H.Z. Model-free finite-time terminal sliding mode control with a novel adaptive sliding mode observer of uncertain robot systems. *Asian J. Control* **2022**, *24*, 1437–1451. [CrossRef]
12. Boukattaya, M.; Mezghani, N.; Damak, T. Adaptive nonsingular fast terminal sliding-mode control for the tracking problem of uncertain dynamical systems. *ISA Trans.* **2018**, *77*, 1–19. [CrossRef]
13. Zhang, S.J.; Cao, Y. Cooperative Localization Approach for Multi-Robot Systems Based on State Estimation Error Compensation. *Sensors* **2019**, *19*, 3842–3852. [CrossRef] [PubMed]
14. Wang, Y.; Li, B.; Yan, F. Practical adaptive fractional-order nonsingular terminal sliding mode control for a cable-driven manipulator. *Int. J. Robust Nonlinear Control* **2019**, *29*, 1396–1417. [CrossRef]
15. Hao, S.; Hu, L.Y.; Liu, P. Second-order adaptive integral terminal sliding mode approach to tracking control of robotic manipulators. *IET Control Theory Appl.* **2021**, *15*, 2145–2157. [CrossRef]
16. Van, M.; Mavrovouniotis, M.; Ge, S.Z. An adaptive backstepping nonsingular fast terminal sliding mode control for robust fault tolerant control of robot manipulators. *IEEE Trans. Syst. Man Cybern. Syst.* **2019**, *49*, 1448–1458. [CrossRef]
17. Shao, X.Y.; Sun, G.H.; Xue, C. Nonsingular terminal sliding mode control for free-floating space manipulator with disturbance. *Acta Astronaut.* **2021**, *181*, 396–404. [CrossRef]
18. Zhang, S.; Cao, Y. Consensus in networked multi-robot systems via local state feedback robust control. *Int. J. Adv. Robot. Syst.* **2019**, *16*, 1–7. [CrossRef]
19. Wang, Y.M.; Han, F.L.; Feng, Y. Hybrid continuous nonsingular terminal sliding mode control of uncertain flexible manipulators. In Proceedings of the IECON Proceedings (Industrial Electronics Conference), Dallas, TX, USA, 29 October–1 November 2014; pp. 190–196.
20. Fang, H.R.; Wu, Y.; Xu, T.; Wan, F.X. Adaptive neural sliding mode control of uncertain robotic manipulators with predefined time convergence. *Int. J. Robust Nonlinear Control* **2022**, *32*, 223–238. [CrossRef]
21. Mourad, K.; Magdi, S.M. Robust-QSR γ-dissipative sliding mode control for uncertain discrete-time descriptor systems with time-varying delay. *IMA J. Math. Control Inf.* **2018**, *35*, 735–756.
22. Charfeddine, S.; Boudjemline, A.; Ben Aoun, S.; Jerbi, H.; Kchaou, M.; Alshammari, O.; Elleuch, Z.; Abbassi, R. Design of a fuzzy optimization control structure for nonlinear systems: A disturbance-rejection method. *Appl. Sci.* **2021**, *11*, 2612. [CrossRef]
23. Abbassi, A.; Ben Mehrez, R.; Bensalem, Y.; Abbassi, R.; Kchaou, M.; Jemli, M.; Abualigah, L.; Altalhi, M. Improved Arithmetic Optimization Algorithm for Parameters Extraction of Photovoltaic Solar Cell Single-Diode Model. *Arab. J. Sci. Eng.* **2022**, *47*, 10435–10451. [CrossRef]
24. Yang, G.C.; Yao, J.Y.; Dong, Z.L. Neuroadaptive learning algorithm for constrained nonlinear systems with disturbance rejection. *Int. J. Robust Nonlinear Control* **2022**, *32*, 6127–6147. [CrossRef]
25. Yang, G.C.; Yao, J.Y.; Dong, Z.L. Multilayer neuroadaptive force control of electro-hydraulic load simulators with uncertainty rejection. *Appl. Soft Comput.* **2022**, *130*, 109672. [CrossRef]
26. Zhou, X.; Zhao, B.; Liu, W. A compound scheme on parameters identification and adaptive compensation of nolinear friction disturbance for the aerial intertially stabilized platform. *ISA Trans.* **2017**, *67*, 293–305. [CrossRef] [PubMed]
27. Si, Y.J.; Song, S.M. Adaptive reaching law based three-dimensional finite-time guidance law against maneuvering targets with input saturation. *Aerosp. Sci. Technol.* **2017**, *70*, 198–210. [CrossRef]
28. Sun, H.Q.; Zhang, S.J.; Quan, Q.L. Trajectory Tracking Robust Control Method Based on Finite-Time Convergence of Manipulator with Nonsingular Fast Terminal Sliding Mode Surface. *J. Control Sci. Eng.* **2022**, *2022*, 2271804. [CrossRef]

Article

Dynamic Control of a Novel Planar Cable-Driven Parallel Robot with a Large Wrench Feasible Workspace

Sergio Juárez-Pérez [1,†], Andrea Martín-Parra [1,†], Andrea Arena [2,†], Erika Ottaviano [3,*,†], Vincenzo Gattulli [2,†] and Fernando J. Castillo-García [1,†]

1 School of Industrial and Aerospace Engineering, University of Castilla-La Mancha, Av. Carlos III, 45071 Toledo, Spain
2 Department of Structural and Geotechnical Engineering, Sapienza University of Rome, Via Eudossiana 18, 00184 Rome, Italy
3 Department of Civil and Mechanical Engineering, University of Cassino and Southern Lazio, Via G. Di Biasio 43, 03043 Cassino, Italy
* Correspondence: ottaviano@unicas.it
† These authors contributed equally to this work.

Abstract: Cable-Driven Parallel Robots (CDPRs) are special manipulators where rigid links are replaced with cables. The use of cables offers several advantages over the conventional rigid manipulators, one of the most interesting being their ability to cover large workspaces since cables are easily winded. However, this workspace coverage has its limitations due to the maximum permissible cable tensions, i.e., tension limitations cause a decrease in the Wrench Feasible Workspace (WFW) of these robots. To solve this issue, a novel design based in the addition of passive carriages to the robot frame of three degrees-of-freedom (3DOF) fully-constrained CDPRs is used. The novelty of the design allows reducing the variation in the cable directions and forces increasing the robot WFW; nevertheless, it presents a low stiffness along the x direction. This paper presents the dynamic model of the novel proposal together with a new dynamic control technique, which rejects the vibrations caused by the stiffness loss while ensuring an accurate trajectory tracking. The simulation results show that the controlled system presents a larger WFW than the conventional scheme of the CDPR, maintaining a good performance in the trajectory tracking of the end-effector. The novel proposal presented here can be applied in multiple planar applications.

Keywords: cable-driven parallel manipulator; dynamic control; wrench feasible workspace; dynamics model; vibration control

Citation: Juárez-Pérez, S.; Martín-Parra, A.; Arena, A.; Ottaviano, E.; Gattulli, V.; Castillo-García, F.J. Dynamic Control of a Novel Planar Cable-Driven Parallel Robot with a Large Wrench Feasible Workspace. *Actuators* **2022**, *11*, 367. https://doi.org/10.3390/act11120367

Academic Editors: Marco Carricato and Edoardo Idà

Received: 1 November 2022
Accepted: 5 December 2022
Published: 7 December 2022

Publisher's Note: MDPI stays neutral with regard to jurisdictional claims in published maps and institutional affiliations.

1. Introduction

Parallel manipulators are a type of robot in which the end-effector is controlled by and connected to driving rigid links, called limbs. These robots are designed for achieving high stiffness and high payload capacity at the end-effector [1]; therefore, the weight of the links and their maximum strokes may considerably limit the velocities and the feasible workspace.

In order to overcome these evident limitations, cable-driven parallel robots have been introduced in the last two decades. In these systems, lightweight flexible cables actuate the end-effector. Each cable is wound around a winch, which is driven by an actuator. The winch can provide from millimetres up to several metres of cable, not only enabling large workspaces, but also assuring rather limited inertia, if compared to the rigid links' counterparts, and high velocities at the end-effector. CDPRs also provide the possibility of implementing deployable and reconfigurable topologies, which may expand the application of robotics in new environments, e.g., search and rescue operations [2,3], motion aiding systems [4,5], large buildings' maintenance and construction [2], entertainment [6], manufacturing or heavy handling [7], and service [8], but also considering classical high-speed pick and place manipulators in [9,10]. Applications related to the large working volume are the NIST Robocrane [11] and the five-hundred-metre Aperture Spherical Telescope (FAST), as reported in [12].

Despite these advantages, the current development of CDPRs seems to be limited mainly by the consequences deriving from the unilateral nature of cables; they can only pull the end-effector, but not push it. For operating a CDPR, the cable tensions must be kept positive, e.g., tensed during motion. If a cable or more cables become slack, the end-effector cannot be controlled properly for following a prescribed trajectory or exert the required wrench to perform a given task. This occurrence may introduce relevant performance and safety issues, since the control of the end-effector may be lost [13]. Nevertheless, the use of CDPRs seems very promising and has already been suggested in several different operation fields, such as heavy handling and industrial manufacturing, but also medical rehabilitation, home assistance, or sport shooting.

In the future, a wide use of CDPRs is expected thanks to their lightweight structure, which makes them energy efficient, modularity and reconfigurability, which makes them flexible and easy to transport, and finally, the potentially high dynamics and payload capacity, which makes them effective in a wide range of industrial applications.

In the analysis and design of CDPRs, but also for the operating issues and control, it is of great importance considering the number of active cables and their collocations; therefore overconstrained CDPRs have been defined as robots having a number of actuated cables higher than the EE degrees-of-freedom (DoFs) so that the cables can pull each other. Underconstrained CDPRs are those for which the cables cannot control all the DOFs of the robot; usually, under this category, the cable suspended robots fall; they are defined as CDPRs for which all active cables are suspending the EE, and the bottom of the robot is free from cables [14,15]. In [16], a point-to-point trajectory model was proposed for suspended CDPRs.

A large amount of research on CDPRs was initially devoted to workspace determination by referring to the so-called wrench closure workspace (WS) [17], i.e., a pose belongs to the WS if the static equilibrium is satisfied by positive tensions or tensions bounded by given limits of wrench feasible WS [18]. Therefore, limits were defined as the highest load capacity for the upper bound and cables' slackness avoidance, for the lower bounds, modelling an ideal cable, neglecting mass and elasticity.

Advanced models for the kinetostatic analysis consider the mass and elasticity of the cables to take into account sag effects in positioning capabilities [19–22]. In particular, direct and inverse kinematic models have been proposed.

Nevertheless, taking into account many practical applications, such as those mentioned above, a static or quasi-static assumption is rather limited since dynamic conditions are involved. In particular, vibrations are not simply induced by rapid end-effector velocity changes, wind disturbance, and/or friction of the cables around pulleys [23]. In applications requiring high performances, especially dynamic performances, e.g., [9,24], or in the presence of wind, e.g., [25,26], vibrations are an issue since they can affect the positioning accuracy of the end-effector and yield fluctuations around a desired nominal end-effector trajectory.

Dynamic models of CDPRs were initially developed neglecting the contribution of the cables, i.e., the cables were assumed as mass-less and non-elastic elements, as reported in [27]. Natural frequency is widely used in the literature as an index for the dynamic stiffness evaluation of CDPRs [23,28]. Several works proposing the stiffness analysis of CDPRs have been introduced by considering cable elasticity, but negligible cable mass [29]. In [30,31], cables were modelled as mass-less elastic springs and additionally taut strings, for analysing the contribution of the cables' flexibility in the axial and transverse directions to the vibrations of the moving mass constituting the end-effector. The kinematics and dynamics of the IPAnema3 cable robot were proposed in [32] by modelling cables as spring–damper elements with variable spring and damping rates due to the change in effective lengths. The above-mentioned model presents limitations, especially in the case of CDPRs with heavy and/or long-span cables. In fact, the axial cable stiffness is not the only source of the static stiffness. Sag-introduced stiffness should also be considered. As actual trends, research on CDPRs is focusing on the dynamic stiffness for its importance to have better stability, positioning accuracy, and flexibility [33]. The stiffness of a vibrating cable depends on the mechanical properties, the configuration of the cables, environmental effects, the initial cable tension, and cable motion [34].

Most of the reported works in the literature consider both cable elasticity and cable mass in two-dimensional inclined cables for dynamic analyses of the CDPRs with free vibration [34,35], i.e., without external forces acting on the system. However, for some applications, the forced vibrations are the result of external excitements/disturbances.

The vibration analysis for CDPRs with a large workspace was proposed in [36,37], where the cables were modelled with distributed mass and axial stiffness. The resulting solving equations were solved using different numerical techniques, e.g., converting to ordinary differential equations using the finite element method [36] and the assumed-mode method [37].

This paper presents a novel planar CDPR design based on adding passive carriages to the frame, providing the ability to passively reconfigure the distal anchor points of the robot [38]. The mechanical scheme together with the workspace gain and another kinematics issued were presented in [39].

Some works present active reconfigurable cable-driven parallel robots (RCDPRs) (e.g., [40]) or variable-structure cable-driven parallel robots (VSCR) (e.g., [41]. In RCDPRs, the position of the distal anchor points can be controlled by actuators, since in VSCRs, the distal anchor points are relocated by adding circumventing obstacles. In this sense, the scheme of this work pretends to increase the feasible workspace of the robot without adding more actuators.

The previous work [39] concluded that the addition of a passive carriage notoriously increases the feasible workspace of the robot, but on the contrary, the end-effector suffers a loss of stiffness along the horizontal axis, and non-desirable vibrations appear on the end-effector during its manoeuvres.

This work continues the work in [39]. The objective is to develop a dynamic control that allows the end-effector positioning

For this purpose, this paper is focused on the following:

(a) To present the dynamics model of the robot, which includes the dynamics model of the passive carriages. The addition of these carriages makes the system have five degrees-of-freedom and four actuators (underconstrained), but only three of them shall be controlled (fully constrained).

(b) To develop a suitable trajectory generator. In this sense, the distal anchor points [38] of the robot are passively reconfigurable, and trajectory planning for the underconstrained scheme shall be applied. Our proposal is based the trajectory generator on the statically feasible workspace [42].

(c) To implement a dynamic control that tackles the problems of positive tension and vibrations' rejection, owing to the low stiffness that the robot presents in the horizontal axis. In this sense, a novel control approach based on the addition of a control signal offset together with conventional PID controllers are also presented.

At the end, our final objective is to obtain a planar CDPR with a large workspace with an accurate end-effector position control.

This paper is organised as follows: Section 2 describes the novel proposal and presents the nomenclature. Section 3 details the mathematical model of the robot, including the dynamics model, required to design a proper control approach, which is presented in Section 4. Sections 5 and 6, respectively, show the simulation and experimental results, and finally, Section 6 discusses these results and summaries the main conclusions.

2. System Description

2.1. Overall Description

This paper is focused on a planar fully constrained CDPR [38]. The end-effector has three degrees-of-freedom, and its pose is commanded by four cables. At the same time, each cable is driven by a motor (placed at the frame) that rolls in or out the cable into/from a winch. A scheme of this conventional planar CDPR is shown in Figure 1a.

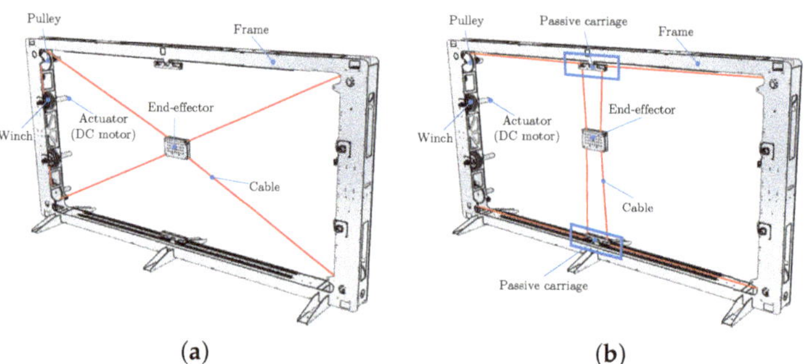

Figure 1. Scheme comparison. (**a**) Conventional scheme; (**b**) new proposal.

Our proposal is based on the addition of passive carriages over the conventional scheme. The end-effector has the same three degrees-of-freedom, and the motors drive the length of the four cables in the same way as in the conventional scheme. Nevertheless, now, each cables is led to a pulley placed in a carriage that can freely move over a linear guide. Note that Cables 1–2 and 3–4 are attached to the lower and upper carriages, respectively. The scheme of the proposal is shown in Figure 1b. Our previous work [39] presented the proposal, reporting the workspace gain in comparison to the conventional scheme as the main advantage.

2.2. Nomenclature

In a preliminary way, Figure 2 shows the basic nomenclature, where for both schemes:

- The end-effector pose is defined as $\mathbf{Q}_e = [x_e, y_e, \delta_e]$.
- The tension of the cables is T_i for $i = 1$ to 4.
- The angle of the cables is θ_i for $i = 1$ to 4.
- The angle of the motors is defined as α_i for $i = 1$ to 4.

Finally, for the novel proposal, the horizontal coordinate of the upper carriage is denote as x_{uc} and for the lower carriage as x_{lc}.

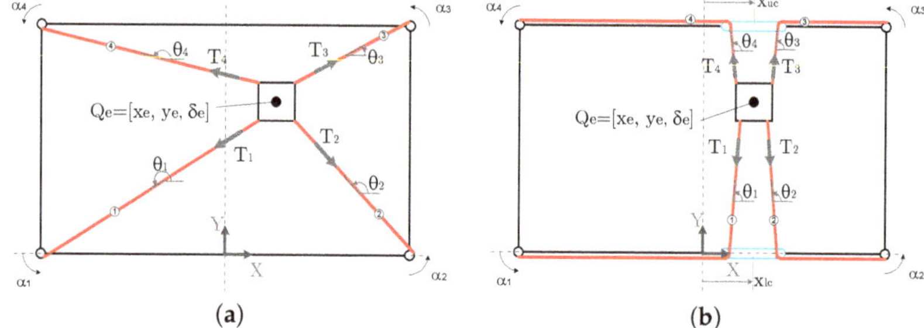

Figure 2. Scheme for nomenclature. (**a**) Conventional scheme; (**b**) new proposal.

2.3. Workspace and Stiffness Comparison

Assume that both schemes are in the vertical plane and gravity has an influence on the Y axis force equilibrium. Under a static point of view, the workspace can be easily obtained by establishing an end-effector pose and checking if a tension combination $T = [T_1, T_2, T_3, T_4]^T$ exists that allows a force/torque equilibrium at the end-effector (for the conventional scheme) and at the end-effector and both carriages (for the novel scheme). This tension combination must remain inside of the allowed limit $[T_{min}, T_{max}]$ as any CDPR (see [38] for more details).

The equilibrium equations for the conventional scheme can be written as:

$$\mathbf{A_s}(\mathbf{Q}_e, x_{uc}, x_{lc})\mathbf{T} + \mathbf{W}_e = 0 \tag{1}$$

where $\mathbf{A_s}$ is a geometrical matrix that depends on the parameters of the robot and the end-effector pose (e.g., [38]) and $\mathbf{W}_e = [0, -m_e \cdot g, 0]^T$ is the external force array, with m_e the end-effector mass and g the gravity acceleration. Only for illustrative purposes, for several end-effector orientations ($\delta_e = 0°$, $\pm 1°$, $\pm 3°$, $\pm 5°$, $\pm 7°$, and $\pm 9°$), the feasible workspace, defined as the set of points that fulfil (1), is represented for both schemes, conventional and the new proposal, in Figure 3, where W and H are the width and the height of the frame, respectively. These results are detailed in [39].

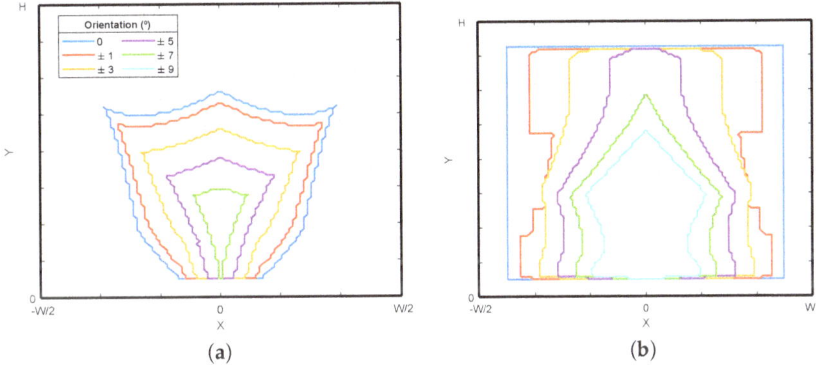

Figure 3. Workspace comparison. (**a**) Conventional scheme; (**b**) new proposal.

This workspace gain makes the proposal suitable to apply planar CDPRs, which require a complete workspace with regard to the frame size. Nevertheless, on the contrary, the proposed scheme presents lower stiffness along the horizontal axis, which yields a non-desirable vibration during the robot manoeuvres. The stiffness matrix of a CDPR, defined by Pott [43], is the sum of the cable and the geometric stiffness matrices. The cable stiffness matrix depends on the cables' stiffness and the robot structure matrix $\mathbf{A}_e^T$ (further discuss), which in turn depends on the robot geometry and end-effector's/carriages' position. Furthermore, the geometric stiffness matrix depends on the cable force distribution and the $\mathbf{A}_e^T$ variation with end-effector pose. In order to evaluate how the novel design is affected, the directional stiffness index defined by Moradi [44] is used. Employing the parameters from Table 1, we are able to compare the stiffness of the new CDPR with that of a conventional design. For illustrative purposes, the robot frame is swept on equally spaced points to determine the end-effector non-oriented position. Applying the method proposed by Pott in [45], the force distribution is obtained within the cable force limits of $[10, 100]$ N. Figure 4 compares the stiffness map obtained under a force applied in the x axis positive direction for both schemes. A significant decrease of the stiffness can be observed when applying forces in the x axis. More explanation about these results can be found in [39].

As a conclusion, the proposal here presents a much larger feasible workspace with regard to the conventional scheme, but, on the other hand, the horizontal stiffness is much lower, producing non-desirable vibration during the robot movements.

In this sense, the next section introduces the mathematical model of the system, including the dynamic model, in order to design a control approach that allows rejecting the negative effects of the low horizontal stiffness of the proposal.

Table 1. Model parameters.

Subsystem	Parameter	Value
Frame	H	0.8 m
	W	1.2 m
Carriages	w_{cl}	0.2 m
	w_{cu}	0.2 m
	m_c	3 kg
	b_c	0.2 Ns/m
Motor/winch set	J	4.61×10^{-4} Kgm^2
	b_m	7.69×10^{-3} Nms
	r	0.04 m
End-effector	m_e	1 kg
	w	0.1 m
	h	0.1 m
	I_e	2×10^{-3} Kgm^2
Controllers	ω_c	40 rad/s
	φ_m	60°
	K_p	0.6357
	K_d	0.0121
	K_i	0.55

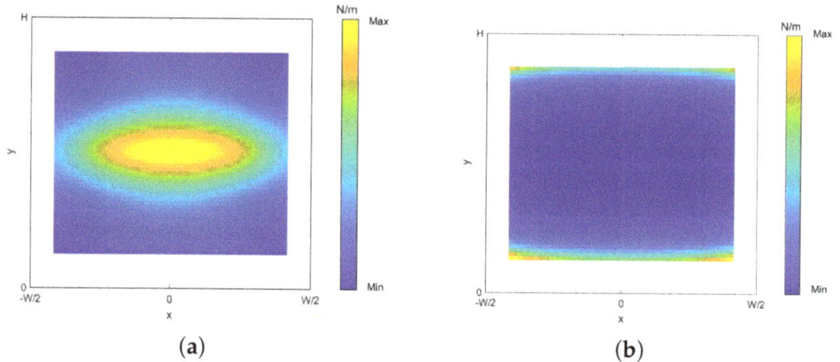

Figure 4. X axis stiffness comparison. (**a**) Conventional scheme; (**b**) new proposal.

3. Mathematical Model

3.1. Kinematic Model

In this section, the inverse kinematic model is presented. This model will be used to implement the dynamics model and in the control approach of the robot. The inverse kinematic determines the actuator position, α_1 to α_4, to obtain a desired end-effector pose $\mathbf{Q}_e = [x_e, y_e, \delta_e]^T$.

First of all, to complete the nomenclature of Section 2.2, it is required to define some still missing geometrical parameters of the robot: end-effector width, (w), end-effector height (h), upper carriage width (w_{uc}), and lower carriage width (w_{lc}).

It must be pointed out that the geometry model was obtained by taking into account the passive pulley radius effect on the cable lengths [38]. The carriages' widths, w_{cu}, w_{cl}, are considered to be the distance between the passive pulleys' centres; the cable goes vertically from the pulley centre to the winch, although the contact point in the pulley from the end-effector depends on the positioning between the carriage and the end-effector.

The frame is defined by the proximal anchor points (PAPs), $A_i \in R^2 (i = 1, \ldots, 4)$, with respect to the World reference system κ_0 shown in Figure 5. On the other hand, the end-effector geometry is defined by the distal anchor points (DAPs), $b_i \in R^2 (i = 1, \ldots, 4)$, in the local reference system κ_e with respect to the end-effector. Figure 5 is included to clarify all vectors used in the kinematics model and the points of interest.

The end-effector pose, $\mathbf{Q}_e = [x_e, y_e, \delta_e]^T$, is a combination between its position, $\mathbf{q}_e = [x_e, y_e]^T$, and orientation, δ_e. In this sense, the DAP with respect to κ_0 can be obtained as

$$\mathbf{B}_i = \mathbf{q}_e + \mathbf{R}_z(\delta_e) \cdot \mathbf{b}_i, \tag{2}$$

where

$$\mathbf{R}_z(\delta) = \begin{bmatrix} \cos(\delta) & -\sin(\delta) \\ \sin(\delta) & \cos(\delta) \end{bmatrix} \tag{3}$$

is the rotation matrix around the z axis. The position of the passive carriages with respect to κ_0 is defined as x_{uc} and x_{lc} for the upper and lower carriage. Taking this into account, the position of the passive pulley centres with respect to κ_0 is defined as

$$c_i = x_c + v_{ci}, \tag{4}$$

with $x_c = x_{lc}$ for $i = 1, 2$, $x_c = x_{uc}$ for $i = 3, 4$ and v_{ci} being the vector that goes from the carriage centre to the ith passive pulley centre.

As is known from earlier, parallel robots have two possible kinematic formulations. The inverse kinematic model obtains the cable lengths, $\mathbf{L} = [L_1, \ldots, L_4]^T$, from an arbitrary end-effector pose:

$$\mathbf{L} = \varphi^{IK}(\mathbf{Q}_e, x_{lc}, x_{uc}) \tag{5}$$

The cables are considered massless, so that they are formulated as straight lines connecting the previously defined geometric points. Considering the scheme in Figure 5 and the effect of the passive pulleys' radius, r_p, the total cable length yields $L_i = L_{i,1} + L_p + L_{i,2}$, with:

$$\begin{aligned} L_{i,1} &= |A_{i,x} - c_{i,x}| \\ L_{i,2} &= \sqrt{BC_i^2 - r_p^2} \\ L_p &= \beta_u r_p, \end{aligned} \tag{6}$$

where $BC_i = ||B_i - c_i||$, $|| \cdot ||$ being the two-norm of the vector $\cdot$ and β_u the angle formed by the cable rolled in the passive carriage pulleys, which can be obtained by means of c_i and B_i as:

$$\beta_u = \cos^{-1}\left(\frac{L_{i,2}}{BC_i}\right) + \cos^{-1}\left(\frac{\text{sign}(c_{i,x} - A_{i,x}d_x)}{BC_i}\right), \tag{7}$$

$d_x = B_{i,x} - c_{i,x}$, $\text{sign}(x)$ being the sign function. The actuators angle for an arbitrary end-effector and carriages' pose can be obtained by:

$$\alpha = \frac{\mathbf{L} - L_0}{r_w}, \tag{8}$$

where $\alpha = [\alpha_1, \ldots, \alpha_4]^T$ is the array of the joint coordinates, L_0 is the initial cable lengths defined for an initial end-effector and carriages' pose/position, $\mathbf{Q}_{e0}, x_{lc0}, x_{uc0}$, and r_w is the combined winch radius taking into account any gearbox connected to the actuators.

On the contrary, the forward kinematic consists of determining the end-effector pose, $\mathbf{Q}_e$, for a given set of joint coordinates, α:

$$\mathbf{Q}_e = \varphi^{FK}(\alpha, x_{lc}, x_{uc}) \tag{9}$$

Due to its much more complicated development and owning to the fact that the forward kinematics has infinite solutions [38], only the inverse kinematics is fully developed and applied for the control scheme.

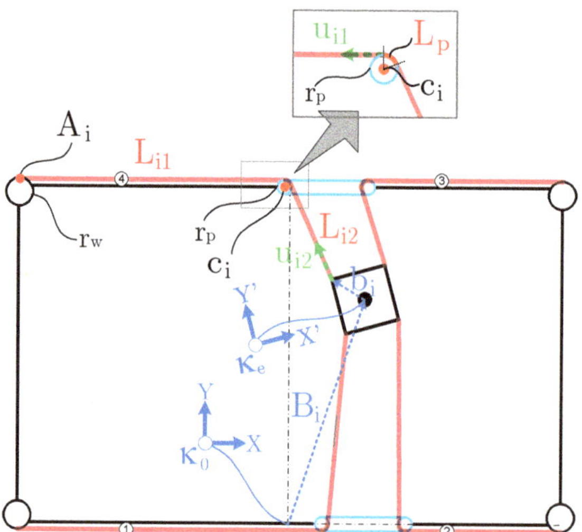

Figure 5. Nomenclature: arrays and points of interest.

The inverse kinematic model developed here is used to implement the dynamic model presented in the following section and for the control scheme detailed in Section 4.

3.2. Dynamic Model

While the kinematics and static models relate the end-effector and carriage positions to the cable lengths and tensions, the dynamics of the system describes its behaviour. The input variables to the system are the torque exerted by the motors, τ_i with $i = 1, \ldots, 4$, and the output is the end-effector pose, $\mathbf{Q}_e = [x_e, y_e, \delta_e]^T$.

Due to the novel design, the dynamics of the motor sets, end-effector, and carriages must be considered to develop a representative model. The motor and end-effector dynamics is equal to the one used for a conventional CDPR design. The end-effector dynamics is represented by the following expression:

$$\mathbf{M}_e \ddot{\mathbf{Q}}_e = \mathbf{A}_e^T (\mathbf{Q}_e, x_{lc}, x_{uc}) \mathbf{T} + \mathbf{W}_e, \tag{10}$$

where

$$\mathbf{M}_e = \begin{bmatrix} m_e & 0 & 0 \\ 0 & m_e & 0 \\ 0 & 0 & I_e \end{bmatrix}, \tag{11}$$

m_e being the end-effector mass, I_e the end-effector rotational inertia, $\mathbf{T} = [T_1, \ldots, T_4]^T$ the cable tensions' vector, $\mathbf{W}_e = [0, -m_e g, 0]^T$ the external wrench vector, which is only subjected to its own weight, and $\mathbf{A}_e^T$ the structure matrix, which yields:

$$\mathbf{A}_e^T = \begin{bmatrix} \mathbf{u}_{12} & \cdots & \mathbf{u}_{m2} \\ \mathbf{R}_z(\delta_z)\mathbf{b}_1 \times \mathbf{u}_{12} & \cdots & \mathbf{R}_z(\delta_z)\mathbf{b}_m \times \mathbf{u}_{m2} \end{bmatrix}, \tag{12}$$

$\mathbf{u}_{i,2} = \mathbf{R}_z(\beta_u)\mathbf{e}_x$ being the unity vector containing the directions in which the cable forces are exerted on the end-effector, with $\mathbf{e}_x$ the unity vector in the x axis direction.

The motor sets' dynamics can be expressed as:

$$\tau = \mathbf{J}\ddot{\alpha} + \mathbf{b}_m \dot{\alpha} + r_w \mathbf{T}, \tag{13}$$

where $\mathbf{J}$ is the rotational inertia matrix, $\alpha = [\alpha_i, \ldots, \alpha_m]^T$ is the array of joint coordinates, $\mathbf{b}_m$ is the viscous friction coefficient, and r_w is the winch radius, respectively.

On the other hand, the carriages' dynamics are yielded as:

$$m_c \ddot{x}_i^c + b_c \dot{x}_i^c = \mathbf{A}_c(Q_e, x_{lc}, x_{uc})\mathbf{T}, \tag{14}$$

where m_c is the carriage mass, $\ddot{x}_i^c$ is the x direction acceleration of the carriage i, and

$$\mathbf{A}_c = \begin{bmatrix} \mathbf{u}_{11,x} - \mathbf{u}_{12,x} & \mathbf{u}_{21,x} - \mathbf{u}_{22,x} & 0 & 0 \\ 0 & 0 & \mathbf{u}_{31,x} - \mathbf{u}_{32,x} & \mathbf{u}_{41,x} - \mathbf{u}_{42,x} \end{bmatrix} \tag{15}$$

is a $2 \times m$ matrix that contains the x direction cables' exerted forces on the carriages, $\mathbf{u}_{ii,x}$ being the x component of the ii cable section direction.

Equations (9), (13), and (14) constitute the dynamic model of the system. They can be expressed by means of end-effector/Cartesian coordinates (16) or by means of the joints' coordinates (17).

$$\begin{aligned} \mathbf{M}_e \ddot{Q}_e &= \mathbf{A}_e^T(Q_e, x_{lc}, x_{uc})\mathbf{T} + \mathbf{W}_e \\ \tau &= \mathbf{J}\ddot{\varphi}^{IK}(Q_e, x_{lc}, x_{uc}) + \mathbf{b}_m\dot{\varphi}^{IK}(Q_e, x_{lc}, x_{uc}) + r\mathbf{T} \\ m_c \ddot{x}_i^c + b_c \dot{x}_i^c &= \mathbf{A}_c(Q_e, x_{lc}, x_{uc})\mathbf{T} \end{aligned} \tag{16}$$

$$\begin{aligned} \mathbf{M}_e \ddot{\varphi}^{FK}(\alpha, x_{lc}, x_{uc}) &= \mathbf{A}_e^T(\varphi^{FK}(\alpha, x_{lc}, x_{uc}), x_{lc}, x_{uc})\mathbf{T} + \mathbf{W}_e \\ \tau &= \mathbf{J}\ddot{\alpha} + \mathbf{b}_m\ddot{\alpha} + r\mathbf{T} \\ m_c \ddot{x}_i^c + b_c \dot{x}_i^c &= \mathbf{A}_c(\varphi^{FK}(\alpha, x_{lc}, x_{uc}), x_{lc}, x_{uc})\mathbf{T} \end{aligned} \tag{17}$$

Dynamic Model (16) is widely used, instead of (17), owing to three reasons: (a) the CDPR's forward kinematics is hard to develop; (b) actuators use to be more easily sensorised than end-effectors; (c) the joint coordinates' control is easier to implement than Cartesian coordinates' control [38]. For these reasons, the following section assumes Dynamic Model (16) for the control approach.

4. Control Approach

4.1. Control Scheme

One of the main control goals for a CDPR is to design a control scheme for end-effector positioning. This control approach shall achieve that the end-effector pose $Q_e = [x_e, y_e, \delta_e]^T$ follows a trajectory reference $Q_e^* = [x_e^*, y_e^*, \delta_e^*]^T$, but maintaining all cables' tension in the allowed range $[T_{min}, T_{max}]$.

Assuming that the measurable variables are the angular position of the motors, under an overall point of view, the CDPR can be controlled on Cartesian coordinates ($Q_e = [x_e, y_e, \delta_e]^T$) or on joint coordinates ($\alpha = [\alpha_1, \alpha_2, \alpha_3, \alpha_4]^T$); this make reference to the variable that is used in the feedback loop of the control scheme. Figure 6 represents both control schemes.

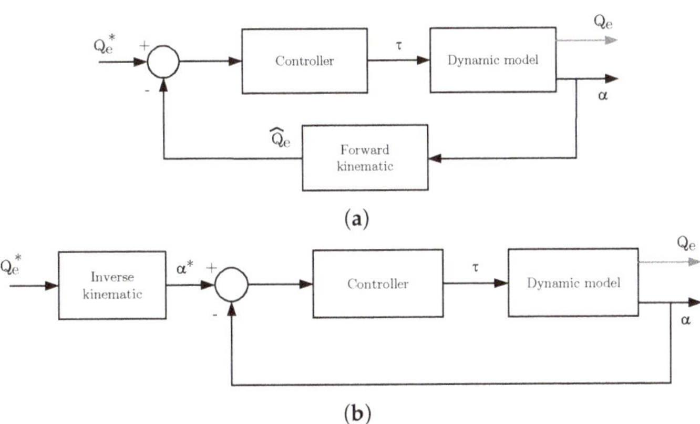

Figure 6. Overall control approaches. (**a**) Cartesian coordinates' space control; (**b**) joint coordinates' space control.

Note that Cartesian coordinates' control (Figure 6a) requires the estimation of the end-effector pose, $\hat{\mathbf{Q}}_e$, by means of the forward kinematics. In addition, the controller block receives a 3×1 error signal and must output a 4×1 control signal, and therefore, the controller block is a coupled 4×3 matrix. This makes the control tuning difficult, and an uncoupling technique must be applied to obtain the controllers' parameters (e.g., [46]).

On the contrary, the joint coordinates' space control requires the inverse kinematics, for the joints' reference generator, $\boldsymbol{\alpha}^*$, and the controller block is formed by a 4×4 diagonal matrix, while SISO controllers can be tuned for each motor. For this reason, the joint coordinates' space control was selected for the end-effector positioning control.

4.2. Trajectory Generation

The trajectory generation in our proposal is one of the main problems to be solved. Note that, for joint space control, the inverse kinematics is required, but it has infinite solutions. We can obtain a given $\mathbf{Q}_e^* = [x_e^*, y_e^*, \delta_e^*]$ with the infinity position of the lower and upper carriage position, x_{lc} and x_{uc}, but most of then are not compatible with the dynamic model.

In this sense, together with the $\mathbf{Q}_e^*$ generation, it is required to also generate x_{lc}^* and x_{uc}^*. In particular, the trajectory generator proposed here is based on the statically feasible workspace [42] and follows these steps:

(a) Generation of the desired end-effector trajectory $\mathbf{Q_e}^*$.
(b) For a given $\mathbf{Q_e}^*$, the tension value of one cable is fixed to T_o^*; in our case, T_1 was arbitrary set.
(c) For the established values of $\mathbf{Q_e}^*$ and $T_1 = T_o^*$, the static equilibrium problem is solved to determine x_{lc}^* and x_{uc}^*. To solve this problem, the *Levenberg–Marquardt* algorithm is applied [47].

The simulation results in Section 5 show that this trajectory-generation technique allows reaching a high-performance trajectory tracking of the end-effector.

4.3. Positive Tension Problem and Vibration Rejection

Finally, we cannot forget the problem of maintaining cables' tension within the allowed tension range $[T_{min}, T_{max}]$. Owing to the fact that, in our proposal, the cable angles θ_i remain very close to $\pm\pi/2$ rad in almost all workspaces, a simple but effective modification of the control scheme shown in Figure 6b can be applied for maintaining the cables' tension within the allowed range. Figure 7 presents this simple modification.

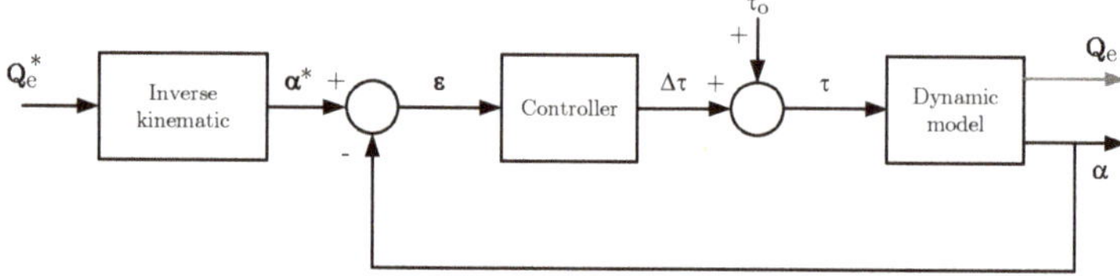

Figure 7. Final control approach for end-effector positioning.

It consists of adding a control signal offset, τ_o, to the control signal generated by the controller block, $\Delta\tau$. In this way, a new control signal is introduced to the system:

$$\tau = \Delta\tau + \tau_o \tag{18}$$

where τ_o can be set, in a first approach, equal for all cables to a fixed value within the allowed tension range $\mathbf{T}_o^* \in [T_{min}, T_{max}]$ multiplied by the winch pulley radius, r_w. The final control signal can be therefore written as:

$$\tau = \Delta\tau + r_w \mathbf{T}_o^* \tag{19}$$

Finally, if a conventional PID controller is used for the end-effector positioning, the final control signal for commanding the CDPR is:

$$\boldsymbol{\tau}(t) = \mathbf{K}_p \boldsymbol{\varepsilon}(t) + \mathbf{K}_d \frac{d\boldsymbol{\varepsilon}(t)}{dt} + \mathbf{K}_i \int_0^t \boldsymbol{\varepsilon}(\sigma)d\sigma + r_w \mathbf{T}_o^* \tag{20}$$

$\varepsilon = \boldsymbol{\alpha}^* - \boldsymbol{\alpha}$ being the error signal and $\mathbf{K}_p$, $\mathbf{K}_d$, and $\mathbf{K}_i$ diagonal matrices with the proportional, derivative, and integral constants of the controller, respectively. This controller constant can be obtained, for example, by tuning the regulator by means of the frequency domain or time domain techniques (e.g., [48]).

Without this offset control signal, τ_o, a controller block to control the position of the joints can be easily found, but the tension of the four cables can reach negative values, so, as a consequence, this is not a valid solution in the real prototype. Actually, it is required to ensure not only that all tensions remain positive, but also $T_i > T_{min}$ for $i = 1$ to 4, T_{min} being the minimum tension allowed to avoid the sagging effect in the cables.

In this sense, many authors have proposed solving an optimisation problem, which results as the control signal to the motors, to guarantee that all tensions remain in the allowed range $[T_{min}, T_{max}]$ [49]. Another extended tendency is to develop a force distribution algorithm [45], which can be applied for example to generate the trajectory tracking together with the control signal law [50]. Under our point of view, in the control of cable-driven parallel robots, the control scheme concept is more important that the particular controller technique (PID, sliding control, fuzzy controller, etc.).

After checking that the proposed control scheme can control the end-effector position and all tensions remain inside of the allowed range, we established the value of T^* to minimise (or reject) the x axis vibration of the end-effector.

Although the inherent stiffness is not controlled by the cable tension, it has a significant impact on the system stiffness [51]. What is the system stiffness variation when T^* changes? Can we, therefore, obtain a value of T^* that rejects the x axis vibration of the end-effector, allowing a proper trajectory tracking? To answer this question, we changed the T^* value from T_{min} to T_{max}, and we executed horizontal rapid trajectories (worst case) and determined the amplitude and frequency of the end-effector when it reaches the steady state. Assuming an illustrative tension range of $[50, 200]$ N for the cables' tension, the result after executing the same trajectory, with the same PID controller, but different T^* values, is shown in Figure 8.

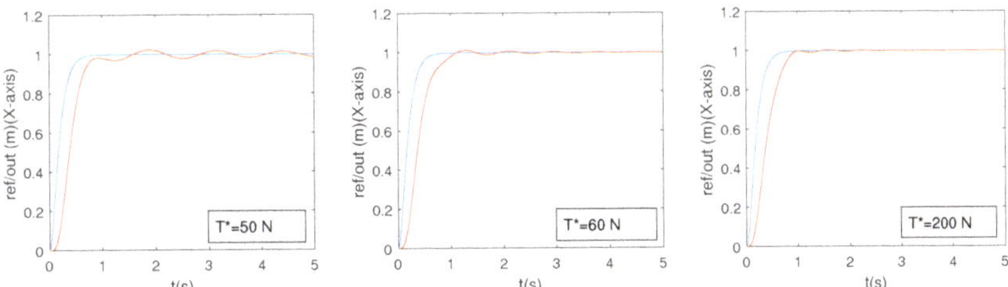

Figure 8. Effect of T^* on the vibration rejection.

In this sense, it is important to mention that the higher T^*, the higher the frequency of the vibration is and the lower is the amplitude. In this sense, the simulation results of the following section show that the non-desirable vibration of the end-effector owing to its low horizontal stiffness can be easily rejected by increasing the T_o^* value within the $[T_{min}, T_{max}]$ range.

4.4. Controller Tuning

The PID controller has three parameters to be tuned, K_p, K_i, and K_d. As Dynamic Model (16) is non-linear, some linearisation technique must be used for tuning the controller.

The most-applied linearisation techniques are: feedback linearisation and removing the non-linear part near an equilibrium point. If we look at some works where PID control is used (e.g., [52]), most of them finally *select* values of the controller gains to ensure the stability, but no linear technique is described to analytically obtain the gains.

In our case, owing to the cables' angle changing much less than the conventional scheme during end-effector movement (see θ_i in Figure 2), the non-linear effect is lower than in the conventional CDPRs, and we can directly remove the non-linear term of the dynamic model to assume an LTI model, which corresponds to the model of a DC motor, and then, apply linear tuning methods for the ID controller, in our case the frequency domain technique.

Assuming the corresponding linear transfer function of one joint, $G(s)$, the PID controller was tuned by a simple frequency domain technique based on defining the gain crossover frequency, ω_c, and phase margin, φ_m, of the equivalent open-loop transfer function system. The complex equation to obtain the controllers' parameters is [48]:

$$R(j\omega_c)G(j\omega_c) = -e^{j\varphi_m} \tag{21}$$

The controller parameters can be therefore easily obtained with this tuning equation:

$$K_p = \Re\{\frac{-\cos(\varphi_m) - j\sin(\varphi_m)}{G(j\omega_c)}\} \tag{22}$$

$$K_d = \frac{1}{\omega_c}\left(-\frac{K_i}{\omega_c} + \Im\{\frac{-\cos(\varphi_m) - j\sin(\varphi_m)}{G(j\omega_c)}\}\right) \tag{23}$$

where $\Re\{\}$ and $\Im\{\}$ are the real and imaginary parts and K_i can be selected to minimise the influence of the non-linear term of the dynamics model, for example minimising the tracking error by means of simulations.

5. Simulation Results

5.1. Preliminaries

This section validates the control approach by executing several simulated trajectories using Matlab® and Simulink®. A fixed sample time of 1 ms was set while using the ode4 (Runge–Kutta) solver. Table 1 summarises the parameters used for the simulations.

5.2. Control Approach Validation

One of the main drawbacks of the new design is the loss of stiffness along the x direction. This leads to non-desirable end-effector vibrations while executing manoeuvres. In order to validate the proposed control, several trajectories were executed while monitoring the end-effector and carriage positions and the cable tensions. Dynamic Model (16) is a second-order system, and the fifth Bezier trajectories were implemented to ensure smooth trajectories while avoiding abrupt changes in the control signal values, which can lead, once again, to non-desirable end-effector vibrations.

For illustrative purpose, a set of trajectories was executed with zero end-effector orientation. In addition, the last trajectory was executed while orienting the end-effector, proving that the control is capable of avoiding non-desirable vibrations, even with more complex manoeuvres. Note that, for all cases of study, the end-effector tracks the reference during all the trajectory. The average trajectory tracking error, e_a, is estimated by:

$$e_a = \frac{\sum_{i=1}^{N} ||\mathbf{Q}_e^* - \mathbf{Q}_e||_2}{N} \tag{24}$$

N being the number of samples of the simulation:

- Case I: Figure 9 shows a diagonal trajectory from initial pose $\mathbf{Q}_e^0 = [-0.2, 0.3, 0]$ to $\mathbf{Q}_e = [0.2, 0.5, 0]$. The average error (24) of the executed trajectory is 1.45×10^{-4} m-rads.

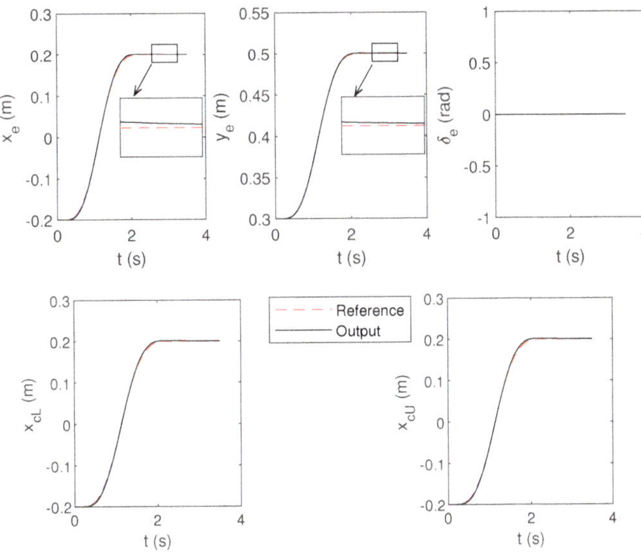

Figure 9. Case I results.

- Case II: Figure 10 shows a reorientation trajectory from initial pose $\mathbf{Q}_e^0 = [0, 0.4, 0]$ to $\mathbf{Q}_e = [0, 0.4, 6]$. The average error (24) of the executed trajectory is 4.02×10^{-4} m-rads.
- Case III: Figure 11 presents a diagonal trajectory from initial pose $\mathbf{Q}_e^0 = [-0.2, 0.3, 0]$ to $\mathbf{Q}_e = [0.2, 0.5, -6]$. This trajectory introduces a change in the end-effector orientation while executing the diagonal movement. The average error (24) of the executed trajectory is 2.17×10^{-4} m-rads.

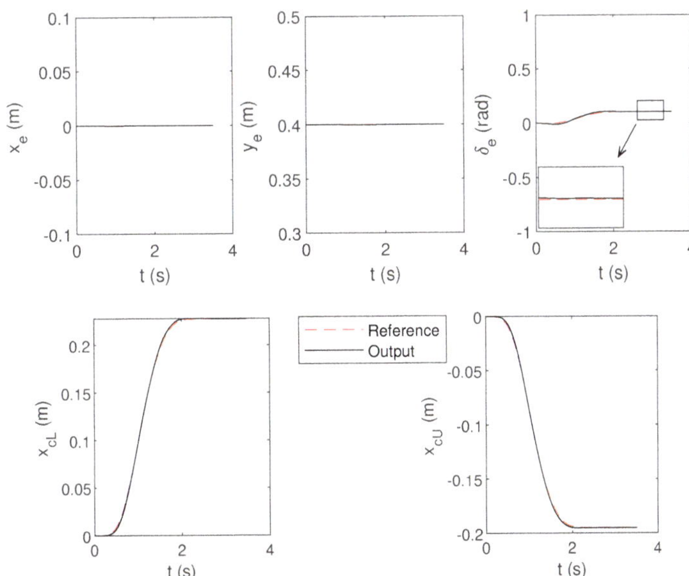

Figure 10. Case II results.

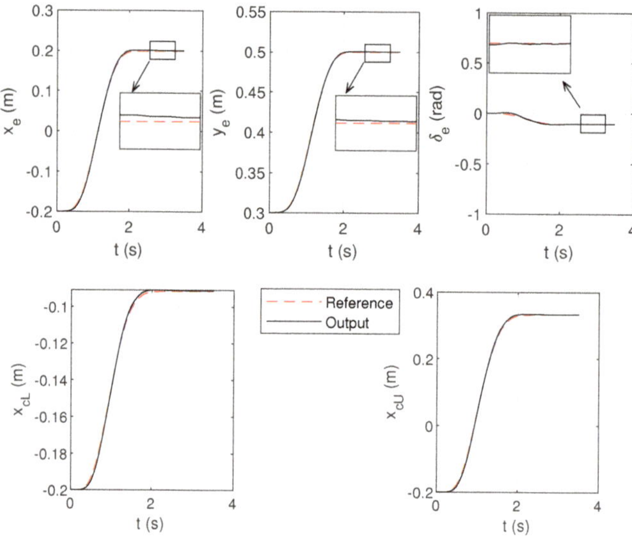

Figure 11. Case III results.

These three cases were simulated for an allowed tension range, $[10, 100]$, and the reference tension, T_o^*, to determine that the control signal output has been fixed to 50 N. Figure 12 represents the tension of the cables for Cases I, II, and III. Note that the cables' tensions remain within the allowed range of tensions.

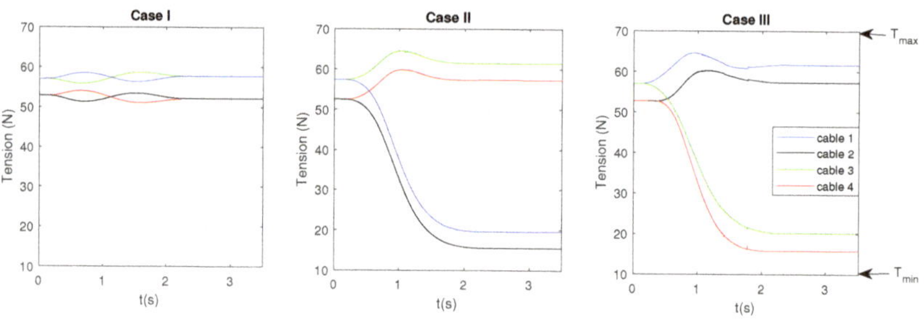

Figure 12. Cable tensions: Cases I, II, and III.

5.3. WFW Increase Validation

As mentioned previously, this novel design increases the WFW in comparison to the conventional CPDR scheme. In this sense, more frame area could be covered while executing manoeuvres. It should be pointed out that the oriented WFW of the novel design is also bigger than the one of a conventional CDPR (see Figure 3).

In order to validate this, a multi-line trajectory near the WFW limits shown in Figure 3b with different end-effector orientations was executed. In particular, the case of studies are as follows:

- Case IV: The orientation of the end-effector is fixed at $\delta_e = 0°$. The multi-line trajectory tracking results are shown in Figure 13—Case IV. The average error (24) of the executed trajectory is 3.87×10^{-4} m-rads.
- Case V: The orientation of the end-effector was fixed to $\delta_e = -5°$. The trajectory tracking results are shown in Figure 13—Case V. The average error (24) of the executed trajectory is 1.98×10^{-4} m-rads.

- Case VI: The orientation of the end-effector was fixed to $\delta_e = 9°$. The trajectory tracking results are shown in Figure 13. The average error (24) of the executed trajectory is 4.81×10^{-4} m-rads.

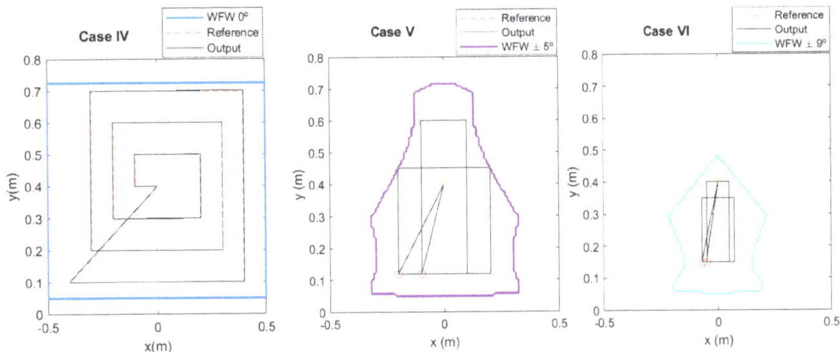

Figure 13. Results of Cases IV, V, and VI.

In Figure 13, the circle in green represents the initial position of the trajectory and red ones the final positions.

It is notable that, in the three cases of study, the selected trajectory covers a great area of its corresponding maximum WFW (see Figure 3). In addition, the cable tensions for Cases IV, V, and VI also remain within the allowed range of tensions. The results obtained in this section allowed the authors to validate both the control approach and the WFW increase.

6. Experimental Results

6.1. Preliminaries

The proposal here is experimentally validated by means of the prototype shown in Figure 14, where the main functional elements are: ① frame; ② end-effector; ③ carriages; ④ linear guides; ⑤ winches; ⑥ DC motors. In Figure 14, the cable path is coloured in red for illustrative purposes. The main parameters of the robot are summarised in Table 2.

The joint coordinates of the motors, Maxon RE40 DC Motors, are controlled by means of servo amplifiers, ESCON 70/10, which receive the control signal from the control equipment. This control device is a Ni MyRio 1900 real-time controller, which is able to maintain a sample time of 1 ms during all the manoeuvres. The controllers' parameters were tuned using a gain crossover frequency $\omega_c = 40$ rad/s and a phase margin $\varphi_m = 60°$, which yield the controller parameters: $K_p = 0.6891$, $K_i = 0.57$, and $K_d = 0.0102$.

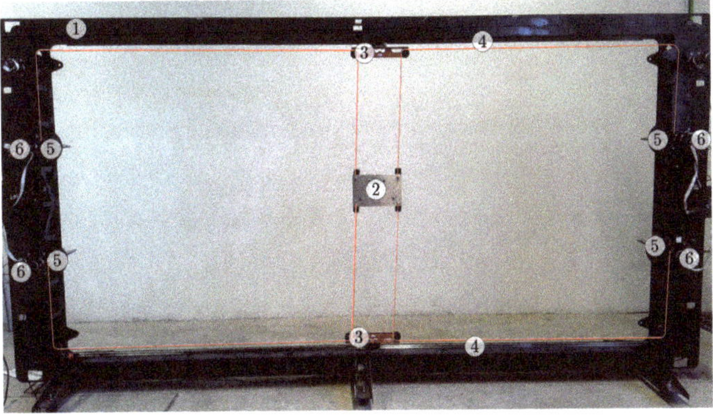

Figure 14. Prototype.

Table 2. Prototype parameters.

Subsystem	Parameter	Value
Frame	H	1.2 m
	W	2.0 m
Carriages	w_{cl}	0.21 m
	w_{cu}	0.21 m
	m_c	1.26 kg
	b_c	0.2 Ns/m
	r_c	0.01 m
Motor/winch set	J	4.61×10^{-4} Kgm^2
	b_m	7.69×10^{-3} Nms
	r	0.06 m
End-effector	m_e	1.8 kg
	w	0.18 m
	h	0.11 m
	I_e	2×10^{-3} Kgm^2

The end-effector and the carriages' poses were obtained by recording the movement with an IP-evo 4K camera with a resolution of 3264×2448 and a frame rate of 30 frames per second. The videos were processed offline with Matlab^{TM} to detect the position of a reflector located on the points of interest.

6.2. Trajectory Tracking

In order to experimentally validate the proposal, a horizontal trajectory (worst case) was tested with a length of 0.500 m, maintaining the orientation at $0°$. Figure 15 shows the results for $T^* = 50$ and 85 N in order to illustrate the influence of the control signal offset, τ_o, in the trajectory tracking.

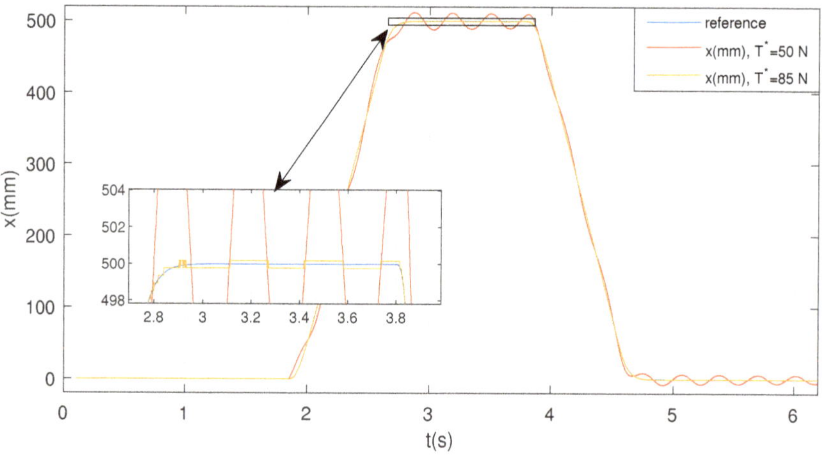

Figure 15. Trajectory tracking of horizontal manoeuvres with different values of T^*.

Note how a greater T^* lowers the amplitude of the end-effector vibration, but a compromise between the T^* value and control signal saturation must be stated to avoid the actuator saturation. Figure 16 represents the tracking error for both cases. Note that the maximum error for $T^* = 50$ N is about 16 mm, since for $T^* = 85$ N, it is about 0.5 mm.

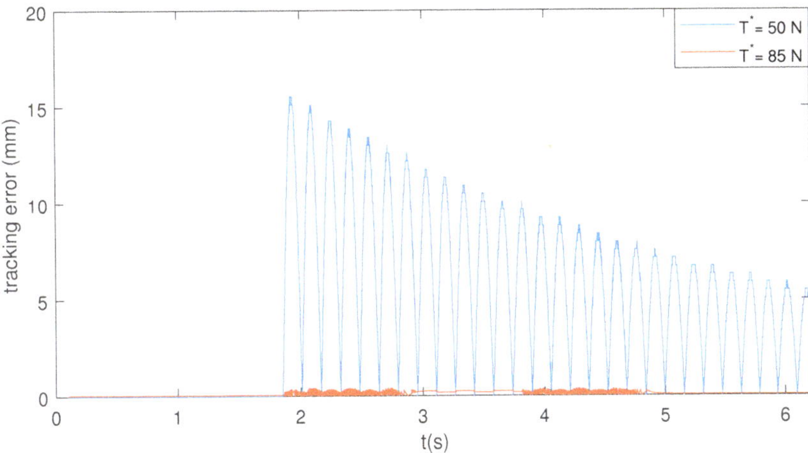

Figure 16. Tracking error of horizontal manoeuvres with different values of T^*.

7. Conclusions

This paper proposed a new dynamic control technique for a novel CDPR design. This novel design is based in the addition of passive carriages to the robot frame of a 3DOF CDPR. The main goal of the new concept is to increase the robot WFW because of the smaller change in the cable directions.

As the main disadvantage of the new design is the loss of stiffness in the x direction, one of the goals of the new dynamic control technique was to minimise the vibrations originated in this direction, ensuring an accurate trajectory tracking.

The kinematic and dynamic equations of the robot were presented in order to select an appropriate control approach for the implemented dynamic model.

The control strategy presented tackles with the trajectory generation, the positive tension problem, and the vibration rejection. This control approach is based on the addition of an offset control signal.

In order to validate the dynamic control, several tracking trajectories for the end-effector were simulated. The average error always remained lower than 1 mm, and a good avoidance of vibrations was accomplished. In addition, to assess the improvement of the WFW while validating the novel control, a set of different trajectories was used for several end-effector orientations. The results showed how the new design together with the proposed control allowed the CDPR to manoeuvre in a much larger area with good dexterity.

For experimental validation, a horizontal trajectory (worst case) was executed for two values of T^* to illustrate how the offset control signal value can reject the vibration of the end-effector along the horizontal axis.

Author Contributions: The authors contributed equally to this work. All authors have read and agreed to the published version of the manuscript.

Funding: Authors wish to thank the University of Castilla-La Mancha, Spain for the financial support provided by the post- and pre-doctoral grants 2020-PREDUCLM-16080 and to the regional government Junta de Comunidades de Castilla La Mancha (European Regional Development Fund) by the grant SBPLY/21/180501/000238.

Data Availability Statement: Not applicable.

References

1. Merlet, J.P. *Parallel Robots*; Springer Science & Business Media: Berlin/Heidelberg, Germany, 2005; Volume 128.
2. Bosscher, P.; Williams, R.L.; Tummino, M. A concept for rapidly-deployable cable robot search and rescue systems. In Proceedings of the International Design Engineering Technical Conferences and Computers and Information in Engineering Conference, Long Beach, CA, USA, 24–28 September 2005; Volume 47446, pp. 589–598.
3. Tadokoro, S.; Verhoeven, R.; Hiller, M.; Takamori, T. A portable parallel manipulator for search and rescue at large-scale urban earthquakes and an identification algorithm for the installation in unstructured environments. In Proceedings of the 1999 IEEE/RSJ International Conference on Intelligent Robots and Systems, Human and Environment Friendly Robots with High Intelligence and Emotional Quotients (Cat. No. 99CH36289), Kyongju, Republic of Korea, 17–21 October 1999; Volume 2, pp. 1222–1227.
4. Ottaviano, E.; Ceccarelli, M.; De Ciantis, M. A 4–4 cable-based parallel manipulator for an application in hospital environment. In Proceedings of the 2007 Mediterranean Conference on Control & Automation, Athens, Greece, 27–29 June 2007; pp. 1–6.
5. Gallina, P.; Rosati, G. Manipulability of a planar wire driven haptic device. *Mech. Mach. Theory* **2002**, *37*, 215–228. [CrossRef]
6. Skycam. Available online: http://www.skycam.tv/2022 (accessed on 10 October 2022).
7. Gueners, D.; Chanal, H.; Bouzgarrou, B.C. Design and implementation of a cable-driven parallel robot for additive manufacturing applications. *Mechatronics* **2022**, *86*, 102874. [CrossRef]
8. Castelli, G.; Ottaviano, E.; Rea, P. A Cartesian Cable-Suspended Robot for improving end-users' mobility in an urban environment. *Robot.-Comput.-Integr. Manuf.* **2014**, *30*, 335–343. [CrossRef]
9. Kawamura, S.; Choe, W.; Tanaka, S.; Kino, H. Development of an ultrahigh speed robot FALCON using parallel wire drive systems. *J. Robot. Soc. Jpn.* **1997**, *15*, 82–89. [CrossRef]
10. Gonzalez-Rodriguez, A.; Castillo-Garcia, F.; Ottaviano, E.; Rea, P.; Gonzalez-Rodriguez, A. On the effects of the design of cable-Driven robots on kinematics and dynamics models accuracy. *Mechatronics* **2017**, *43*, 18–27. [CrossRef]
11. Albus, J.; Bostelman, R.; Dagalakis, N. The NIST robocrane. *J. Robot. Syst.* **1993**, *10*, 709–724. [CrossRef]
12. Duan, B.; Qiu, Y.; Zhang, F.; Zi, B. Analysis and experiment of the feed cable-suspended structure for super antenna. In Proceedings of the 2008 IEEE/ASME International Conference on Advanced Intelligent Mechatronics, Xi'an, China, 2–5 July 2008; pp. 329–334.
13. Tho, T.P.; Thinh, N.T. An Overview of Cable-Driven Parallel Robots: Workspace, Tension Distribution, and Cable Sagging. *Math. Probl. Eng.* **2022**, *2022*, 2199748. [CrossRef]
14. Abbasnejad, G.; Carricato, M. Direct geometrico-static problem of underconstrained cable-driven parallel robots with n cables. *IEEE Trans. Robot.* **2015**, *31*, 468–478. [CrossRef]
15. Castelli, G.; Ottaviano, E.; González, A. Analysis and simulation of a new Cartesian cable-suspended robot. *Proc. Inst. Mech. Eng. Part J. Mech. Eng. Sci.* **2010**, *224*, 1717–1726. [CrossRef]
16. Idà, E.; Bruckmann, T.; Carricato, M. Rest-to-rest trajectory planning for underactuated cable-driven parallel robots. *IEEE Trans. Robot.* **2019**, *35*, 1338–1351. [CrossRef]
17. Gouttefarde, M.; Gosselin, C.M. Analysis of the wrench-closure workspace of planar parallel cable-driven mechanisms. *IEEE Trans. Robot.* **2006**, *22*, 434–445. [CrossRef]
18. Gouttefarde, M.; Daney, D.; Merlet, J.P. Interval-analysis-based determination of the wrench-feasible workspace of parallel cable-driven robots. *IEEE Trans. Robot.* **2010**, *27*, 1–13. [CrossRef]
19. Kozak, K.; Zhou, Q.; Wang, J. Static analysis of cable-driven manipulators with non-negligible cable mass. *IEEE Trans. Robot.* **2006**, *22*, 425–433. [CrossRef]
20. Riehl, N.; Gouttefarde, M.; Krut, S.; Baradat, C.; Pierrot, F. Effects of non-negligible cable mass on the static behavior of large workspace cable-driven parallel mechanisms. In Proceedings of the 2009 IEEE International Conference on Robotics and Automation, Kobe, Japan, 12–17 May 2009; pp. 2193–2198.
21. Merlet, J.P. The kinematics of cable-driven parallel robots with sagging cables: Preliminary results. In Proceedings of the 2015 IEEE International Conference on Robotics and Automation (ICRA), Seattle, WA, USA, 26–30 May 2015; pp. 1593–1598.
22. Ottaviano, E.; Arena, A.; Gattulli, V. Geometrically exact three-dimensional modeling of cable-driven parallel manipulators for end-effector positioning. *Mech. Mach. Theory* **2021**, *155*, 104102. [CrossRef]
23. Du, J.; Bao, H.; Cui, C.; Yang, D. Dynamic analysis of cable-driven parallel manipulators with time-varying cable lengths. *Finite Elem. Anal. Des.* **2012**, *48*, 1392–1399. [CrossRef]
24. Kawamura, S.; Kino, H.; Won, C. High-speed manipulation by using parallel wire-driven robots. *Robotica* **2000**, *18*, 13–21. [CrossRef]
25. Zi, B.; Duan, B.; Du, J.; Bao, H. Dynamic modeling and active control of a cable-suspended parallel robot. *Mechatronics* **2008**, *18*, 1–12. [CrossRef]
26. Bruckmann, T.; Sturm, C.; Lalo, W. Wire robot suspension systems for wind tunnels. In *Wind Tunnels and Experimental Fluid Dynamics Research*; IntechOpen: London, UK, 2010; pp. 29–50.
27. Oh, S.R.; Agrawal, S. A reference governor-based controller for a cable robot under input constraints. *IEEE Trans. Control. Syst. Technol.* **2005**, *13*, 639–645. [CrossRef]

28. Ma, O.; Diao, X. Dynamics analysis of a cable-driven parallel manipulator for hardware-in-the-loop dynamic simulation. In Proceedings of the 2005 IEEE/ASME International Conference on Advanced Intelligent Mechatronics, Monterey, CA, USA, 24–28 July 2005; pp. 837–842.
29. Du, J.; Ding, W.; Bao, H. Cable vibration analysis for large workspace cable-driven parallel manipulators. In *Cable-Driven Parallel Robots*; Springer: Berlin/Heidelberg, Germany, 2013; pp. 437–449.
30. Diao, X.; Ma, O. Vibration analysis of cable-driven parallel manipulators. *Multibody Syst. Dyn.* **2009**, *21*, 347–360. [CrossRef]
31. Behzadipour, S.; Khajepour, A. Stiffness of cable-based parallel manipulators with application to stability analysis. *J. Mech. Des.* **2006**, *128*, 303–310. [CrossRef]
32. Tempel, P.; Miermeister, P.; Lechler, A.; Pott, A. Modelling of kinematics and dynamics of the IPAnema 3 cable robot for simulative analysis. *Appl. Mech. Mater.* **2015**, *794*, 419–426. [CrossRef]
33. Yuan, H.; Courteille, E.; Deblaise, D. Static and dynamic stiffness analyses of cable-driven parallel robots with non-negligible cable mass and elasticity. *Mech. Mach. Theory* **2015**, *85*, 64–81. [CrossRef]
34. Wu, Q.; Takahashi, K.; Nakamura, S. Formulae for frequencies and modes of in-plane vibrations of small-sag inclined cables. *J. Sound Vib.* **2005**, *279*, 1155–1169. [CrossRef]
35. Zhou, X.; Yan, S.; Chu, F. In-plane free vibrations of an inclined taut cable. *J. Vib. Acoust.* **2011**, *133*, 031001. [CrossRef]
36. Du, J.; Cui, C.; Bao, H.; Qiu, Y. Dynamic Analysis of Cable-Driven Parallel Manipulators Using a Variable Length Finite Element. *J. Comput. Nonlinear Dyn.* **2014**, *10*, 011013. [CrossRef]
37. Meunier, G.; Boulet, B.; Nahon, M. Control of an overactuated cable-driven parallel mechanism for a radio telescope application. *IEEE Trans. Control Syst. Technol.* **2009**, *17*, 1043–1054. [CrossRef]
38. Pott, A. *Cable-Driven Parallel Robots: Theory and Application*; Springer: Cham, Switzerland, 2018.
39. Martin-Parra, A.; Juarez-Perez, S.; Gonzalez-Rodriguez, A.; Gonzalez-Rodriguez, A.G.; Lopez-Diaz, A.I.; Rubio-Gomez, G. A novel design for fully constrained planar Cable-Driven Parallel Robots to increase their wrench-feasible workspace. *Mech. Mach. Theory* **2023**, *180*, 105159. [CrossRef]
40. Youssef, K.; Otis, M.J.D. Reconfigurable fully constrained cable driven parallel mechanism for avoiding interference between cables. *Mech. Mach. Theory* **2020**, *148*, 103781. [CrossRef]
41. Rushton, M.; Khajepour, A. Planar variable structure cable-driven parallel robots for circumventing obstacles. *J. Mech. Robot.* **2021**, *13*, 021011. [CrossRef]
42. Trevisani, A. Planning of dynamically feasible trajectories for translational, planar, and underconstrained cable-driven robots. *J. Syst. Sci. Complex.* **2013**, *26*, 695–717. [CrossRef]
43. Pott, A. Geometric and Static Foundations. In *Cable-Driven Parallel Robots*; Springer: Berlin/Heidelberg, Germany, 2018; pp. 45–117.
44. Moradi, A. *Stiffness Analysis of Cable-Driven Parallel Robots*; Queen's University: Kingston, ON, Canada, 2013.
45. Pott, A. An improved force distribution algorithm for over-constrained cable-driven parallel robots. In *Computational Kinematics*; Springer: Dordrecht, The Netherlands, 2014; pp. 139–146.
46. Lieslehto, J. MIMO controller design using SISO controller design methods. *IFAC Proc. Vol.* **1996**, *29*, 1152–1156. [CrossRef]
47. Ranganathan, A. The levenberg-marquardt algorithm. *Tutoral LM Algorithm* **2004**, *11*, 101–110.
48. Ogata, K. *Modern Control Engineering*; Pearson: Upper Saddle River, NJ, USA, 2010; Volume 5.
49. Zhang, B.; Shang, W.; Cong, S.; Li, Z. Coordinated dynamic control in the task space for redundantly actuated cable-driven parallel robots. *IEEE/ASME Trans. Mechatron.* **2020**, *26*, 2396–2407. [CrossRef]
50. Yuan, H.; Courteille, E.; Deblaise, D. Force distribution with pose-dependent force boundaries for redundantly actuated cable-driven parallel robots. *J. Mech. Robot.* **2016**, *8*, 041004. [CrossRef]
51. Cui, Z.; Tang, X.; Hou, S.; Sun, H. Research on controllable stiffness of redundant cable-driven parallel robots. *IEEE/ASME Trans. Mechatron.* **2018**, *23*, 2390–2401. [CrossRef]
52. Khosravi, M.A.; Taghirad, H.D. Robust PID control of fully-constrained cable driven parallel robots. *Mechatronics* **2014**, *24*, 87–97. [CrossRef]

MDPI
St. Alban-Anlage 66
4052 Basel
Switzerland
www.mdpi.com

Actuators Editorial Office
E-mail: actuators@mdpi.com
www.mdpi.com/journal/actuators